Modern Communications
Receiver Design and Technology

For a listing of recent titles in the
Artech House Intelligence and Information Operations Series,
turn to the back of this book.

Modern Communications Receiver Design and Technology

Cornell Drentea

ARTECH HOUSE

BOSTON | LONDON

Library of Congress Cataloging-in-Publication Data
A catalog record for this book is available from the U.S. Library of Congress.

British Library Cataloguing in Publication Data
A catalogue record for this book is available from the British Library.

ISBN-13: 978-1-59693-309-5

Cover design by Vicki Kane

© 2010 Artech House
685 Canton Street
Norwood, MA 02062

10 9 8 7 6 5 4 3 2

Contents

CHAPTER 17

CHAPTER 18

CHAPTER 19

CHAPTER 20

CHAPTER 21

CHAPTER 26

Conclusions

Foreword

Any author who attempts to comprehensively cover the many, many aspects of receiver design and their complex interrelationships is up to a gigantic job. To even know that certain aspects exist often takes decades of experience in not only the design and analysis of radio receivers, but also, most importantly, in the lab bench experience gained in building actual receiver hardware. Cornell Drentea is an excellent author for this task.

Readers who are new to the topic of radio communications receivers will find the early chapters of this book of most value. As the experience of the reader grows, the advanced material in subsequent chapters will make more sense. This book provides a comprehensive overview of how RF system requirements flow into designs as well as hands-on use of state-of-the-art system design tools. It is ultimately an advanced study of practical receiver systems and circuit design.

One aspect of this book that I particularly like is Mr. Drentea's placing of much of this material in a historical context. It is important, if one is to make true progress in any field, to understand how that field of work has evolved into its current state. One of my favorite quotations is, "Those who cannot remember the past are condemned to repeat it" (Santayana). We are fortunate that Mr. Drentea has a tremendous amount of experience and remembers how he got it. For those of us who cannot remember the past because we are new to a topic we must study its past to build a useful foundation. This leads to an appropriate paraphrase of Santayana: Those who do not study history are doomed to repeat it. I commend Mr. Drentea for providing this reference.

Because the subject is so vast, Mr. Drentea chose to emphasize certain technical areas that cannot usually be found in one single publication and combine them with facts and experiences of his own in such a way as to provide a complete and accelerated learning reference. I find this to be a well thought out book with lots of material good for advanced engineering students as well as for practicing hardware and system design engineers. The format and approach are logical and easy to follow. This book is complete and in depth where it matters. It flows well and it has a lot of material that is not easily (or possibly) found anywhere else.

This work should be good for anyone considering an RF signal processor (radio) concept. A very useful reference, this book is particularly helpful to an RF systems engineer or operations planner in EW, reconnaissance, or a similar field dealing with a wide variety of signals, environments, and operating conditions. It gives powerful tools for the trade-off of requirements against potential solution approaches.

This book provides an excellent knowledge base of well-interlocked topics. It reflects the passion the author has put into what he does best. I appreciate the level of effort it took to put together a course of this magnitude. I was especially impressed with the presentation of this complex information content, and the great spirit that is put into it.

Earl McCune, Jr., Ph.D.
RF Communications Consulting
Santa Clara, California
August 2010

Preface

This book is built on the success of my previous book, *Radio Communications Receivers*, first published in 1982 and republished in 1984.

While keeping the same basic format, the new book attacks the topics of all types of receivers, not just communications receivers. It focuses on the high probability of intercept receivers. The material has been greatly expanded with many new and complex topics to reflect today's state-of-the-art in RF and anticipate tomorrow's systems and technologies. Special attention was given to treat the subjects from a physics and mathematical point of view as well.

As such, the size of the book has doubled to combine new and complex topics into an authoritative scientific and engineering manual for all RF system and circuit designers involved in a wide range of receiver applications and over a wide frequency range.

High probability of intercept means different things to different people, and no book can be everything to everyone. Although this book was written in the spirit of sharing information, the subject of high probability of intercept receivers is so vast that this book cannot cover every aspect in full depth. This book is designed to logically interconnect chapters such that you can get a complete picture of what can be done in receiver design from all points of view: from historic implementations to today's modern EW microwave implementations.

The book treats the subject of receiver design broadly and in depth. However, the book is not intended as a how-to book, but rather as a tool to quickly educate the new or more experienced engineer entering the field of RF engineering. Its main purpose is to treat all topics in concert and keep a balance between theory and practice.

This book will make a conscious attempt to stay away from the typical large collection of disconnected chapters backed by multiple pages of equations. Instead, it will concentrate on the interactions of topics of RF design while using the necessary mathematics and present an actual design implementation of a high probability of intercept, high dynamic range, fully synthesized coherent transceiver. This transceiver project introduces the reader to a practical high performance receiver design, which has numerous other applications at much higher frequencies.

Modern Communications Receiver Design and Technology is an advanced work, as well as a beginner's book. One can use it as a comprehensive course in RF, a historical and technology book, or a highly technical reference book.

Intended for the electrical engineering student, the new engineer, the senior practicing RF engineer, and the receiver construction aficionado, this book combines the necessary theory, mathematics, and practical circuitry in such a way as to make the learning process fun.

Modern Communications Receiver Design and Technology combines more than 40 years of RF engineering experience into a single authoritative source. If you are a beginner, after reading this book you should have a good understanding of what is involved in the design and development of a modern receiver or transceiver. If you are an experienced engineer, you may find this book to be a good reference that can help you in critical times. If you are a constructor, many of the circuits presented in this book are state-of-the-art proven designs that work well. Although this book is not a construction handbook, the practical ideas shown in it can be easily carried into new designs. I hope you will enjoy my book.

To the Memory of My Father

It is a beautiful Sunday morning in September 1946. I am going to be six years old in another week. With a pair of pliers in one pocket and a screwdriver in the other, I am patiently waiting for my parents to leave home. They are going to visit relatives and will not return until late in the afternoon. I can finally take the radio apart to see what the little people inside look like, though something tells me that this cannot be true ... but, how does it work?

"Good-bye, Mama! Bye, Dad. I'll be good, I promise!"

The plan is to take the radio all apart and put it back together before they get back at five o'clock. If everything works right, I should have enough time to find all there is to know about this magic box that my dad got last Christmas. I know that I've been told not to touch it, but I really have to find out what is inside. If my dad ever found out about this, I would be in big trouble. It is a heavy radio to move, but it's on the table now. I better start working. What comes out first?

It is two o'clock and by this time I have lots of nuts and bolts lying around the table. I also have these funny-looking objects made out of glass with things inside them ... and no little people yet. There are also some tweaking things that I cannot take out, but something tells me not to touch them. So that is where the sound comes from, this big round-looking thing with paper inside ... and a magnet? But why? I hear noises—it can't be; it's only two-thirty. They're back. The door opens.

"What did you do?" I hear my father's voice as I was unveiling another gray area of the radio, with the pliers in my right hand and the screwdriver in my mouth. Quick thinking tells me not to lie at this point.

"Well, I wanted to know how it works, Dad, but I'll put it back together, I promise. I promise." I feel tears coming out of my eyes.

"Very well, you put it back together, and if it still works, I will forgive you. But it has to work."

I am scared. Will it still work? I know I was very careful to remember where everything came from, but I didn't plan on having so many little parts. Where did this come from? Oh yes, here. And this funny-looking bottle, it doesn't want to go back ... oh yes, it has a key there.

It's seven o' clock and I have three screws left. I don't know what to do with them; I really don't remember where they came from. Will it work without them? I have to take a chance.

"I am ready, Dad," I call as I am hiding the screws in my pocket. The big test is here. Oh, God, help me. The set is plugged in and suddenly the little people are back again. What a relief!

Although I never found out where the three little screws went, I later went to school and learned a lot more about how things worked and became an engineer. My father has long since passed away, but I will never forget my first contact with a radio. This book is dedicated to the memory of my father, who taught me the laws of self-discipline and perseverance.

Acknowledgments

I wish to express my gratitude to the many professional colleagues, friends, and companies who have contributed towards the enrichment of this book. In particular I wish to thank: Dr. Earl McCune for his careful review of the entire book, and his valuable suggestions, in-depth technical discussions and writing the foreword; Robert Zavrel for his careful review of the entire book, his comprehensive comments, and the writing of the Introduction; Stephen Heald of Tait Electronics, New Zealand for the many outstanding discussions, suggestions, and reference contributions; Lee R. Watkins for his in-depth discussions about filter design and valuable materials contributions; Phillip N. Eide for his discussions and input on power supply design and materials contributions; Constantin Popescu for his valuable help in putting the Star-10 project together; Jerry Scrivano for his help during our wonderful trip together to the Arecibo Radio telescope in Puerto Rico; Colin Horrabin for our comprehensive discussions on the performance of the H-mode mixer; Dana Whitlow and Tony Acevedo of the Arecibo Observatory, National Astronomy and Ionospheric Center in Puerto Rico, operated by Cornell University, for their materials contributions and interesting scientific discussions during my visit to the observatory; Larry Wolfgang and Steve Ford of the ARRL organization for being instrumental in making all my QEX materials available for publication; Hans Zahnd of ADAT, Switzerland for the many outstanding discussions and software-defined radio materials contributions; Kirt Blattenberger of RF Cafe for his suggestions and encouragement throughout the writing of this book; Milan Hudecek of WinRadio, Australia for his materials contributions and the in-depth discussions regarding software-defined radios; Randy Burcham for his help with the Star-10 transceiver; Mike Lindsay of Raytheon for his comments, suggestions, encouragement, and help with creative graphics throughout the writing of this book; John Richardson of Wenzel Associates for his valuable input in the area of super-regenerative dividers and materials contributions; Joe Jensen of Hughes Research Laboratories (HRL) for his materials contributions and constructive reviews; Tom Ashburn for his contributions and review of the EW receivers section materials; Jacqueline Hansson of IEEE Intellectual Property Rights Office for her expedient dealing with release requests; Franck Darde of AA Opto-Electronic Group, France, for his acousto-optic reference contributions; Danny Fung, David Seidel, Lute Maleki, and Carolyn Daeseleer of OEwaves, Inc. for their reference material contributions and reviews; and Dave Ames of Stanford Research Systems, Larry Steckler of *Electronics Now* magazine, Thomas Sanders and Donald P. Havens of Rockwell Collins—Filter Products, Steve Raymer of the Museum of Broadcasting (Minneapolis, Minnesota), Nang-Haung Sheng of Euvis, RCA, Brent Campbell and

Ace J. Blackburn of Z-Communications, and Kipp Schoen of Picosecond Laboratories for their material contributions.

Finally, I would like to thank my wife, Dominique Drentea, who took over all my duties and responsibilities and allowed me to devote all my time to writing. Without her devoted love, care, contributions, understanding, and support, this book would never have been written.

Introduction

Early in 2010 I received an e-mail from my old friend Cornell Drentea asking if I could review a new text he was writing on RF engineering and provide a short introduction. Without hesitation I agreed. What followed was a literary, technical journey into the depths of the RF engineering art, which I might say is unprecedented. There are many outstanding texts on RF engineering in general and many more on specific topics. Few if any provide a novel-like narrative meant to be read cover to cover. Although not intended to be just a historic text or a cookbook, its utility as a wealth of reference material becomes clear upon reading.

Cornell not only provides the answers to how and what, but also answers the often neglected why. The book begins with a short history of radio and the key inventions that have led to today's state-of-the-art. Then he weaves history into state-of-the-art engineering, taking the reader along a fascinating and very instructive journey. This reading should be recommended for a wide audience: advanced radio amateurs, engineering students, seasoned practicing engineering professionals, and anyone wishing to learn the nuts and bolts of radio engineering in one cohesive flowing narrative backed by advanced analysis tools and a thorough mathematical treatise where required.

Cornell combines world-class technical expertise with his own unique writing style and the art of RF design. In this book, you won't find page after page of complex nonderived equations. Rather, Cornell distills very complex topics into plain English to the greatest possible extent without diluting the depth of coverage. Of course the narrative also combines the necessary mathematics of the trade, but avoids endless and unnecessary details. This can only be successfully accomplished by someone who has "been there and done that." It also requires world-class mentoring and teaching skills. The combination of these skills and expertise has culminated in this wonderful new text. Clearly this is the result of a labor of love from a wonderful teacher.

Therefore, I am pleased and honored to introduce this seminal new work and encourage anyone reading this introduction to read this text cover to cover.

Robert J. Zavrel, Jr.
CEO and Founder
Plum Valley Systems LLC
Elmira, Oregon
August 2010

CHAPTER 1

Introduction to Receivers

Whenever thinking of radio, we usually think of one man: Guglielmo Marconi. However, radio, as a technology, resulted from the work of many people. The following is a list of some of the many men and women involved in the development of radio:

Robert Adler
Ernest F. W. Alexanderson
Edwin H. Armstrong
Jones J. Berzelius
Edouard Branly
George Campbell
John Carson
Arthur A. Collins
Frank Conrad
William Crookes
Jacques and Pierre Curie
Amos E. Dolebear
R. L. Drake
William D. Duddel
H. H. Dunwoody
Thomas A. Edison
Albert Einstein
Robley Evans
Michael Faraday
Reginald A. Fessenden
John A. Fleming
Lee De Forest
Ben Franklin
L. Alan Hazeltine
Oliver Heaviside

Heinrich R. Hertz

Christian Huygens

Karl Jansky

Arthur E. Kennelly

Irving Langmuir

Heddy Lamar

Oliver J. Lodge

James C. Maxwell

G. M. Minchin

Samuel F. B. Morse

Greenleaf W. Pickard

Alexander Popov

William H. Preece

Theodore Roosevelt

David Sarnoff

Nikola Tesla

Jules Verne

Figure 1.1 Communications pioneers—a group of distinguished scientists visiting RCA's experimental Trans-oceanic Communications Station at New Brunswick, New Jersey in 1921. From left to right: (starting fourth from left) David Sarnoff, Thomas J. Hayden, Dr. E. J. Berg, S. Benedict, Albert Einstein, John Carson, Dr. Charles P. Steinmetz, Dr. Alfred N. Goldsmith, A. Malsin, Dr. Irving Langmuir, Dr. Anthony W. Hull, E.B. Pillsbury, Dr. Saul Dushman, R. H. Ranger, and Dr. G. A. Campbell. (Photo courtesy of RCA.)

T. L. Wadley

Clemens Winkler

One must also not forget the many dedicated ham radio operators around the world. Some of these men were science fiction dreamers. Their contribution to this invention was that of stimulating the other's imaginations. Still others were scientists and mathematicians. Their role was to pave the road for future developments. Some were politicians and businessmen, and some were inventors and technical practitioners, or what we usually refer to as engineers (see Figures 1.1 and 1.2).

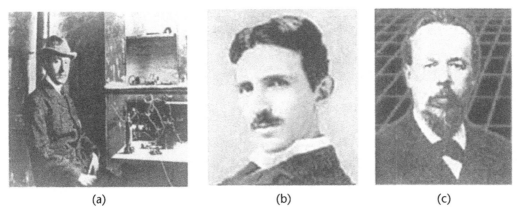

(a)   (b)   (c)

Figure 1.2 Who invented radio? (a) Italian Guglielmo Marconi and his famous receiver at St. John's, Newfoundland, December 21, 1901, where the first transatlantic signal was heard. He performed his first RF tests in 1895 in London, allegedly copying Tesla's invention. Marconi's first wireless patent application was filed in England on June 2, 1896. (Photo courtesy of RCA.) (b) Croatian of Istro-Romanian origin Nikola Tesla demonstrated the first transmissions of intelligent RF signals in 1893 in the U.S. and obtained first U.S. patent #645,576. He used a 5-kW spark transmitter. Despite winning disputes with Marconi who claimed he never read Tesla's technical papers, he is not usually thought of as the father of radio. However, Tesla invented radio.(c) Russian Alexander Popov entered the wireless field through his development of a receiver that detected thunderstorms on May 7, 1895. He conceived the idea of using the Branly coherer to pick up natural "damped" waves generated by lightning. On March 24, 1896, Popov demonstrated sending and receiving wireless signals across an 800-foot distance.

CHAPTER 2

The History of Radio

The following is a brief history of the development of radio.

2.1 The Coherer

The radio receiver appeared as a consequence of an invention called the coherer (Figure 2.1). Edouard Branly, of France, discovered that a glass tube with two silver electrodes, filled with loose iron particles, will conduct DC electricity better in the presence of so-called Hertzian waves generated with conventional spark generators.

2.2 The First Radio Receiver

Branly would not see the real application for this device, but Marconi developed a receiver on this principle in 1895, apparently copying an earlier invention of Nikola Tesla. He applied for a patent in June of 1896. According to the records, Croatian Nikola Tesla demonstrated the first transmissions of intelligent RF signals in 1893 in the United States and obtained the first U.S. patent #645,576 for this invention. He used a 5-kW spark transmitter. Despite winning disputes with Marconi who claimed to never have read Tesla's technical papers, he is not usually thought of as the father of radio. However, Tesla is the inventor of radio.

A Russian scientist by the name of Alexander Popov in 1895 claimed to have created a storm receiver, which rang a bell every time an electrical storm would approach within several miles. This receiver had no antenna and was of limited range.

In 1895, Marconi discovered that by adding two wires with large metal plates at the ends to his receiver and/or transmitter, the range could be increased considerably. He called these wires "catch wires," and soon buried one of these wires in the ground, while elevating the other, thus discovering the HF antenna system as we know it today.

It appears that the two inventors were in correspondence for a while. While Popov did not see any further application for his receiver, the enterprising Marconi perfected the machine and demonstrated its usefulness. He noticed that once conducting, the coherer would stay in that state. Thus, he invented the decoherer.

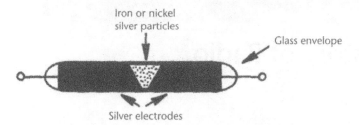

Figure 2.1 The coherer was invented by Edouard Branly in 1891.

2.3 The Decoherer (Practical Coherer/Decoherer Receivers)

The decoherer was nothing more than an electrical bell, slightly modified and connected in series with the coherer and/or the receiver relay as shown in Figure 2.2. Every time a signal was received, the coherer would be set in an "on" state, triggering the relay and, therefore, the electrical bell, which in turn would knock on the coherer, resetting it for a new signal. Although practical receivers of this type required a decoherer, they were typically referred to as coherer receivers.

Marconi perfected this receiver and produced several versions of it. He left Italy and went to England where he improved the coherer, producing equipment that established maritime communications for the Royal Navy. By 1899, he established communications across the English Channel, and two years later he sent the letter "S" over the Atlantic Ocean, by using a similar receiver [see Figure 1.2(a)]. He realized the limitations of the coherer receiver, but it wasn't until 1906 that any improvement came about.

2.4 Galena Crystal Discovery, the Fleming Valve, and the Audion

General H. H. Dunwoody of the U.S. Navy discovered the crystal detector in 1906 (see Figure 2.3). This produced a new type of receiver, which was more sensitive: the crystal, or galena, receiver (see Figure 2.4). The crystal receiver set didn't last long as a commercial receiver type because of Fleming's valve (1904) and the De Forest audion (1906).

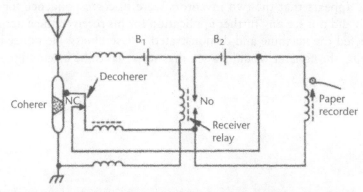

Figure 2.2 A typical coherer receiver from 1901 had an automatic decoherer.

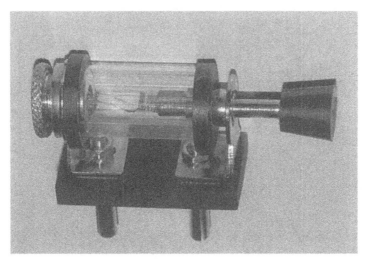

Figure 2.3   A typical commercial galena crystal detector has a well-designed "cat whisker arm."

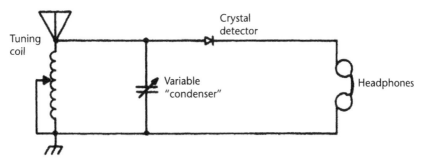

Figure 2.4   A typical 1906 crystal (galena) receiver.

As a result of these inventions, new and even more sensitive receivers evolved from the crystal receiver as shown in Figure 2.5.

2.5   The Audion and the Regenerative Receiver

A new generation of radio receivers was born in 1906: the regenerative sets. Edwin Howard Armstrong is responsible for the invention of the regenerative receiver. He understood that the Audion circuit used in Figure 2.5(b) was not satisfactory. In this circuit, the Audion's grid was connected in a tuned circuit to the receiver's antenna, while the plate was connected in series with the headphones and the anode battery to ground. The circuit was closed through the electron-emitting filament, which in turn was activated by the filament battery.

Armstrong found that some alternating current was produced in the plate headphone circuit (called the wing circuit) where it wasn't expected and attempted tuning it in much the same manner as the antenna-to-grid circuit, thus coupling some of the amplified output signal back to the grid circuit inductively, as shown in Figure 2.6. This method was called regeneration, and the "tickling" level adjustment shown in the figure provided for a threshold, allowing for large amplification

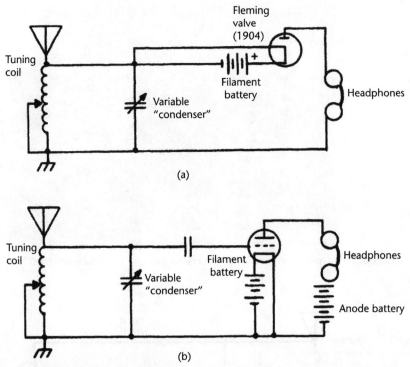

Figure 2.5 The evolution of the crystal (galena) receiver was a result of the Fleming valve and the Audion. (a) The Fleming valve replaced the crystal detector in this 1907 receiver. (b) This receiver uses De Forest's Audion in a detector/amplifier mode.

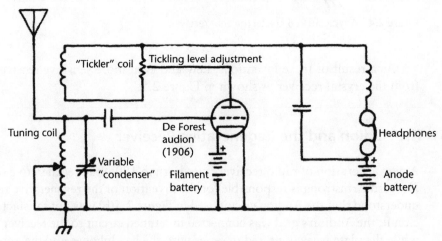

Figure 2.6 A typical regenerative receiver using De Forest's Audion.

not previously possible with any other method. This newly discovered feedback amplifier circuit made practical long distance reception a reality for the first time.

2.6   The Audion and the Local Oscillator

Both Armstrong and De Forest noted that when pressing the Audion to higher amplification (by increasing the tickling level, over the threshold point) in a regenerative set, an audible hissing would result. While De Forest dismissed this fact as an "irritating noise that hindered proper operation," Armstrong went on to prove that the Audion was not only a receiving device, but an oscillator of electromagnetic waves, which would serve later as the basis for the local oscillator (LO) to be used in the superheterodyne receiver.

2.7   The Audion and the Tuned Radio Frequency (TRF) Receiver

Another form of radio receivers characteristic of this era was the tuned radio frequency (TRF) receiver. The TRF receiver was nothing more than a chain of individually tuned amplifiers. This radio took advantage of the Audion valve in an amplifying mode. Frequently, many dials were present on these early radios and it took patience to tune one of them. Better versions employed complex mechanical tracking for slaving several variable capacitors to the motion of a single knob in order to provide identical tuning as shown in Figures 2.7, 2.8, and 2.9. These radios were popular until 1928.

However, this concept is still being used today at much higher frequencies. In its modern implementation, called the crystal video receiver it is used for high probability of intercept applications such as electronic countermeasures (ECM), primarily for detection of low-duty cycle pulses or CW signals over ultrawideband frequency ranges and in relatively close proximity to the emitters. Its purpose is near instantaneous reception of many signals over widebands without the need to sweep a local oscillator, which takes time.

Figure 2.7 Early 1920s Atwater-Kent (AK-42) tuned radio frequency (TRF) receiver implementation using belts and variable capacitors to provide a triple tracking filter function combined with tubes for amplification. (Courtesy of the Museum of Broadcasting, Minneapolis, Minnesota.)

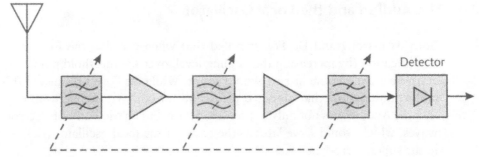

Figure 2.8   Typical TRF receiver has a triple tuned circuit.

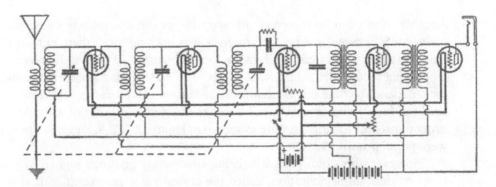

Figure 2.9   Schematic diagram of a typical TRF receiver of the 1920s.

A typical crystal video receiver consists of a multiplexer or a power divided filter bank, which splits a wide input frequency range into several broad contiguous bands, which in turn are further filtered, logarithmically amplified, and detected. Amplitude and video detection occurs at the baseband in each of the bands using a crystal square law detector. We will discuss this type of high probability of the intercept receiver along with others in more detail in Chapter 26.

2.8   Early Progress in Radio Receivers

As a result of the early Audion invention, amplitude modulation (AM) was born, and before long the spectrum was crowded with voices and music; electromagnetic interference (EMI) finally began.

The type of radios described earlier were quite sensitive, but had very poor selectivity. Many stations were received at the same time. It is interesting to look at some statements made during this time, statements that probably define the radio receiver as we know it today.

"Imagination is better than knowledge," said Albert Einstein, and scientists and engineers certainly had imagination during this time in history.

William Crooks, an English physicist who missed the discovery of X-rays, en-visioned the progress of radio receivers. He said in 1892: "More delicate receivers which will respond to wavelengths between certain defined limits and be silent to all others remain to be discovered."

David Sarnoff (see Figure 2.10) a self-educated technical genius, entrepreneur, and a ham radio operator famous for receiving the *Titanic* distress message, said in 1916: "The receiver can be designed in the form of a simple radio music box and arranged for several different wavelengths, which should be changeable with the pulling of a single switch or pressing of a single button." This vision is shown in Figure 2.11. The realization of such selective receivers did not come true until 1918, when Professor Armstrong invented the superheterodyne receiver.

Figure 2.10   David Sarnoff receiving traffic at the John Wanamaker store in New York City in 1912. Note the Cunard Line photo on the wall. He stayed at his post for 72 hours to report the *Titanic* disaster, demonstrating the importance of radio and generating new interest in radio technology. Sarnoff later became the president of RCA. (Photo courtesy of RCA.)

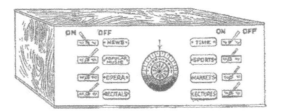

Figure 2.11   David Sarnoff's vision of a futuristic receiver (1916). (*From:* [1].)

Reference

[1]   Dellinger, J. H. , and L. E. Whittemore, *Lefax Radio Handbook*, Washington, D.C.: Radio Laboratory, U.S. Bureau of Standards, 1922.

CHAPTER 3

The Superheterodyne Receiver

The superheterodyne is a type of radio receiver that uses a process in which the incoming signal is mixed with a local oscillator (LO) signal in a manner generating new signal components, which are equal to the sum and the difference of the original frequencies. One of these products is designated as the intermediate frequency (IF) and is passed by a tuned circuit, which rejects the undesired products as well as the original incoming frequency while still maintaining the information contained in it. This low level signal (caused by the natural losses along the propagation path between the transmitting point and the receiver's antenna) is further amplified before detection takes place. The process of mixing will be discussed in detail in Chapter 15.

In the amplitude modulation (AM) mode, the audio frequencies riding on the carrier are detected through the process of rectification, directly from the IF, while Morse code signals (CW) or single sideband signals (SSB) can be recovered by another mixing process, which produces audible heterodynes with the help of a beat frequency oscillator (BFO).

In the case of frequency modulation (FM) where the carrier frequency moves around a center point at the rate of the audio frequency, a frequency discriminator or a phase-locked loop are used for detection. The final step after detection is audio amplification.

In the Armstrong superheterodyne, the audion tube performed several functions as shown in Figure 3.1. It first acted as a local oscillator (LO), which produced high frequency currents of a different frequency than the signal to be received. It was used again as a mixer when this energy was heterodyned with the incoming frequency from the antenna, producing the intermediate frequency (IF), which was amplified again by the audion. Amplitude modulation (AM) detection was performed with the help of the Fleming valve, while final audio amplification used the audion tube again. Today there are several types of superheterodynes, and their mixing schemes dictate their terminology, as shown in Table 3.1.

3.1 Single Conversions

A single conversion receiver can downconvert from the received frequency or upconvert. Both downconvert and upconvert can be used in a single conversion superheterodyne, when the range to be covered is so wide that the intermediate frequency falls somewhere in the middle of it, as in the case of many general coverage HF

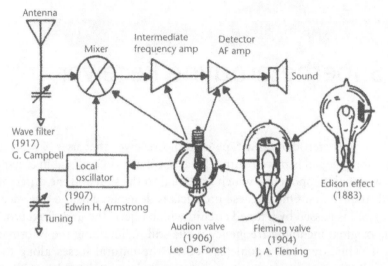

Figure 3.1 The impact of earlier inventions on the superheterodyne receiver, which was invented in 1918 by Edwin H. Armstrong.

Table 3.1 Superheterodyne Receiver Types

1. Single conversions	Downconversion;
	Upconversion;
	Combinations of both.
2. Multiple conversions	Double conversion;
	Triple conversion;
	Quadruple conversion.
3. Direct conversion	Conversion to baseband.

radio receivers with IFs at 5.5 MHz, 9 MHz, or 10.7 MHz. There are advantages and disadvantages in these schemes, as we will see later in Chapter 16.

3.2 Multiple Conversions

Another superheterodyne type is the double conversion, and as a growth from the double conversion, the triple and quadruple conversions.

3.3 Direct Conversion (Zero IF)

The last and the simplest form of superheterodyne is the direct conversion type. We will analyze all these types of radio receivers, and examine their advantages and disadvantages. Branching from this concept, we also look at the direct sampling no conversion approach, which takes advantage of modern A/D, DSP, and D/A techniques.

CHAPTER 4

Implementing Single Conversion Superheterodynes

The most popular superheterodyne system ever realized was a single conversion type with an IF of 455 kHz. The block diagram in Figure 4.1 shows a typical single conversion superheterodyne, which tunes the frequency range from 500 kHz to 30 MHz. The IF amplifier is tuned to 455 kHz.

The incoming signal anywhere between 500 kHz to 30 MHz passes through a tunable filter called the preselector. The preselector has the role of filtering out all frequencies coming from the antenna, except for the one that is needed. This filtered signal goes to the *mixer*, which can take several forms, where it combines with the frequency coming from the local oscillator (in our case a variable frequency oscillator) and produces an IF signal at 455 kHz.

As you have probably already noticed, this is a downconversion scheme because the IF falls below all of the frequencies to be received within the range of 500 kHz to 30 MHz. The variable frequency oscillator (VFO) LO is operating 455-kHz above the incoming frequency at all times. For example, if we want to receive 500 kHz, the VFO must generate 955 kHz, and if we want to receive 30 MHz, the VFO should be 30.455 MHz.

As a result of the interaction between the VFO and the signal, there are two signals (referred to as *products*) coming out of our mixer: the sum and the difference (other products are generated as a result of the mixing and they will be reviewed later in this book). The higher product is rejected by the IF filter tuned to 455 kHz.

Note that no matter where we tune the receiver, the IF is always the same, making filtering and processing of the signal identical for all RF input frequencies to be received.

This is the major idea behind the superheterodyne approach. Mixers will be discussed in detail in Chapter 15. Unlike the previously used tunable filters in the TRF receiver, any input frequency can simply be reduced to a common denominator, the IF, and can be processed through a single fixed frequency filter with a bandwidth matching the information bandwidth and a set of amplifiers, which are optimized at a single frequency (the IF) and easier to handle. This is why the superheterodyne has lasted until now as the best design.

The 455-kHz signal, which also contains the information that came from the antenna, is filtered and amplified in the IF amplifier and is finally detected and amplified again at audio frequency before being delivered to the speaker. This scheme remains, to date, a basic approach of the single conversion downconvert

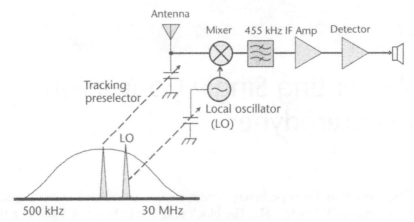

Figure 4.1 Simple block diagram of a general coverage single conversion superheterodyne receiver. Note that the preselector tracks together with the local oscillator (LO).

superheterodyne receiver. Note the tracking mechanism of the scheme (Figure 4.2). In this type of radio, it is imperative that the preselector and the VFO track together. Because of the large RF input bandwidth of six octaves, it becomes very difficult to provide proper preselection with a single filter.

In a basic single conversion HF receiver covering 500 kHz to 30 MHz, the incoming signal goes through separate filters, which are switched in and out for different ranges, one at the time, allowing for several bands to be covered in steps. The VFO is also switched in similar bands, which track exactly with the preselector bands.

The tuning element of the preselector, the variable capacitor, is mechanically ganged with the VFO capacitor. They track together, so that when the preselector peaks at some frequency, the VFO operates exactly 455 kHz above that frequency, as shown before. For example, if the preselector is tuned to exactly 1 MHz, its capacitor is mechanically ganged with the VFO capacitor, so that the VFO produces exactly 1.455 MHz. The rest of the circuitry remains the same, as described previously. However, we have a new problem, the image. This is shown in Figure 4.3. Tracking variable capacitors used in such receivers are shown in Figure 4.4.

A practical five-band receiver could be implemented as shown in the schematic diagram in Figure 4.5. In this design, the signal from the antenna goes to switch S1-1, which couples it with any one of the five preselector tuned circuits, corresponding to the five bands in this case. Switch S1-2, which is mechanically ganged with S1-1, picks up the output of the preselector range chosen. Together with the variable capacitor, which is always in the circuit, the chosen range tracks the exact frequency to be received, which is exactly 455-kHz below the frequency of the VFO.

The VFO capacitor is mechanically coupled with the preselector capacitor. S1-3A and S1-4 are also mechanically connected with S1-1 and S1-2 in order to provide the right ranges for the VFO to match the preselecctor. The VFO signal couples through C18 into Q2 where it combines with the proper preselected signal and produces a difference frequency of exactly 455 kHz for any selected receiver frequency between 500 kHz and 30 MHz.

The signal is directed through a ceramic filter, which eliminates all the unwanted mixing products and keeps only the 455-kHz signal. This signal is then routed

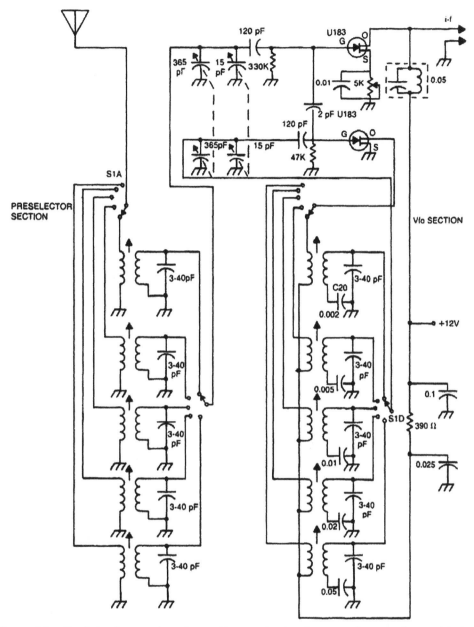

Figure 4.2 Classic implementation of a tracking mechanism for a preselector and a variable frequency oscillator, showing a complex band selector in a standard 455-kHz IF receiver.

through a chain of amplifiers; demodulated; fed through a volume control into an audio amplifier made of Q1, Q2, Q3, and Q4 (circuit board #2); and finally moves into the speaker.

By now, some readers have probably wondered why 455 kHz was the choice for the IF. Is there anything special about this frequency? There are three reasons for this choice:

- It was right below the broadcast band.

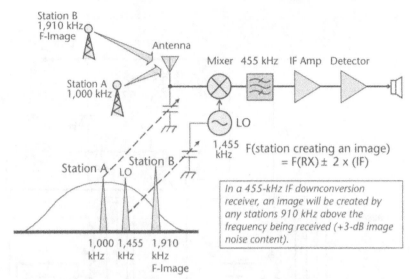

Figure 4.3 Tracking system: a 455-kHz IF, downconvert AM superheterodyne receiver. The image is introduced by the low frequency IF.

Figure 4.4 Modern variable capacitors featuring as many as four tracking sections have been used in AM and HF superheterodyne receivers.

- It was low enough in frequency to allow adequate selectivity to be obtained with conventional inductors and capacitors (L/C).
- It was empirically found that using this IF frequency, the image was kept out of a standard AM receiver.

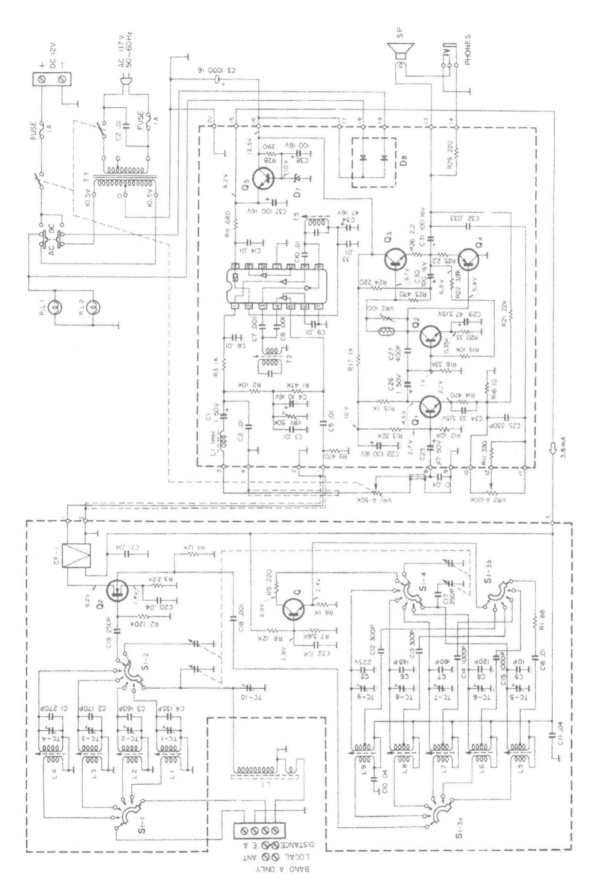

Figure 4.5 Modern solid-state implementation of a general coverage 455-kHz IF, single conversion superheterodyne receiver.

4.1 The Image Problem

The users were all very happy with their radios and could hear a lot of stations with their general coverage receivers. Some of these stations were one and the same (received in several places on the dials). Even though this phenomenon was discovered during the 1920s, it wasn't until the 1950s until manufacturers did something about it.

Here is what the 1926 edition of the *Radio News Superheterodyne Book* [1] says in regard to this problem: "There are many other instances where slight improvements may be made. Two straight-line frequency condensers, mounted on one shaft, for the oscillator and tuner circuits are recommended. This not only simplifies tuning, but eliminates the objectionable feature of every superheterodyne—the reception of all stations on more than one setting of the oscillator dial." This phenomenon is called *image*. It is the result of the mixing process, which will be discussed in Chapter 15.

It has been established that the incoming signal frequency from the antenna meets with the local oscillator to produce the intermediate frequency for further amplification. The local oscillator could be below or above the incoming frequency by the intermediate frequency value. Usually the oscillator operates above the incoming frequency signal, as in our case, and it would oscillate, for our example, at 14.545 MHz to convert a 14.090-MHz signal to an IF of 455 kHz as shown in Figure 4.6.

Another station, such as WWV at 15.000 MHz, would also mix with the 14.545-MHz local oscillator producing the 455-kHz IF. If the front-end selectivity of the receiver is relatively poor, and the 15.000-MHz station is strong, it will also be heard with the receiver tuned to 14.090 MHz and interfere with the 14.090 MHz signal; both signals will show up concurrently in the 455-kHz IF.

To dramatize this, consider the case in Figure 4.6 where station A broadcasts at 14.090 MHz with 100 watts of power, and station B broadcasts at 15.000 MHz

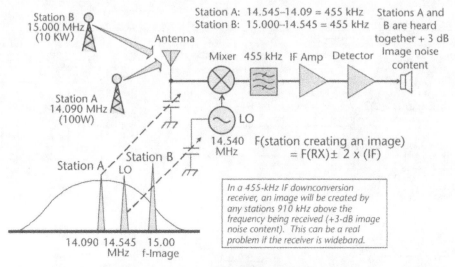

Figure 4.6 Image problems in a wideband general coverage receiver with a 455-kHz IF. Stations A and B are heard together plus a 3-dB image noise content.

with a power of 10 kilowatts. The relatively wide preselector is set at 14.090 MHz as shown. The VFO oscillates at 14.545 MHz. Station A combines with the VFO producing 455 kHz, and this signal is processed further in the IF chain and amplifier and is detected and heard in the speaker. Station B, at 15.000 MHz, falls within the passband of the preselector, although it is somewhat attenuated by it. The preselector is of the type discussed previously. It involves one or two tuned circuits, and station B is powerful enough to make up for the off-frequency rejection of the preselector as shown in our exaggerated case here. Station B at 15.000 MHz also mixes with the 14.545-MHz local oscillator producing the 455-kHz IF signal, which is processed along with the signal from station A and is heard at the same time, thus interfering with the real signal.

The image is twice the intermediate frequency above the received signal if the oscillator frequency is above the desired signal frequency or twice the intermediate frequency below the signal frequency if the oscillator is below the desired signal frequency. We can then say that in a typical 455-kHz receiver, images will be created by any stations 910-kHz above the frequency being received.

From a mathematical point of view, image is a third order mixer product. A product is not a harmonic but a sum or difference of n- and m-order harmonics of both the RF signal as well as the LO signal. An in-band product, together with the desired signal, one heard concurrently in the IF. Image is caused by a signal removed in frequency from the desired wanted signal by two times the IF frequency. If precautions are not taken, both signals are heard concurrently within the receiver's IF. Consequently, image is a direct function of the IF frequency choice. This relationship is shown in Figure 4.7.

Another way of defining image in a superheterodyne receiver is the contribution of unwanted energy in the IF from a possible RF emitter removed from the

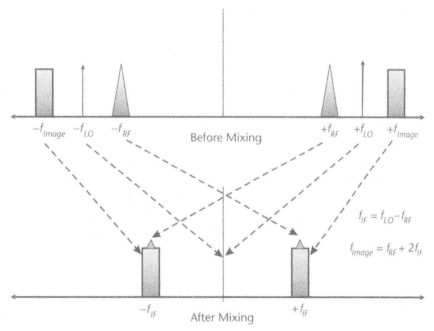

Figure 4.7 The mathematical explanation of image.

frequency being received by two times the IF frequency (+ or − depending on the superheterodyne scheme used). This energy can be another RF station and/or noise present at the image frequency, which contributes or degrades the receiver's noise figure or its minimum discernable signal (MDS) by 3 dB. Once in the IF, the image component cannot be removed by any means. Rejecting image has to be resolved before the first mixer or in the first mixer circuitry.

An RF receiver or transceiver system should be designed to reject image over its total RF input bandwidth. This is relatively simple in narrowband (NB) RF systems, but it can be very complex or sometimes impossible in wideband (WB) systems depending on the RF input bandwidth requirements and the IF choice.

The wider the RF front end bandwidth is, the more complex the image rejection mechanism will become (as we will see later in Chapter 14). If the IF is not properly chosen when compared with the RF input bandwidth, the image can be in-band and rejecting it can require the front end to become extremely complex. For instance, in a professional wideband microwave ECM receiver covering from 2–20 GHz, and using a typical down convert scheme with an IF at 450 MHz, the preselector can physically occupy up to half the receiver's physical volume in order to keep the image rejection consistent over the entire RF coverage. Such a receiver would use complex switched preselectors to guarantee equal image rejection over the entire RF coverage. This, in turn, complicates the design of the receiver and makes for a bulky package.

It can be seen from this explanation that using the 455-kHz IF from our example is limited to relatively narrow RF bandwidth applications such as in an AM receiver (see Figure 4.8), but when used for wideband applications, it imposes complicated preselection schemes.

An example of an ultrawideband HF receiver (4 octaves RF) with a 455-KHz IF was previously shown in Figure 4.6. A similar situation exists in the case of using a 200-kHz IF, which was historically utilized in AM radios before the 455-kHz IF was adopted.

If $F_{image} = F_{RF} +/- 2F_{IF}$, then the image of a superheterodyne AM radio with a 200-kHz IF will be in the AM band without a chance of being rejected by the preselector. This is shown in Figure 4.9.

To make the point of how complex wideband image reject mechanisms can be, the schematic diagram of a commercial general coverage HF receiver with an IF of 455 kHz is shown in Figure 4.10. Because of the wide RF coverage requirements and the low 455 kHz IF, the preselector has to be switched in several bands for equal image rejection throughout the entire RF range. Such a radio would require a very complex design as shown. This particular example requires a six-wafer switch times six poles per each wafer, ganged together in a complicated assembly. This cumbersome preselector and the switched local oscillator mechanism allows for only 30 dB of image rejection throughout the entire coverage.

Improving preselector selectivity would mean increasing the number of tuned circuits with additional complex mechanical switching.

Some attempts have been made to minimize the switching of multipole preselectors by using combinations of wide-range variable capacitors with magnetically or electromagnetically tuned ferrite coils. This property was further developed in a preselector that uses opposing polarity magnets with a toroidal ferrite coil sandwiched between them as shown in Figure 4.11.

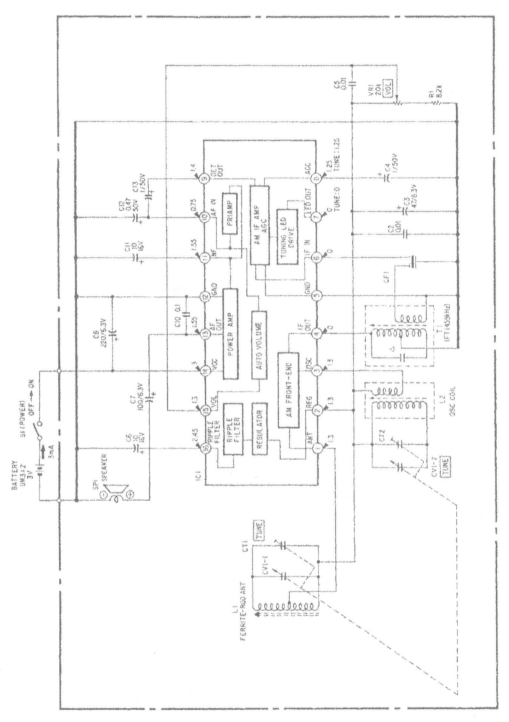

Figure 4.8 Modern implementation of a classic 455-kHz IF, AM receiver.

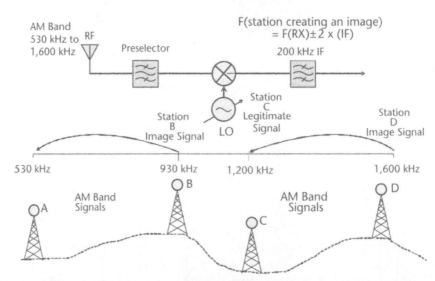

Figure 4.9 Impact of image in an AM superheterodyne receiver using a 200-kHz IF. If listening to station A at 530 kHz, station B at 930 kHz will be heard at the same time. Conversely, listening to station C at 1,200 kHz, station D at 1,600 kHz will be heard at the same time.

A set of four magnets polarized as shown are the stator of the device, while another set of four magnets are mounted in the rotor. The ferrite toroidal coil (part of the preselector filter) is mounted between the stator and rotor and forms the first pole of the preselector. The second pole is made of another toroidal ferrite coil which is identical with the first one and is mounted on the other side of the rotor, which is now common to both poles. A third set of magnets polarized the same way as the first part of the stator completes the arrangement.

The specially arranged magnetic field created by the rotor can then permeability tune both toroidal ferrite coils together, making for an extremely wide tuning nonswitching preselector. The number of poles could be increased providing for a sharper response. This preselector was designed to cover from 0.5 to 30 MHz by only switching the capacitors across once. Several problems occur with such an approach. Too much magnetic field will drive ferrites into saturation and the point of saturation is not always predictable since it depends on temperature. Also, unless the preselector is magnetically shielded, AC magnetic fields from power supply transformers and other sources can easily modulate the ferrite coils. In reality, this preselector would become impractical.

4.2 Upconverting—The Rule of 35%

The real fix for suppressing the image problem is to choose the IF at a much higher frequency. Mathematically, this proves to be preferred. If the frequency of a strong station creating an image equals $F_{RX} + 2F_{IF}$, and the range to be covered by the receiver is for example, 500 kHz to 30 MHz, a rule of thumb is that the intermediate frequency should be chosen above the highest frequency to be received (in our case 30 MHz) by about 35% (or higher), which in this case happens to be 41 MHz as shown in Figure 4.12. This is easy to establish mathematically if we use the formulas

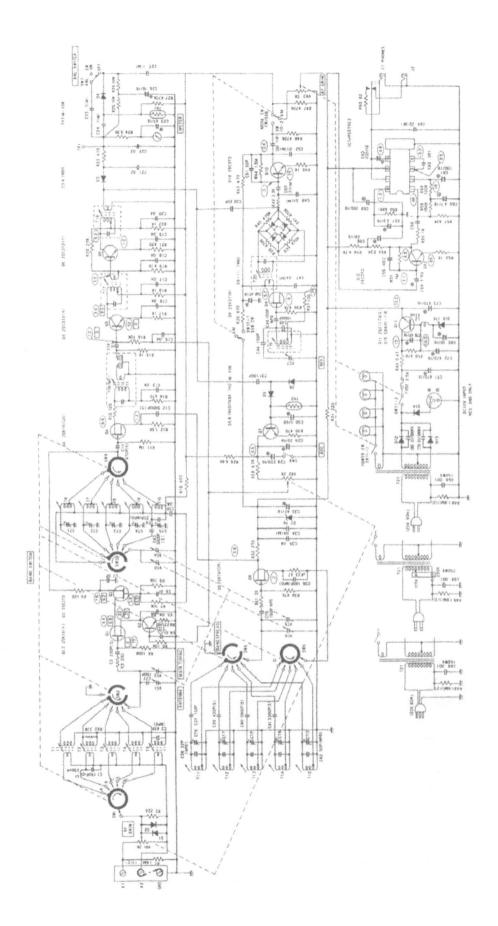

Figure 4.10 Example of a wideband (4 octave) preselected receiver with a 455-kHz IF. The mechanism uses six wafers by six poles for each wafer, ganged together with the switched LO bank. This cumbersome preselector and the local oscillator mechanism allow for only 30 dB of image rejection throughout the coverage.

(a)

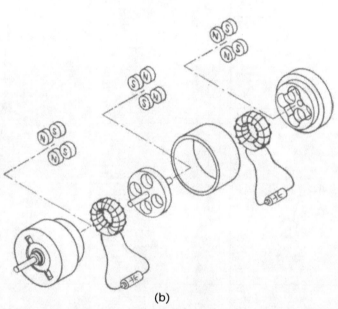

(b)

Figure 4.11 (a) Two-pole permeability-tuned preselector. (b) Arrangement of parts in the permeability-tuned preselector. Opposing magnets are positioned in front and behind the inductors in such a way that when turning the rotor, they impact the permeability of the ferrite material in the toroid coils, which in turn change inductance of the two coils over a wide range. A wideband preselector is therefore realized.

from Figure 4.7, but there was another element that impacted this design—selectivity. It wasn't always possible to produce relatively narrow IF filters compatible with the information bandwidth at the upconvert IF frequencies because of a phenomenon called percentage bandwidth. Thus, multiple conversion superheterodyne receivers were born, which used the upconvert approach with additional conversions and subsequent lower-frequency IFs, which allowed for narrower filters to be realized. We will discuss multiple conversion superheterodynes in Chapters 5 and 6. The upconvert superheterodyne receiver combined with additional downconverters is the preferred way of designing receivers today even at microwave frequencies.

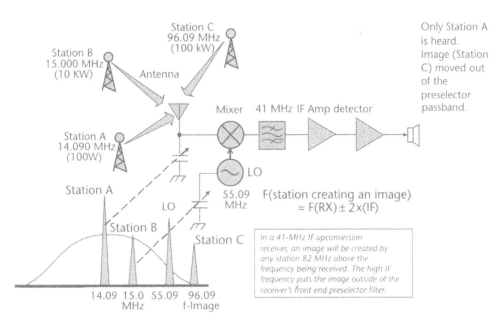

Figure 4.12   Image with a 41-MHz IF wideband. The upconversion receiver is 82 MHz above the received frequency (96.090 MHz) and can be easily filtered out at the preselector filter with considerably less hardware. The former image causing station at 15 MHz is no longer a problem with this arrangement. This upconvert receiver obeys the rule of 35%.

Later in Chapter 26, we will discuss a patented receiver scheme that covers 2 to 20 GHz in an upconvert system obeying the 35% rule with a 26-GHz first IF.

4.3   Selectivity and IF Filters

We have seen that the main advantage of the superheterodyne concept is to be able to process RF input signals through a common frequencies denominator, the IF. Selectivity at IF is achieved by using electrical filters implemented through using various techniques. Depending on what center frequency is chosen for the IF, such filters can take many forms and use various topologies.

In a single conversion downconvert superheterodyne receiver, the IF filter bandwidth is usually known as the ultimate bandwidth element as it reflects the needed requirements for the type of modulation received (e.g., modulated video, pulsed video, voice, music, or data). In a multimode receiver, various IF filters can be selected or cascaded to achieve the desirable ultimate information bandwidth for each and every modulation requirement. However, the image problem dictates higher IFs where narrowband ultimate rejection filters may not be possible. Thus, multiple conversion receivers result.

In multiple conversion receivers, the first IF bandwidth can be wider than the ultimate bandwidth. This is either by default (ultimate bandwidth filters cannot be achieved at the higher upconvert IF frequencies) or by design, where the later IF conversions can tune inside the first wide IF providing large steps of the RF frequency spectrum that can be processed instantaneously in bulk and in real time using Fourier transformers such as Bragg cells and parallel DSP signal processing. We will discuss these receivers in Chapter 25.

The realization of IF filters with proper ultimate information bandwidths depends on the IF frequency used, which in turn depends on the receiver's system design and the technologies dictated by it.

Many IF frequencies have been used in the past. First IFs of 900 MHz through 1.5 GHz have been used extensively in satellite receivers. Radio astronomy IFs can range from 700 MHz to 4 GHz featuring instantaneous bandwidths of 2 to 4 GHz at center frequencies of 8 to 12 GHz. These IFs have been implemented in receivers covering the RF spectrum at up to several hundred GHz.

A 4-GHz-wide IF centered at 26 GHz has been introduced in [2, 3]. Even wider first IF filters are being contemplated. An 8-GHz-wide first IF with a very low noise performance has been reported in radio astronomy receivers. It presents a substantial design and development case. The impact of the noise figure of a wideband approach on the overall receiver system performance has to be carefully analyzed before it can be implemented successfully.

We will discuss these types of receivers in much more detail later on in this book, especially in Chapter 25.

The discussion of IF filter realizations at various frequency choices calls for an understanding of an important concept: the concept of percentage bandwidth. We will now divert our discussion to this topic.

So far, we have determined that in a superheterodyne receiver, the process of mixing allows the incoming signals to generate new signal components, which are equal to the sum and the difference of the original frequencies. One of these products is designated as the intermediate frequency (IF) and is passed by a tuned circuit intended to reject the undesired products and reproduce only the original incoming information. The advantage of the superheterodyne approach is that the IF filter bandwidth and processing remain the same for any of the RF input signals, thus simplifying the receiver design.

The superiority of the superheterodyne receiver is evident. It evolved to where it is today and will remain with us for a long time despite rumors that total digital sampling receivers will replace it soon. Although digital conversion and DSP have come a long way at HF, VHF, UHF, and beyond, designing fully digital receivers at microwave frequencies will remain a stretch goal for some time. The superheterodyne approach will remain the choice of receiver designs beyond 2 GHz. This is because progress in A/D and D/A technology beyond this frequency has been limited. Going up in frequency in these devices usually limits the number of realizable bits, which in turn impacts dynamic range. We will discuss direct sampling receivers in Chapters 10 and 25. Let's now turn to the percentage bandwidth concept.

4.4 Defining Baseband and Broadband: The Concept of Percentage Bandwidth

The concept of percentage bandwidth plays an important role not only in IF filter design, but also in designing the right RF system and choosing the proper technologies for it. As we discussed, using higher first IF frequencies can present certain advantages from an information bandwidth handling point of view. In synthesizers, using well above the required frequency generators and dividing these signals down digitally can improve phase noise performance and, consequently, the signal-to-noise-ratio or dynamic range in superheterodyne receivers.

To better understand what can be done at the higher IF frequencies, we will first turn to the concept of baseband versus broadband and the definition of percentage bandwidth, which is used by RF system and filter designers to determine how much information can be packed in a certain IF frequency bandwidth and what filter approximation technique, technology, and topology can be used to implement it. We will begin with a simple concept of baseband.

In any communication system, baseband is how information starts and ends up. It has a finite limited bandwidth, which usually extends from zero hertz to several megahertz and cannot normally be transmitted through space because of physics properties and regulatory limitations.

Consider a 100-kHz hypothetical computer baseband signal as shown in Figure 4.13. It can be seen from this example that such a signal, when transmitted or received over a twisted copper wire (such as in a telephone link, LAN, or DSL link), occupies 100% of the medium's bandwidth at any given time. Only half-duplex baseband signals can be transmitted or received at a given time over such a medium. Consequently, the percentage bandwidth of such a system is said to be 100% because it occupies 100% of the available bandwidth.

The percentage bandwidth is defined as follows:

$$\%BW = \frac{2[f_H - f_L]}{f_H + f_L} \times 100 \tag{4.1}$$

or

$$\%BW = \frac{bw_{total}}{c_f} \times 100 \tag{4.2}$$

where f_H and f_L are the upper and lower edges of the signal, respectively.

However, modulating the 100-kHz baseband computer information over a 100-MHz carrier produces an RF signal, which contains the original baseband information but represents only 0.1% of the 100-MHz frequency bandwidth. This translates into the ability to pack more than one baseband at this frequency leaving room for additional signals to be packed side by side when using a superheterodyne approach in a process called frequency division multiplexing (FDM). Several 100-kHz basebands can be transmitted and received at the same time over a medium such as a coaxial cable network or through the air. This is the concept

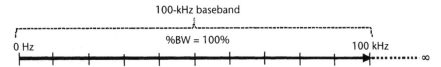

Figure 4.13 Baseband occupies 100% of the available bandwidth. Only one transmission can be made over the medium allowing only half-duplex communications.

of broadband communications. This process is also called percentage bandwidth modulation. This concept is shown graphically in Figure 4.14.

It can be seen from this discussion that at 100 MHz, there is room for at least four (and many more) 100-kHz-wide baseband signals to be modulated on a carrier for a total of 400 kHz, which in turn can allow for the four computers in our example to communicate full duplex wirelessly over a simple local area network (LAN). The percentage bandwidth at 100 MHz for the entire necessary bandwidth would be 0.4%.

Figure 4.15 shows the hypothetical network implementation of such a system. Our case uses no guard bands to simplify the example. In a practical implementation, electrical filters would be used at the center RF frequencies to protect from adjacent channel crosstalk.

It can be seen from this discussion that going to higher frequencies allows even more basebands to be modulated on the RF frequencies. Extremely small percentage bandwidths are possible at light frequencies, thus increasing communication networking capacity tremendously. This is shown in Figure 4.16. The IF filters at light frequencies would take the form of colored filters. This is how extremely large bandwidths can be multiplexed over broadband fiber-optic networks. These systems can be either wavelength-division multiplexed or frequency-division multiplexed as shown. Our example uses laser diodes, which are temperature tuned to emit at slightly different light frequencies. A baseband signal modulates each diode and the modulated frequencies are transmitted simultaneously. At the receiver end the composite light is mixed with that from another laser, which functions as a local oscillator (LO), producing a distinct color for each channel. These frequencies are converted to electrical signals by a photodiode and are electronically sorted by filters. Frequency-division multiplexing allows channels to be packed

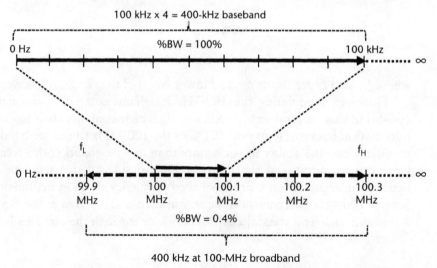

Figure 4.14 The process of modulating baseband over broadband (mixing) allows for four baseband signals (or more), each 100 kHz wide, to communicate in a full duplex mode. The 100% percentage bandwidth of the baseband information (top) translates into a 0.1% percentage bandwidth at 100 MHz and can allow for many signals to communicate in a full-duplex system at this frequency. The total percentage bandwidth of the four communicating computers in the example is 0.4%, which is much better than the 100% percentage bandwidth at baseband. This process is also called frequency division multiplexing (FDM).

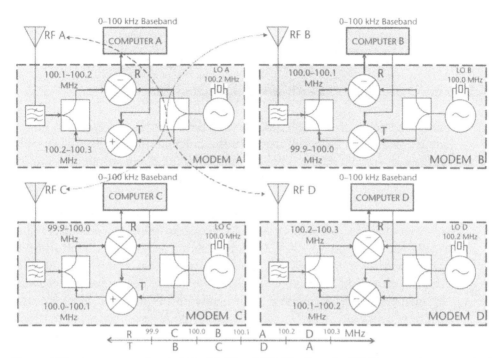

Figure 4.15 A percentage bandwidth of 0.4% at 100 MHz allows for FDM between four networked computers. Full duplex communications between A and D, and B and C, can be achieved using baseband data modulated over RF at 100 MHz. Each computer has a total baseband receive and transmit bandwidth of 200 kHz for full-duplex communications. The baseband data can be modulated in various ways over RF including FSK or spread spectrum. No guard bands have been considered in this example for simplicity.

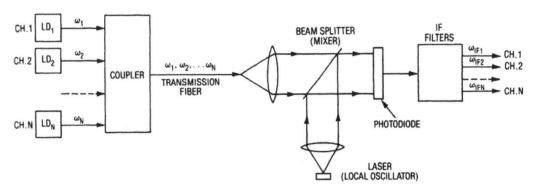

Figure 4.16 Extremely small percentage bandwidths are possible at light frequencies. Laser diodes are temperature tuned to emit at slightly different frequencies. A baseband signal modulates each diode and the modulated frequencies are transmitted simultaneously. At the receiver, the composite light information is mixed with the light from a laser used as an LO. Distinct IFs are produced for each channel. These frequencies are then converted to electrical signals by a photodiode and are electronically sorted by filters. Large data bandwidths can be frequency-division multiplexed over broadband fiber-optic networks.

more closely than wavelength-division multiplexing because electrical separation at radio frequencies is more selective than optical separation, which suffers from optical temperature variations due to the very high frequencies involved. Closer spacing of signals allows for more channels to be transmitted over optical fibers, which exhibit the lowest loss.

4.5 Percentage Bandwidth and Filter Design

The percentage bandwidth concept discussed above also applies to the design of analog and digital filters. It uses a simple calculation that gives a normalized measure of how realizable a filter can be. As pointed out in our previous examples, when frequency goes up, the absolute bandwidth may also increase and its percentage bandwidth may change.

Filter designers define percentage bandwidth by using (4.2) where the bandwidth of the filter at the 3-dB points is divided by the center frequency times 100%.

It can be seen from these definitions that if we design a bandpass filter with a center frequency of say, 10 GHz and a 3-dB bandwidth of 1 GHz, the percentage bandwidth will be 10%. Conversely, if we design a bandpass filter with a center frequency of 100 GHz with a 3-dB bandwidth of 10 GHz, the percentage bandwidth will also be 10%. The major point of these examples being that we can pack more information at the higher center frequency. Also, the percentage bandwidth remains constant as the bandwidth capacity increases drastically with frequency.

In filter design, percentage bandwidth has to be greater than 1/10 of the Q of the elements used at the particular center frequency for a filter to be realizable. Filter designers today use normalized tables to determine how narrow or wide a filter can be produced at various center frequencies depending on these rules. The Q of filter elements varies with frequency and impacts the technology used in the design of the filter. In order to have a filter response as close as possible to the theoretical curves, the filter element Q should be greater than ten times the Q of the filter, which is the reciprocal of the percentage bandwidth. For instance, a 10% bandwidth has a filter Q of 10 (1/0.1). Therefore the element Q should be ≥ 100 in order to approximate the theoretical curves.

We can see from this discussion how the shape factor is degraded as the Q ratio gets smaller. This is shown in the Figure 4.17, which is an example of a fifth-order Butterworth filter design table.

4.6 The Seven-Layer ISO-OSI Model

A good knowledge of the ISO-OSI model will help in understanding the implications of percentage bandwidth in hardware implementation.

This book is not intended to discuss in-depth communications networking. However, a good knowledge of protocols will help designers understand where receiver and filter technologies fit in the big picture. Although not directly connected with receivers, we will briefly discuss the ISO-OSI communications model.

In 1978, the International Organization for Standardization (ISO) introduced a high-level methodology of classifying the many aspects of end-to-end communications. This is the seven layer Open System Interconnection (OSI) model, a concept for developing communications between heterogeneous devices. This initially applied only to data communications as used in local area networks (LAN), but has

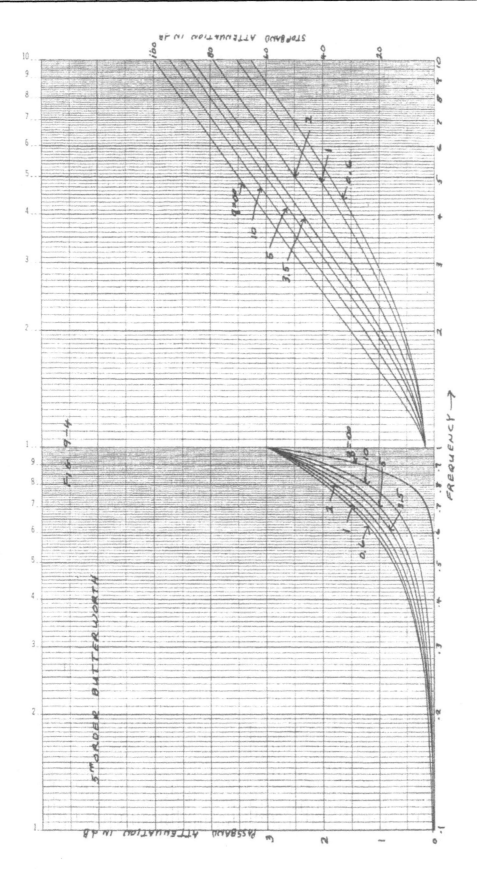

Figure 4.17 Modeling a fifth-order Butterworth filter design table. In order to have a filter response as close as possible to the theoretical curves, the filter element Q should be greater than 10 times the Q of the filter, which is the reciprocal of the percentage bandwidth. The shape factor is degraded as the Q ratio gets smaller. (*From:* [4]. Courtesy of Lee R. Watkins.)

been recently adopted by the RF and wireless industry in order to define the many facets of communications networking of radios or RF modems. The model has since been adopted by the Institute of Electrical and Electronics Engineers (IEEE) and is reflected in the IEEE-802 and other standards.

The ISO-OSI model is shown in Figure 4.18. This standard ensures compatibility between networked equipment and components manufactured by many different companies, with only a minimum amount of work required of the user.

The idea is that in order to provide reliable end-to-end communication between two modems or radios, information has to flow from the top left application layer of the A terminal, down through the remaining layers, to the physical layer, and then over the network medium (which can include the propagation medium and repeaters), into the right column through the physical, data link, and network layers and all the way up, again to the application layer of the B terminal. For full duplex operation, the process is reversed from the top of the B terminal and up to the top of the A terminal.

It can be seen that the A and B layers are parallel and growing in sophistication as they go up with the top 5, 6, and 7 layers at the software level of the communications architecture. The baseband and broadband concepts discussed earlier, including the hardware associated with them and their consequential design, fit on the primitive layers 1, 2, 3 of the ISO-OSI model.

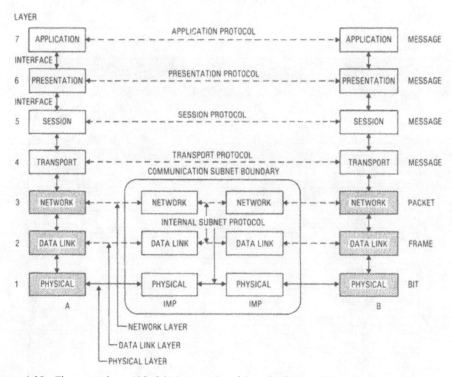

Figure 4.18 The seven-layer ISO-OSI (International Standardization Organization—Open Systems Interconnection) model. All receiver hardware discussed in this book fits on the bottom three layers known as the "primitive" layers.

4.7 IF Filters, an Introduction—History of Filter Design

A key area in the progress of radio receivers has been the invention of the electrical filter. The first lumped element electrical filter was developed in 1915 by George Campbell in the United States and Karl Wagner in Germany. It used the transmission line theory. A better filter design procedure was developed by Otto Zobel of Bell Telephone Laboratories in 1921. In 1923 he published his image parameter filter design procedure in [5]. His methodology was based on the m-derived matching sections concept, which addressed matching impedance to fixed resistive sources and loads for the first time. Zobel's design procedure lasted for the next 30 years, until the early 1950s. His work was responsible for many filter design terms such as *cutoff frequency*, *impedance*, and *m-derived sections*.

An improved design methodology known as the modern filter design or the insertion-loss design procedure was created during the 1940s and early 1950s. This methodology synthesized networks to produce a desired filter response and gradually made Zobel's image parameter procedure obsolete. It was found to be more versatile than the Zobel method, especially because it allowed the production of filter networks with many different and desired response characteristics using a minimum of components. This, in turn resulted in the development of new filter approximations by the Russian mathematician Pafnuti Chebyshev, the German mathematician Wilhelm Cauer, and the British engineer Stephen Butterworth.

During the 1950s, laborious and complex mathematics prevented the progress of filter design, but with the development of new digital computers, it became possible to calculate and publish tables of normalized designs for the more popular approximations. In 1958 the publication of normalized Cauer parameter (also known as the elliptic function) designs [6] gave professional filter designers the tools to more quickly design these types of networks.

In the meantime, Telefunken in Germany published R. Saal's normalized tables of Chebyshev and Cauer designs with an explanation of the design procedure in German. This was later translated in English and it became the most authoritative reference source for the professional filter designer until the 1960s. In 1963, P. Geffe published the classic book *Simplified Modern Filter Design*, also known as the "little blue book" [7].

Since then, several other books on the subject of filter design were published, with Anatoli Sverev's 1967 *Handbook of Filter Synthesis* [8] being the most widely used to this day. Butterworth, Chebyshev, and Cauer-elliptic approximations, as well as several other filter approximations, are still with us today. We will discuss next the various analog IF filter implementations and approximations. Digital signal processing (DSP) and digital filters will be discussed in Chapter 24. They also use the same approximations as the analog filters.

Certain rules apply to choosing IF filter bandwidths in receivers. Ideally, IF filters in single conversion superheterodyne receivers should reproduce with high fidelity the information bandwidth to be received. This kind of bandwidth is known as the ultimate bandwidth. It would provide the best possible signal-to-noise ratio and facilitate the highest dynamic range and highest probability of intercept. However, this is not always possible as we discussed earlier. As we increase the IF frequency to avoid image, narrow filters (see Section 4.2) having the same bandwidth as the information bandwidth may become unrealizable. This is because of the percentage bandwidth phenomenon and its relationship with filter realization at these frequencies. The result is then multiple conversion receivers with second and third IFs at lower frequencies where the percentage bandwidth can allow for the narrow ultimate bandwidth to be achieved.

In addition, perfectly square response filters with ideal characteristics would be desirable. However, such filters do not exist except maybe in mathematical equations. Practical IF filters are somewhat of a compromise. Sometimes they have to be relatively wider than the information bandwidth in order to allow for the least distortion.

IF filter bandwidth generally depends on real life design factors such as ripple, phase and group delay, linearity, and shape factor requirements. In simple applications such as in superheterodyne AM radios, the bandwidth of IF filters closely follows the information bandwidth. In more critical applications such as those used in various phase modulation RF data communications or in pulsed radar, wider IF bandwidths are the norm in order to guarantee fidelity of data delivery at the cost of signal-to-noise degradation.

Consequently, there are no concrete rules about the choice of IF bandwidth and the filter approximation or technology used. The designer has to fully understand the application and the system design and choose accordingly.

Today, most IF receivers (term used for IF chains) are sampled digitally. Digital sampling of bandpass or baseband IF filters follows the Nyquist theory. This theory states that a signal must be sampled at least twice as fast as the highest frequency component of the IF signal to be received. This means that the sampling rate of an IF in a digital receiver must be at least twice the frequency of the desired signal at the high point of the IF bandwidth.

In typical radio communications receivers, the IF bandwidth varies between about 15 kHz for frequency modulation (FM), to less than 500 Hz for keyed continuous wave (KCW) signals. For convenience, some of these modulation types and IF bandwidths are shown in Table 4.1. Shown in Table 4.2 are additional RF data communications modulation methods, which may require wider IF bandwidths.

It can be seen from these tables that IF filter design with the kind of selectivity and percentage bandwidth required was much easier to accomplish at the lower frequencies such as the 455 kHz. Until the 1950s, we were simply not capable of providing narrow bandwidth IF filters at frequencies much higher than 455 kHz. Technology stumbled along for about 30 years with the 455-kHz scheme, which is still being used in modern consumer and car radio receivers today.

Table 4.1 Some IF Selectivity Requirements for Simple Modes of Radio Communication Using 455-kHz Center Frequency

Modulation Type	IF Bandwidth
Amplitude modulation (AM)	6–8 kHz
Frequency modulation (FM)	≤ 150 kHz
Single sideband (SSB)	2.4 kHz
Independent sideband (ISB)	6.1 kHz
Frequency-shift keying (FSK)	≤ 1 kHz
Keyed continuous wave (KCW)	≤ 0.5 kHz

Table **4.2** Several Other Forms of Modulation Used in Radio Communications Can Require Different IF Bandwidths (Which Imply Higher Frequencies Due to the Percentage Bandwidth)

ASK (OOK)	KCW	WFM	FSK	AFSK	2–32 Tone
BPSK	DBPSK	QPSK	$\pi/4$	DQPSK	M-ary FSK
16QAM	64QAM	GMSK	GFSK	OQPSK	OFDM

Before going any further, it should be mentioned that there is one advantage to the 455-kHz IF receiver: it is relatively simple to implement. However, the image problem caused by the low-frequency IF, especially when used in multioctave HF receivers, remains a problem at present.

We discussed that if going up in frequency in our IFs, the percentage bandwidth versus the frequency of operation is increasingly smaller allowing for wider IF bandwidths to be achieved at much higher IF frequencies. This, in turn, allows for much more data to be processed in bulk through new and faster modulation schemes.

With new filter technologies available, fast scanning bulk microwave receivers, which can rapidly step through ultrawide RF bandwidths in a minimum time are now possible. We will discuss such receivers in detail in Chapter 25.

The trend of IF filter design started with 455-kHz high-Q transformers such as those shown in Figure 4.19(a), and gradually progressed from mechanically coupled coils as shown in Figure 4.20, to mechanical, ceramic, quartz crystal, surface-acoustic-wave (SAW) and bulk-acoustic-wave (BOW) resonators, and then to lumped elements and microwave filters.

At 455 kHz, simple IF transformers have been previously used in conjunction with multistage IF amplifiers to provide selectivity for typical low-cost receiver applications. Their tuning is usually staggered as shown in Figure 4.19(b, c). This technique is still used today in ordinary AM receivers or in some inexpensive radio-controlled RF systems using amplitude-shift keying (ASK) modulation.

These transformers use ferrite materials for coupling and tuning. By mechanically varying the position of the ferrite cores in the IF transformers, different degrees of coupling and, therefore, selectivity are achieved.

The major problem associated with this type of IF filters is their vulnerability to temperature changes and vibration, along with variability of the ferrite material over long periods of time. In addition, the small Qs of the ferrite materials degrade the shape factors of such filters.

Another early method of providing selectivity in IF transformers has been through physically varying the coupling of the primary winding of an IF transformer against the secondary. By changing the distance between the two windings, several degrees of selectivity can be obtained without the use of the unstable ferrite materials. This works best at lower IF frequencies. To illustrate this, we took an IF transformer out of a receiver that uses this approach. Figure 4.20(a) shows how it works. Figure 4.20(b) shows a communications receiver of the past, which uses IF transformer coupling to achieve ultimate selectivity.

A characteristic of such coupled circuits was that the bandwidth or selectivity was largely dependent on the degree of coupling, being greatest with light coupling.

(a)

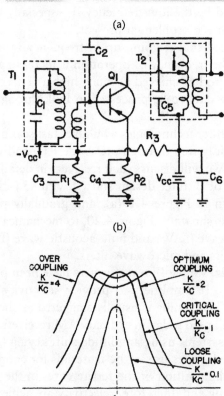

(b)

(c)

Figure 4.19 (a) Modern high-Q RF transformers used in inexpensive commercial AM receivers. (b) Implementation of a staggered-tuned IF stage using tunable ferrite transformers such as those shown in (a). (c) Example of an IF transformer, which is part of a staggered-tuned IF chain using ferrite IF transformers. Achieving desired bandwidth is critical to coupling of the staggered transformers. Over time and temperature changes, this tuning degrades and needs to be redone.

This point became relatively unstable, however. Thus, by switching in different coupling capacitors or precisely varying the distance between the windings, the overall selectivity could be controlled accordingly. As the coupling was increased, a dip in the passband tended to appeared as shown in Figure 4.19(c), thereby changing the passband. Along with this, the response curve spread out from both sides of the initial resonant frequency as can be seen in the Figure 4.19. This problem was usually compensated for by switching in various resonant capacitors for the

(a)

Figure 4.20   (a) A low IF (85 kHz) transformer showing how coupling is achieved by changing the distance between the primary and the secondary transformers. (b) Low IF transformer coupling was used in the Drake 2-C receiver to obtain various degrees of selectivity.

different selectivity positions, in order for the low frequency side of the response curve to remain at a fixed point. A good example of a receiver that used complex L/C mechanical coupling to achieve selectivity was the Collins R-392 shown in Figure 4.21(a).

The R-392 achieved continuous variable bandwidth in the second IF at 455 kHz by using some of the most complex electromechanical implementation ever realized in a receiver to this date. While still considered to be a good receiver even by today's standards, it remains a symbol of the past. Communications receivers of the past have taken full advantage of this method of IF selectivity selection to adapt to various modulation bandwidth requirements. Shown in Figure 4.21(b) are several nostalgic communications receivers of the past. Some of them use the coupling method of controlling the IF bandwidth mentioned here while others use quartz crystal and lumped element IF filters.

4.8   Elements of Modern Filter Design

Modern filter designers use specific topologies for the realization of RF or IF filters. Topology refers strictly to the physical and electrical structure or the order of a network, especially to the manner in which the nodes and filter circuits are physically interconnected to achieve certain desirable results.

The type of filter design used in IF circuits highly depends on the required response (the information bandwidth), which in turn is dictated by the overall receiver system requirements and the information fidelity required. Fidelity is defined as how faithful to the original RF signal the IF chain has to deliver the information, not only from an amplitude point of view, but also from a phase and group delay point of view. Tight filters approaching the ideal square response cannot usually deliver an equal group delay response. As mentioned before, this can be important in certain data communications or in radar applications where equal phase delay

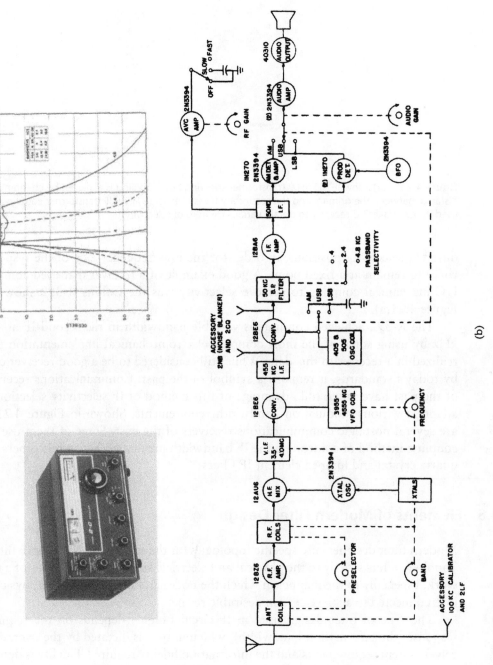

Figure 4.20 (continued)

(a)

Figure 4.21   (a) Several degrees of selectivity were achieved in the Collins R-392 receiver by switching in different mechanically positioned coils in the IF transformers. (b) Several nostalgic communications receivers of the past. Some of them use the coupling method of controlling the IF bandwidth mentioned here while others use quartz crystal and lumped element IF filters.

and group delay are important over the IF filter's passband. In such design cases, the actual bandwidth is purposely enlarged by as much as 30% to 40% to allow near equal time delay response to be achieved over the passband of the IF filter. Enlarging the passband, in turn, reduces the signal-to-noise (SNR) ratio, which has to be considered in the receiver's system design.

Furthermore the IF filter's complexity also depends on the required *shape factor* or the out of the passband attenuation. The bandwidth required can also take into consideration the possible tunability of a second IF (in multiple conversion superheterodyne receivers) inside the first IF with its synthesis implications, and the rejection of in-band and out of band intermodulation products.

We will discuss next the most important elements of modern IF filter design. The materials also apply to other filters as used in a radio receiver, synthesizer, or exciter design. Digital filters DSP implementations will be discussed separately in Chapter 24.

4.9   Passband, Bandwidth, and Stopband

A filter can be designed as a bandpass, bandstop, highpass, or lowpass. Depending on its requirements and percentage bandwidth it can be implemented in any of the approximations, topologies, and/or technologies presented in this chapter.

In a bandpass filter, the *passband* is the frequency range over which the filter is designed to operate. It is customary to define the band edges of such a filter as the frequencies at which the power spectral density is 3 dB below what it is at the

Figure 4.21 (continued)

center of the filter spectrum. This is especially true in narrowband bandpass IFs because their spectrum is generally symmetrical from the center frequency and because the spectral region between the 3-dB points contains approximately 95% of the total spectral energy.

In a bandpass filter such as used in IFs, the passband is expressed by a set of two numbers indicating the range of passed frequencies bounded by the −3-dB attenuation points. This is known as the bandwidth of the filter measured at the −3-dB points.

In a lowpass or a highpass filter, bandwidth also relates to the frequency range of the 3-dB attenuation points; in the lowpass design, the passband extends from 0 Hz to the filter's −3-dB point. In the highpass case, passband has no practical meaning because the bandwidth is infinite, although not realizable. In all cases, this point is known as the −3-dB corner frequency.

Stopband is the frequency range on either side of a bandpass filter, which rejects a range of frequencies attenuated by more than a specified minimum, usually by 60 dB. In a lowpass or highpass filter, the stopband can be on either the high frequency side or the low-frequency side of the filter.

4.10 Shape Factor

Shape factor (SF) is a very important parameter in bandpass filter design. It is defined as the ratio of a filter's 60-dB bandwidth and its 3-dB bandwidth. This is shown in Figure 4.22.

SF relates mainly to the rate at which the attenuation slope decreases from the −3-dB points. A steeper slope means increased filter complexity. For instance, in simple second-order filters, the attenuation slope decreases at the rate of 6 dB per octave from the −3-dB points. This number improves with additional sections, but there is a limit to how many sections can be practically used. The shape factor can be very important when we need high selectivity against adjacent signals. Steeper slopes will be required for better selectivity. This means a higher-order filter design.

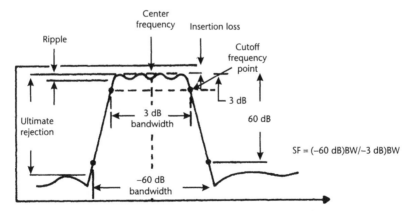

Figure 4.22 Determining shape factor in electrical, mechanical, or piezoelectrical filters. SF is a method of measuring filter performance represented by the ratio between filter bandwidth at −60 dB and −3 dB. The lower the SF number is, the better the filter is.

The intermodulation system design analysis for a receiver can also impact the specification of an IF filter shape factor. It can sometimes reveal the existence of an undesirable IF product located just outside of the filter's passband, which we cannot neglect. This is usually a higher order (e.g., seventh order or higher) product, which will have to be rejected by a certain amount depending on its order and predicted amplitude.

In this case, the attenuation slopes of the IF filter need to be steeper by increasing the number of nodes in the network, or by designing notches that correspond to the specific frequency points in the filter's out of band response in order to alleviate the problem. These are also known as diplexers.

Products in the IF sidebands are common and can change the requirements and especially the order of the IF filter and its shape factor until the undesirable product is out of the picture and does not affect signals in the passband. The new filter's complexity can then impact other design elements such as ripple and insertion loss, phase and group delay, and practical realization. We will discuss an example of how to determine shape factor requirements for a receiver IF filter through the intermodulation system analysis example in Chapter 12.

We have seen that a filter shape factor is defined as the ratio of a filter's 60-dB bandwidth and its 3-dB bandwidth. For instance, a bandpass IF filter with a −3-dB bandwidth of 2 kHz and a 60-dB bandwidth of 4 kHz has a shape factor of 2, and a filter with a 3-dB bandwidth of 2 kHz and a 60-dB bandwidth of 3.2 kHz has a shape factor of 1.6. Shape factors of 1.02 have been realized in receiver IFs using as many as 12 nodes of high quality natural quartz crystals. These filters approach the ideal approximation of a theoretical brick-wall filter.

In conclusion, the smaller the shape factor number is, the more complex the filter becomes and the steeper the filter skirts are. However, there are practical limitations to how complex a filter can be. These limitations depend on such factors as insertion loss, port-to-port isolation, phase and group delay linearity, and so fourth. The receiver designer has a tough job of balancing all these elements in view of all the factors and against the cost.

4.11   Center Frequency and Nominal Center Frequency

We have previously discussed that in a bandpass RF or IF filter, the center frequency is a specific point in the filter's passband, geometrically located half way from the filter's response frequencies at the −3-dB points.

The nominal center frequency of a bandpass filter is the center frequency point used as a reference point for specifying relative levels of attenuation throughout the passband. The nominal center frequency F_o denotes the actual center frequency of a bandpass filter and is described by the following equation:

$$F_o = (f_l \times f_u)^{\frac{1}{2}} \tag{4.3}$$

where f_l and f_u are lower and upper passband limits at the filter's −3-dB points.

The value of F_o will usually be specified as a function of operating temperature range and aging specifications, depending on the filter technology used.

4.12   Attenuation and Insertion Loss

Attenuation in a bandpass filter is expressed by its insertion loss as measured in the passband. It is the power loss expressed in decibels incurred by signals located in the 3-dB passband of the filter.

Absolute attenuation of a filter may be found using a calibrated RF generator and a spectrum analyzer. It is the difference in decibels between a power reading of a direct connection of the source to the matched load and the power reading with the filter inserted in the circuit. The generator is set to deliver the same RF power output in both cases.

Relative attenuation is the power loss expressed in decibels measured relative to a fixed frequency point at the point of minimum loss in the 3-dB passband. Guaranteed attenuation is the power loss expressed in decibels at a favorable fixed frequency point in the −3-dB passband.

Insertion loss can also be defined mathematically as the logarithmic ratio between the power delivered to a load without the filter inserted in the circuit, against the power delivered to the same load with the filter inserted in the circuit. The power insertion loss equation is:

$$IL_{(dB)} = 10\log(P_{L1}/P_{L2}) \tag{4.4}$$

where P_{L1} is the power delivered to the load without the filter inserted and P_{L2} is the power delivered to the load with the filter inserted into the circuit. The voltage insertion loss ratio equation is:

$$IL_{(dB)} = 20\log(V_{L1}/V_{L2}) \tag{4.5}$$

This later equation allows insertion loss to be measured with a simple oscilloscope in terms of voltage.

4.13   Ultimate Rejection

Ultimate rejection is defined as the attenuation of frequencies outside the stopband specification of a filter. Stopband is the frequency range on either side of a bandpass filter, which rejects a range of frequencies attenuated by more than a specified minimum, usually by 60 dB. In a lowpass or highpass filter the stopband can be on either the high frequency side or the low frequency side of the filter. Ultimate rejection is determined by two major factors:

1. The source and load ports lack of isolation causing leakage;
2. Undesired responses outside of the filter's stopband response caused by spurious problems in the piezoelectric filter elements.

Ultimate rejection pertains mainly to the electrical isolation between the source and load ports of the filter as impacted by the layout and construction of the filter. In addition, ground loops may develop if not using separate grounds for the source and load transducers.

All filters have some leakage from the input to the output, both across the filter as a whole, and across the individual poles. In general, each pole normally contributes a –6 dB per octave roll-off to the filter's stopband specification. However, if precautions are not taken or the physical distance between the filter ports is too small (a problem also encountered in high gain IF amplifiers), there can be more leakage at many frequencies outside of the filter's 60-dB stopband. This phenomenon is called "blow by." It is because of this that a filter can have a "spoiled" response outside of its passband response. A bad filter approximates the frequency response characteristic of a piece of wire (it passes all frequencies).

In addition, in quartz crystal filters, there can be many resonant peaks and products outside the response of the quartz elements, far out from its −60-dB stopband. This can combine with the isolation problem explained above to the point that a filter is not a filter anymore because of returns sometimes equal to the passband response. This, in turn can cause problems for the dynamic range of a receiver.

The ultimate rejection of a "practical" filter can also be influenced by:

1. Coupling between every resonator;
2. Residual resistance across the filter elements allowing some energy to pass when not desired;
3. Self-resonance of elements when the frequency is far from the passband. The elements are no longer proper inductors or capacitors. They go into self-resonance and other strange behaviors.

Ultimate rejection of a filter should be specified from the −60-dB stopband point out to a defined frequency point that will not create products or cause intermodulation distortion in the entire RF system. If this is not possible, the ultimate rejection specification should be done in blocks of frequencies for desired ultimate rejection within the certain regions of interest within the spectrum involved. We will discuss frequency planning and designing a system for the least intermodulation distortion in Chapter 12.

In a good receiver intended for SSB or data communications, an ultimate rejection of 120 dB would be recommended since human hearing has a dynamic range of over 100 dB. Unfortunately, this extreme level of rejection is not practical even in digital filters at low baseband audio frequencies, which can usually achieve about 96 dB of digital dynamic range. We will discuss digital filters later in Chapter 24.

Inexpensive IF filters have an ultimate rejection of approximately −40 dB. Well-designed IF filters will exhibit an ultimate rejection of around 80 dB. In some specialized microwave applications, filters with 100-dB ultimate rejection have been realized by using extreme packaging and layout techniques.

4.14 Ripple and Passband Ripple

Ripple in the passband of an RF or IF filter is an important design characteristic. It has a paramount impact on the amplitude, phase, and group delays of the filter, which in turn ensures proper delivery of data through a radio receiver. Ripple in a filter is defined as the difference between the maximum and minimum attenuations within a given passband. Ripple is measured in dB. For instance, if a filter can be designed with steeper skirts, a certain amount of ripple is allowed. We will discuss this in more detail in the automatically switched half-octave front end preselector in Chapter 14. Ideal Chebychev and Cauer-Elliptic functions exhibit equal ripple characteristics, which means that amplitude differences in the passband are equal throughout. Butterworth, Gaussian, and Bessel approximations do not exhibit significant ripple.

4.15 Spurious Response

Unlike lumped elements filters, quartz-based crystal filters can exhibit unpredictable, harmonically related overtones above the desired resonance points. Consequently, almost all crystal filters can have undesirable amplitude and phase irregularities. The spurious responses are generally narrow spikes outside the passband of the filter. Sometimes, in wider bandwidth filters a spurious response can actually occur in the filter passband, affecting its ripple and insertion loss. In some extreme cases, spurs can spoil or reduce the ultimate rejection of an IF filter. In ladder quartz filters, spurs appearing just outside the stopband can actually be cancelled out with proper network-phasing techniques. In these filters, the spurs are generally hard to predict due to the fact that the crystal spurs are a function of crystal lattice irregularities; since the crystals are all different quality and in series, the spurs almost never occur at the same exact frequency from part to part.

In monolithic crystal filters (MCF) experience shows that spurious responses are always above bandpass, and never below. Astute designers will always use two monolithic filter blocks with slightly different bandwidths (in a tandem monolithic configuration) rather than a single higher order assembly, to ensure that the spurious responses of the individual filters do not coincide.

4.16 Linearity

While lumped elements filters can be relatively stable, IF filters manufactured with quartz resonators can be subject to intermodulation distortion depending on the signal levels applied to them. This is particularly true in high-gain IF amplifier sections where IF filters are used or in first conversion IF roofing filters (de facto expression usually referring to upconvert first IF filters intended to limit the information bandwidth to a manageable bandwidth) to filters that are exposed to a wide range of signal levels. We will discuss a high dynamic range IF amplifier using cascaded quartz filters in Chapter 17.

4.17 Intermodulation Distortion (IMD) in IF Filters

Intermodulation distortion occurs not only in active nonlinear devices, but also in piezoelectric quartz-based devices. This happens when such filters are exposed to higher levels of RF energy at several frequencies at once.

These signals can be located in the passband or in the stopband of the filter. As a result, the filter acts in a nonlinear manner causing multiples of harmonics to sum or subtract (mix) according to standard nonlinear product behavior laws such as found in mixers or in linear amplifiers, which behave nonlinearly.

The resulting products (not harmonics but products of harmonics) are called intermodulation products, and can actually show up in the passband of the filter as well as in its stopband.

They are mainly of a third-order and a fifth-order nature, which exhibit some of the highest offending levels. Higher-order products can also be involved.

Out-of-band intermodulation distortion can occur when two relatively high RF signals (typically −20 to −30 dBm in amplitude) in the filter's stopband produce third order and/or fifth order products right in the middle of the filter's passband. This problem is prevalent in superheterodyne receivers when multiple RF adjacent channels outside the system's IF passband are active simultaneously.

The prediction of intermodulation distortion contributions of quartz-based filters to the receiver dynamic range is not fully understood. These predictions are empirical and depend mostly on the quality of the quartz, particularly on its impurities, its surface defects, its resonator manufacturing process, and the nonlinear elastic properties of a particular batch of materials.

This problem is equally important in the design and manufacturing of very low phase-noise quartz crystal oscillators as we will discuss in Chapter 16.

Filter and crystal oscillator designers are always looking for high quality quartz crystals. Most high quality natural quartz crystals are imported from Brazil. Artificially cultured quartz crystals can now be grown in the laboratory but do not exhibit as high of a Q as some of the natural quartz.

4.18   Power Handling Capability

Power handling of a filter is specified as the maximum input power allowable at the input of a filter over a certain period of time. This parameter is usually closely related to the in-band intermodulation distortion specification and applies mainly to quartz networks. Within a certain filter bandwidth and knowing the insertion loss and spurious characteristic, the maximum power handling capability can be empirically estimated. The maximum power delivered to a filter is usually characterized in the form of a pulse at the center of the passband, in which the peak power is much greater than the steady-state power handling capability of the filter.

If exposed to too much RF power, the insertion loss of a filter can change depending on the drive level. As we discussed above, at high RF power levels, quartz resonators become nonlinear. This can also cause the filter insertion loss to change (increase). The phenomenon is little understood but is attributed to the inherent nonelastic piezoelectric properties of the quartz material.

High levels IF signals can actually damage quartz filters if no preventive measures are taken such as good AGCs. In general, most quartz filters have a maximum nondestructive level acceptance of approximately −30 dBm, while much more expensive units can be manufactured, which can withstand +5- to +10-dBm levels for a limited amount of time (a duty cycle of about 10%). If exposed to high levels of RF for a longer time, quartz-based IF filters can be permanently shifted in frequency or can even be damaged or destroyed. The phenomenon is similar to that caused by electromagnetic pulse (EMP) effects on quartz caused by nuclear impacts. A natural phenomenon characteristic to quartz called aging (a property referring to quartz and other piezoelectric materials changing their resonance characteristic over time) can also be accelerated in these situations.

On the other hand, if exposed to low drive levels, quartz resonators can have a hard time maintaining a constant insertion loss (they can act like threshold-triggered devices). This problem can be countered with the application of proper AGC levels and with more stringent manufacturing techniques. Despite these problems, quartz-based IF filters have been manufactured with minimal impact on insertion loss of ±0.005 dB for an approximately 40-dB change in drive levels.

4.19   Settling Time and Rise Time in Filters

Settling time is defined as the time that a filter network takes to settle within a specified overshoot after the input has been excited by a pulsed RF signal. Rise time is defined as the time required for the output of a filter network to ramp up from 10% to 90% of its steady state. The exact value of rise time can be readily determined from filter handbooks. The following equation relating rise time to bandwidth provides a meaningful estimate.

$$T_r = 0.35/f_c \qquad\qquad (4.6)$$

where T_r is the rise time in seconds and f_c is the −3-dB corner frequency in hertz. The rise or fall times can vary as much as four times nominal. They are dependent on the order of the filter and the approximation type selected.

4.20   Phase Delay and Group Delay Distortion

Phase delay or phase linearity of a filter is defined as the deviation of the phase response from a nominal straight line of all signals comprised in the filter's 3-dB bandwidth. This is usually expressed in degrees. Equal phase delay over the passband of an IF can be a very important design parameter in certain applications.

Group delay, also called envelope delay, is the time taken for a narrowband signal to pass from the input to the output of a bandpass filter. It is usually measured in microseconds (μs) and represents the difference between the maximum and the minimum points within the 3-dB passband. It can also be specified at two specific frequencies within the passband, usually as close as possible to the upper and the lower 3-dB edges.

Group delay will usually have response characteristics resembling the letter U inside the 3-dB passband of a filter. The upper points of the U response will tend to peak toward the 3-dB passband points. Consequently, to obtain equal group delay performance in an IF filter, its passband will need to be expanded outwards by about 30% to be able to take advantage of the bottom side of the U shape response, which will remain relatively flat. This is shown in Figure 4.23.

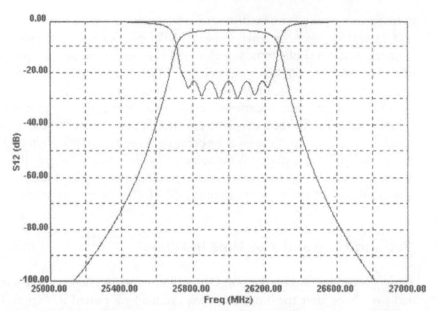

Figure 4.23 Amplitude and phase delay response of a 26-GHz IF filter with a 3-dB passband of 500 MHz. Equal phase delay performance is over 400 MHz of the bandwidth. The phase response looks like the letter U with the ends going up. Only the bottom region of the response can be used to guarantee equal phase and group delay. The 3-dB bandwidth of the filter had to be enlarged purposely by 20% to 30% in order to ensure equal group delay throughout the passband.

Phase and group delays are closely connected with attenuation, ripple and shape factor characteristics, which, in turn, are functions of the type of filter approximations. Group delay is the derivative of the phase response.

In quartz crystal filters, the group delay usually peaks where the amplitude response changes from passband to the transition band, in which case Figure 4.23 may not be representative. Group delay response can be described in relation to the amplitude response of a filter by using the Hilbert transforms (see [8–10], which cover this in detail).

Better shape factors (meaning more complex filters) produce larger delay peaks. For instance, high order filters such as elliptic implementations exhibit large peaks near their 3-dB stopbands. Again, such filters have to be designed even wider to obtain a reasonable equal phase and group delay performance. By contrast, monolithic crystal filters (MCF) and ladder filters have very small group delay distortion characteristics.

Linear phase response refers to a filter, which exhibits a near constant change in degrees per unit of frequency where the plot of phase versus frequency is nearly a straight line. Differential group delay relates to the peak-to-peak difference in group delay between two opposed frequencies in the passband. This method is used to specify group delay performance over a specific passband where the highest and lowest values are used to determine the differential group delay (see Figure 4.23). Differential group delay is also referred to as the delta group delay, group delay flatness, or differential time delay.

4.21   Impedance

Impedance is defined as the amount of opposition presented to the flow of alternating currents (AC) at a particular frequency by a passive or active electronic device such as a filter. Impedance is a combination of ohmic resistance, inductive reactance, and capacitive reactance. It is represented by a complex notation $Z = V/I$, where Z is a complex number given by $Z = R + _jX$. In this expression, R is the real part of the expression (the ohmic resistance), while X is the reactance (the imaginary part of the expression). Impedance is given in ohms.

Filters, by design have impedance, which may or may not match the electrical circuit requirements in which they will be used. In fact, they can have different impedance at the input and the output.

In high-performance receivers, matching of IF filters is usually done at 50 ohms to reduce cable complications and suppress undesired RF radiation. In these designs, IF filters are often matched directly to the mixer parts, which also operate at 50 ohms. In some inexpensive applications, IF filters have impedance values ranging from 200 ohms to 1,000 ohms. They often require additional matching circuits to adapt to active stages. Matching of filters is usually performed with transformers or split capacitor methods.

Matching in filters is shown as terminating impedance, specified as a series source resistance at the input and a load resistance at the output. If circuit impedance matching is not properly addressed, voltage standing wave ratio (VSWR) problems can occur causing inefficient performance (greater insertion loss) and possibly in-band nonlinear spurious problems. In terms of VSWR, it is the ratio of the maximum value of a standing wave to its minimum value, related to the return loss. The return loss expressed in decibels presented by mismatched impedance as a function of VSWR is shown in

$$\text{Return Loss}_{(dB)} = 20\log\big[(\text{VSWR} + 1)/(\text{VSWR} - 1)\big] \qquad (4.7)$$

As an example, the VSWR of a filter device (or any other RF passive or active device) is 1.5:1; it corresponds to a return loss of 20 log [(1.5 + 1)/(1.5 − 1)] = 20 log 5 = 14 dB, which is in addition to the filter's initial insertion loss.

4.22   Vibration-Induced Sidebands

IF filters in modern receivers and transceivers are sometimes exposed to tough vibration, acceleration, and shock conditions. As such, vibration-induced sidebands may appear, especially in quartz crystal filters.

Vibration forces can change the resonance frequency of the quartz elements causing undesirable responses, which, in turn, can change the overall performance of a filter. More often, mechanical vibrations create undesirable sidebands. G-forces can induce similar distortions, and often designers are faced with demanding vibration and shock requirements that make filters ever harder to produce. Vibration-induced sidebands are minimized by damping resonator acceleration sensitivity and by controlling the total mechanical resonance of the network. In these cases, various methods of elastic suspension and element orientation are used internally and/or externally to prevent performance deterioration or filter destruction.

Elaborate test plans are specially designed for a filter application using three axis vibration tables to prove the survivability of a device at several audio frequencies and for certain amounts of time. The filter output is then viewed on a spectrum analyzer, which will show the induced sidebands as offset from the carrier by the frequency of vibration. For most filters, the vibration-induced sidebands are relatively small. However, narrowband quartz IF filters require special attention.

4.23   Modern Filter Approximations

Filter approximation is a terminology used by filter designers to describe the type of filter used in an application. It relates to the design process used to convert certain specifications to particular mathematical transfer functions.

In designing an analog or a digital filter, the designer uses a logical approach of choosing from a repertoire of filter network synthesis and approximation techniques, which in turn allow for optimization of all trade-off solutions. This is not an easy process since performance characteristics exhibit many mutually exclusive conditions. Thus, optimizing for a parameter can result in reduced performance for another. To alleviate this problem, designers also use rules-of-thumb and ample experience to select the proper approximation in a practical filter design.

Filter approximations have been named primarily after their discovering mathematicians, Chebyshev, Butterworth, and Cauer, as previously mentioned. Listed in order of increasing selectivity, the most frequently used modern filter approximations are the Bessel, Butterworth, Chebyshev, and Cauer-elliptic. Although several books could be written on these subjects, we will briefly describe these functions. For a much more in-depth treatise on the subject, the reader is directed to the specialized materials listed in the Selected Bibliography.

4.24   Bessel or Linear Phase

The Bessel response is remarkable in its gradual monotonic rise time, which makes it ideal for passing equal phase and group delay signals without distortion especially about its center frequency. However, its slope characteristics are rather soft, making it a bad choice when superior shape factor is required. The transfer function of this type of filter is derived from a Bessel polynomial. This type of filter exhibits a response similar to that of a Gaussian filter. It has a rather poor insertion loss. The Bessel response is best suited in IFs in receivers dealing with pulses or digital data communications, where rise times and overshoot have to be minimized.

4.25   Butterworth

The Butterworth filter is known for its flat response, minimum ripple, and minimum insertion loss. Selectivity is better than the Gaussian response and is somewhat similar to the Bessel function at the expense of inferior phase and group delay linearity. The stopband is moderately selective and monotonic.

Butterworth filters are relatively insensitive to changes in element values over temperature making them a desirable choice for many applications.

4.26 Chebyshev

This filter approximation is one of the most popular choices in filter design because it offers improved shape factor. It exhibits equal ripple response in the passband. The Chebyshev approximation is characterized by a bell-shaped amplitude response similar to that of a Gaussian or a Bessel response. It exhibits an equiripple phase and group delay response. The selectivity is better than the Bessel or Gaussian responses. The stopband of a Chebyshev filter is highly selective and monotonic.

The Chebyshev approximation is usually a preferred type of filter because it is relatively easy to implement.

4.27 Cauer-Elliptic

This type of approximation is known as having the steepest roll-off for the same number of circuit elements from all types of filter approximations known at the cost of increased ripple in the passband. It is used mainly for harmonic and spurious rejection in synthesizers, transmitters, or automatically switched half-octave preselector banks applications. The complexity of this approximation is higher than the Chebyshev or Butterworth approaches.

4.28 Gaussian

The Gaussian approximation is somewhat similar to the Bessel. It exhibits zero or minimal step and impulse response, so rise times and delays are faithfully reproduced from all the approximation types. Because of this, the Gaussian approximation is almost always chosen in IFs for demanding pulsed radar receiver applications. However, these properties come with a price: They are obtained at the expense of rather poor selectivity, high element count, and limited realization behavior.

The Gaussian to 6 (or 12) dB approximations have a passband response that follows the Gaussian response, but it can be changed to follow a Butterworth response at either the 6- or 12-dB points on its skirts. By doing this, the phase and group delays are improved over the Butterworth approximation along with a better slope attenuation, which is better than the Gaussian approximation. Consequently, the Gaussian 6/12 dB is a good compromise between the Gaussian and Butterworth approximations. However, actual implementation of such a filter can be difficult.

4.29   Synchronously Tuned

This type of realization exhibits the same advantages and disadvantages as the Bessel and Gaussian realizations except that the phase and group delays are better behaved than all other realizations. This type of filter is also rather difficult to implement.

4.30   IF Filter Technologies

We will discuss next the various technologies used to implement the filter approximations described above, as they apply to IF filters in superheterodyne receivers and other applications. Some of these technologies are shown in Figure 4.24.

In this discussion, we will consider previous topics such as the implications of percentage bandwidth, the quality factor (Q), and the upconvert image impacts on choosing these technologies at various center frequencies. We will then close this chapter with the design of a quartz crystal ladder filter intended for an IF centered at 9 MHz.

4.31   Mechanical Filters

The advent of modern IF filters began with Robert Adler of Zenith who in 1946 designed and built a six-element electromechanical filter at 455 kHz (see Figure 4.25). In [9] he wrote:

> Interconnected metal plates that transmit vibrations act as transmission-line type filter. Plates are coupled to electrical circuit by magnetostriction. Filter for 455-kc IF channel of broadcast receiver has very sharp cuttoffs, is small, cheap, easily constructed and efficient.
>
> Intermediate frequency currents, upon entering the filter, are converted into mechanical vibrations of the same frequency. These vibrations then

Figure 4.24   Three types of modern IF filter technology implementation: quartz crystal ladder (top), mechanical (center), and ceramic (bottom).

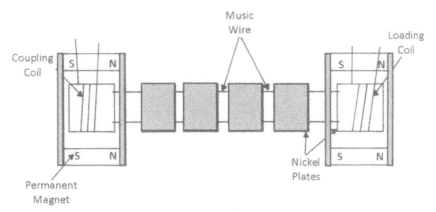

Figure 4.25 Implementation of the first 455-kHz IF mechanical filter by Robert Adler in 1946.

pass through a structure resembling a ladder, consisting of several mechanically resonant metal plates coupled to each other by wires which act as springs.

This structure forms a bandpass filter for mechanical vibrations. Width of the pass-band, as will later be shown, is determined by the design of each individual section and because the several sections are all alike, bandwidth does not depend upon the number of sections. Attenuation outside the band limits, however, increases with the number of sections.

To understand the operation of the filter most easily, let us first consider a familiar electrical filter, composed of inductors and capacitors...

The analogy between electrical and mechanical network elements, with masses substituted for inductors and springs for capacitors, is well known; and it appears quite feasible to build a mechanical filter structure, which is fully equivalent to the electrical filter...by combining masses and springs in analogous fashion.

Adler applied his filter to a 455-kHz IF, AM receiver. He noted [9]:

It was first suspected that such a set would be hard to tune, but tests with a number of lay listeners did not bear this out. Change in tone quality caused by incorrect tuning sets in at two clearly defined points much more abruptly than in conventional receivers; listeners seem to find it quite easy to tune between these points.

Adler made several tests showing clear improvement against adjacent channel interference with his newly invented filter.

Although transformer type L/C filters were still being used, mechanical filters offered superior performance at 455 kHz. Mass production of 455-kHz IF mechanical filters was pioneered later by the Collins Radio Company. By 1952, the first 455-kHz single sideband filters were produced in small quantity by this company. The popularity of such filters continues today when complex mechanical filters with twelve or more resonators can be fabricated with shape factors (SF) as low as 1.25 and passband ripple as low as ±0.1 dB.

Figure 4.26 shows a modern 12-disk mechanical filter produced by Rockwell Collins Filter Products. The theory of operation for this filter is not much different than that of Adler's; however, the technological implementation departs considerably from his implementation.

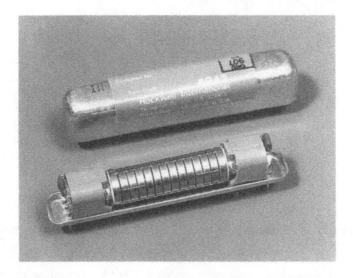

(a)

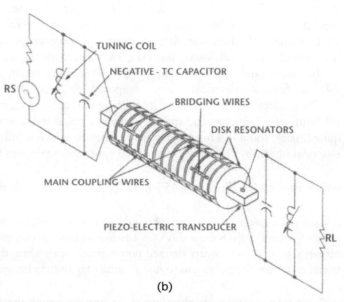

(b)

Figure 4.26 (a) Actual implementation of a modern 12-disk mechanical filter. (Photo courtesy of Rockwell Collins Filter Products.) (b) Internal configuration of the disk-wire mechanical filter equipped with ferrite tuning transducers.

While Adler's filter was using inefficient and lossy magnetostrictive wire transducers, this modern disk-wire mechanical filter uses piezotechnology (PZT) transducers in combination with ferrite cup-cores and negative temperature coefficient capacitors in order to reduce the insertion loss to less than 3 dB, the passband ripple to under 2 dB, and maintain frequency stability over the temperature range of −55°C to +85°C. If the temperature performance is not a requirement, ferrite transducers could be substituted for the PZT transducers. The bridging wires shown in Figure 4.26 allow for phase inversion needed to obtain attenuation poles symmetrically on both sides of the center frequency.

While Adler's filter was produced by laboratory methods, this modern implementation is the result of a highly technological production process, which starts with 6-foot-long cylindrical bars made of a mechanically stable iron-nickel alloy, which was ground to a precise diameter, cut into disks of a certain thickness to resonate slightly below the required frequency, then heat treated and ground within 0.0002 inch for accuracy. The individual disks are then frequency tuned, which is accomplished by drilling out a small amount of metal at their geometrical center. This operation is performed automatically in only seven seconds with the help of a computer-controlled drill press equipped with frequency sensors, which continuously measures and anticipates the resonance of the particular disk and then automatically shuts off the drill motor when the right resonant frequency is achieved. The transducers follow somewhat of the same pattern. In the case of ferrite, iron-nickel oxide powders are wet-mixed and extruded into rods similar to the iron-nickel bars. They are finally fired and cut to size before being attached to the metal disks. The final assembly incorporates the right number of disks, the transducers (which are now attached to the ends), together with coils and magnets.

The bridging and coupling wires are spot welded automatically with the help of a computer-controlled machine, which keeps track of the distance between the disks as well as the precise consistency of the weld. Finally, shock-absorbing rubber grommets are added before the final inspection is performed with the help of network analyzers, and the cover is put on by heat staking or automatic soldering. A more modern implementation of a 455-kHz mechanical filter is the tortional mechanical filter shown in Figure 4.27. Mechanical filters are easy to apply in IF circuits as shown in Figures 4.28.

4.32 Quartz Crystal Filters

Attempts to avoid the image problem in superheterodyne receivers led to using higher frequency IFs especially in HF communications receivers. New IF filter

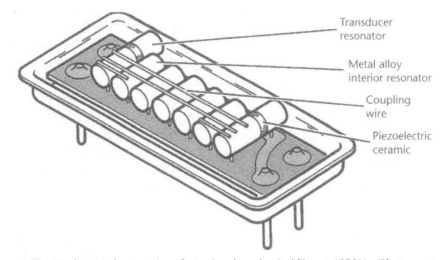

Figure 4.27 Modern implementation of a tortional mechanical filter at 455 kHz. (Photo courtesy of Rockwell Collins Filter Products.)

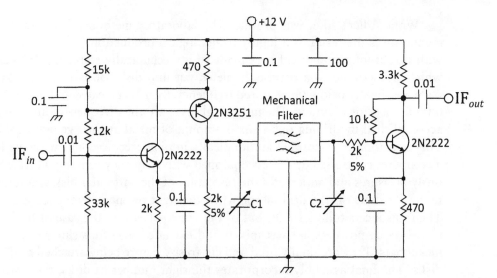

Figure 4.28   Circuit implementation of a 455-kHz mechanical filter.

technologies were introduced in the 1960s with the quartz filter becoming widely used despite newer DSP filter technology. Today's HF superheterodyne communications receiver most likely uses a combination of quartz, mechanical, and DSP technologies in full concert to achieve selectivity.

Because of the very high Q presented by the quartz material, narrow bandpass filters using quartz technology can be realized at IF frequencies ranging from 455 kHz to over 100 MHz (quartz crystals can be manufactured to 100 kHz or less). Ultimate bandwidths for SSB voice or slow data communication such as KCW can be realized at frequencies upwards of 30 MHz. At higher IF frequencies, such as in upconvert HF receivers (e.g., 45, 65, 70, 75, 85, or 115 MHz), narrowband ultimate bandwidths have been limited by the percentage bandwidth laws discussed previously, which in turn led to double and triple conversions to lower frequencies where quartz filters could be used offering superior performance at ultimate information bandwidths. The most popular quartz IF filters have been implemented at 9 or 10.7 MHz. They exhibit 3-dB bandwidths ranging from 200 Hz to 3 kHz depending on the modulation requirements of these receivers.

Modern commercial and military crystal filters involve several individual crystals. The theory and design of such filters is several orders of magnitude more difficult than that of a single crystal. Quartz crystals can exhibit several modes of resonance, resulting in complex interaction of parameters such as center frequency, bandwidth, shape factor, group delay, transient response, phase linearity, insertion loss, temperature and aging, making for exhaustive mathematical analysis, which is usually performed with the help of computers. Networks with as many as thirty-two quartz crystals have been computer modeled and designed by filter engineers.

Quartz crystal filters use quartz elements to achieve high selectivity. Quartz is found in its natural form throughout the world. Because of its crystalline form, its molecules are packed in an even order pattern extending equally in three dimensions. This allows for a wafer of quartz material, with electrodes attached, to generate electricity when mechanically excited, or to vibrate with high-frequency accuracy when an alternating current is applied to it. This is the piezoelectric phenomenon.

There are four basic vibration modes in quartz resonators: flexture, shear, longitudinal, and tortional. These are used to achieve certain goals in the design of filters and oscillators. As a consequence of the quartz piezoelectric phenomenon, surface acoustic wave (SAW) filters and bulk acoustic wave (BAW) filters have also resulted.

Piezoelectricity was discovered by the French scientists, Jacques and Pierre Curie in 1880. Further investigation of piezoelectricity in quartz was performed by Paul Langevin who studied resonators for sonar applications during World War I.

The first quartz crystal oscillator was built in 1917 and patented in 1918 by Alexander M. Nicholson at Bell Telephone Labs. However, his invention was disputed by Walter Guyton, who also claimed the first quartz oscillator. By 1926, quartz crystals were used to control the frequency of AM radio stations. The amateur radio community also helped considerably in the development of these devices as means of controlling frequency in transmitters and in IF filters used in communications receivers.

Natural quartz crystals come in various degrees of quality, which is hard to control. There is considerable variability in the Q of different batches of natural crystals obtained even from the same source. This, in turn, impacts performance consistency and producibility of quartz filters. As a result, it is harder to make quartz IF filters into the megahertz range than to make mechanical filters at 455 kHz where technology and fabrication can be more tightly controlled.

However, quartz crystal technology has improved considerably in the past 30 years. More recently, quartz crystals have been grown artificially maintaining a more consistent quality and purity for the process such that they can now be mass produced.

There are two forms of quartz: the alpha-quartz, which has not been temperature fused, and the beta-quartz, which has been heated to 573°C for stability. Only alpha-quartz is usually used for resonators in IF filters.

Quartz crystal resonators are usually cut in a rectangular shape from a high quality quartz block as shown in Figure 4.29. The various cuts have unique properties, which can be exploited in different applications. For instance, at up to 30 MHz, AT cuts are usually used at fundamental frequencies in filter and oscillator applications. Beyond this frequency, fundamental quartz crystals become very thin and fragile. Consequently, overtone properties of these quartz crystals are being exploited at these higher frequencies. Practical quartz crystals exhibit Qs ranging from 10,000 to over 1 million at frequencies of up to 200 MHz.

A single quartz crystal element exhibits two resonant frequencies, a series resonance point, and a parallel resonant point (also called the antiresonance point). This is shown in Figure 4.30. An electrical model equivalent to a quartz element or a filter using a single quartz element is shown in Figure 4.31.

All quartz crystals can be specified to resonate at either their series or their parallel resonant frequencies. However, it is most common to specify a quartz resonator as a series resonant network as shown in Figure 4.30(a).

A quartz element is modeled through an R, L, C series resonant network as shown in Figure 4.31. Looking at this figure, if a signal generator can be swept

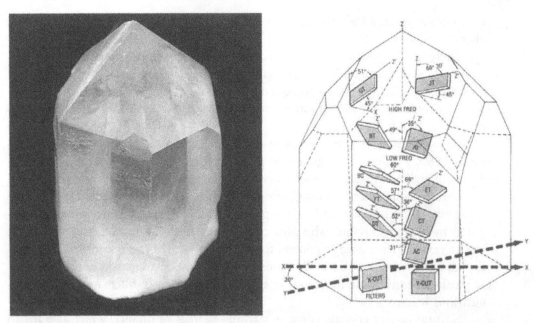

Figure 4.29 Quartz slices are precision-cut from a quality blank (a) at various angles with respect to the axes to yield different resonant qualities for various filter and oscillator applications (b). (*From:* [10]. Reprinted with permission from *Electronics Now*. Copyright Gernsback Publications, Inc.)

slowly over a range of interest, a sharp output would be observed on the spectrum analyzer as we approach the resonant frequency of the crystal.

The behavior of the quartz model from Figure 4.31 can also be expressed mathematically. Looking at Figure 4.31, R_1 is the series resistance, C_1 is the motional capacitance, L_1 is the motional inductance, and C_o is the parallel capacitance (otherwise known as C_p). The series resonant frequency f_s can then be found from:

$$f_s = \frac{1}{2\pi\sqrt{L_1 C_1}} \tag{4.8}$$

This equation is helpful in finding the series resonant frequency of the quartz model from Figure 4.31. Resonance occurs when the inductive and capacitive reactances are equal and cancel at resonance or $X_{L_1} = X_{C_1}$.

The parallel resonant frequency or antiresonance, f_a, can then be found from:

$$f_a = \frac{1}{2\pi\sqrt{L_1\left[\dfrac{C_1 C_o}{C_1 + C_o}\right]}} \tag{4.9}$$

Equation (4.9) uses the parallel capacitance of C_0 and C_1 from Figure 4.31. A crystal is said to operate at its antiresonant frequency, when the impedance is at its maximum and the current flow is at its minimum.

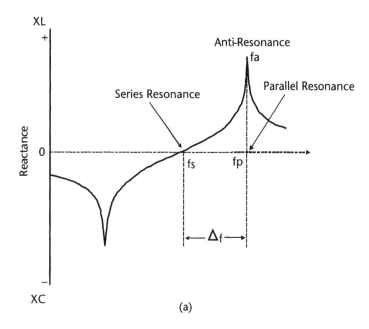

(a)

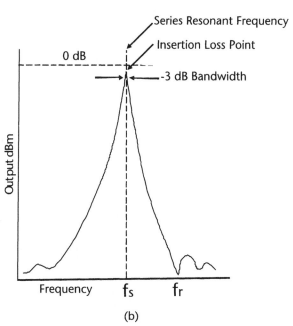

(b)

Figure 4.30 (a) Plot of the reactance versus frequency at the series resonance point of a quartz crystal. (b) Response characteristic of a single quartz crystal filter. The highest response is achieved at the series-resonant frequency (f_s).

As shown in Figure 4.30, the series resonance point f_s is always below the parallel resonance point f_a. Consequently, a quartz crystal is always specified between f_s and f_a. This range of frequencies is known as the parallel resonance area of the crystal. However, there is no difference in the design and implementation of a series

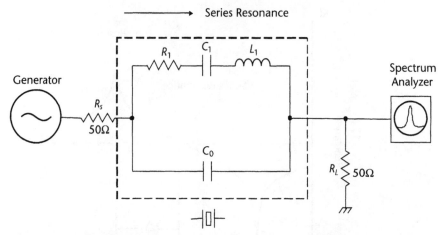

Figure 4.31 A quartz crystal can be modeled and tested as a two-port series resonant electrical network.

resonant quartz crystal from that of a parallel resonant quartz crystal, as they are produced in exactly the same way. The difference is only in how the parallel resonant specification is set somewhat above the series resonant point.

Beyond the 30-MHz limit of the fundamental AT cuts, a quartz crystal element can be made to vibrate at one of its overtone modes, occurring at multiples of the fundamental resonant frequency as shown in Figure 4.32. Only odd-numbered overtones are used since they exhibit the highest amplitude. Such a quartz crystal element is referred to as a third, fifth, seventh, or even ninth overtone crystal.

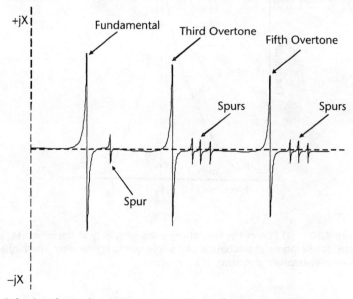

Figure 4.32 Behavior of a single quartz resonator shows a fundamental series resonant component (going up) accompanied by a parallel resonant component (going down), spaced closely together. The frequency response of a quartz resonator also exhibits overtone frequencies (primarily third, fifth, seventh, and ninth), which can be exploited in oscillator and filter design. In addition, various undesirable spurious responses occur as shown. The series resonant component is usually used in filter design.

To accomplish this, a filter network circuit usually includes additional L/C elements to select the desired frequency. However, as we increase the frequency, overtone crystals are not as suitable to design IF filters as AT cuts are. On the other hand, fundamental cuts at the higher frequencies are sensitive to vibration and prone to destruction.

Because of the tight percentage bandwidth and high Q experienced, quartz filters do not lend directly to wideband applications. However, wideband crystal filters have been implemented by using additional inductors and capacitors to contribute additional poles to an L/C design while still using the quartz resonator's response. Percentage bandwidths of up to 10% have been realized.

4.33 Temperature Stability in Quartz Crystal Filters

Temperature stability of quartz IF filters can be very important in receiving systems. While inductors and capacitors can influence the frequency stability of a quartz filter, the crystals are the major contributor to frequency changes over temperature. Since most quartz IF filters are designed using AT-cut crystals, the frequency stabilities of these filters are linked mainly to these piezoelectric elements.

Figure 4.33 shows the temperature curves for fundamental mode AT-cut quartz elements. The frequency change expressed in parts per million (PPM) are shown versus the temperature change from a nominal of 20°C. The designer will select the proper curve to give the minimum deviation over a specified temperature range.

Next, we will design a 9-MHz IF ladder filter for a communications receiver.

4.34 Designing High Performance Quartz IF Filters

We discussed the many aspects of IF filter design, including the various approximations and technologies involved. Contrary to some beliefs, filters are not trivial to design. Many times new designers think of filter design as a simple exercise in computer skills. Nothing could be further from the truth. Although filter design is very much an engineering endeavor, designing high performance filters is considered an art rather than a science. This is because it uses a set of theoretical and empirical skills combined with ample experience in order to get the proper results.

At first glance, there seems to be a lot of computer programs that attempt to do the job easily. A closer look, however, reveals that even with the help of programs, the designer has to do a lot of leg work before a realizable filter can be obtained. This is especially true in narrowband IF quartz filter design, which can take many forms.

This book is not about filter design, but it goes into enough depth to give the reader an appreciation of what it takes to produce a clean realizable network, and not just a paper design, which may not work.

To substantiate this statement, we will now proceed to design an eight-pole quartz ladder filter for a communication receiver. Its −3-dB bandwidth will be 2.4 kHz and its shape factor will be such as to reject a beat frequency oscillator (BFO) leakage by at least 50 dB. The BFO is the local oscillator used to reinsert a carrier into single sideband signals for detection in a communications receiver. Its

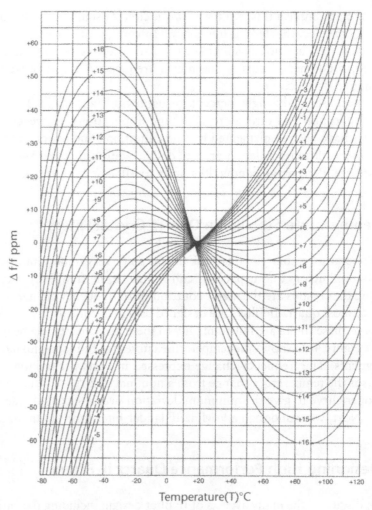

Figure 4.33 Temperature curves for fundamental mode AT-cut quartz elements.

frequency is usually located very close to a filter's passband such that the resulting mixed baseband reflects the bandwidth of the information to be received.

The design goal is to produce a realizable bandpass filter with a good shape factor in order to be selective against adjacent signals as well as to reject the BFO leakage, which could interfere with the received information.

We will choose a quartz crystal ladder approach because it utilizes identical crystals for ease of design. The approximation used will be the Chebyshev 0.1 ripple approach. Although the filter looks like a Cohn (min-loss) filter, this design is not to be confused with it as its ripple and insertion loss are different and much better.

The series elements are AT fundamental-cut quartz crystals of the same exact frequency. For a 9-MHz center frequency filter, the following parameters are found from the quartz crystal manufacturer (note the series resonance (F_s) as being lower than the desired filter center frequency as we previously discussed):

- F_s: Series resonant frequency = 8.998,633 MHz;

- L_s: Motional inductance = 0.00313 μH;
- C_s: Motional capacitance = 0.01 pF;
- R_s: Motional resistance (high and negligible in this example);
- C_p: Parallel case capacitance = 0.035 pF;
- Q_u: Quality factor = 150,000.

These are key parameters usually furnished by the quartz crystal manufacturers. If not available, these parameters have to be found experimentally by actually taking crystals apart and measuring them. Much more on this subject, including equations for the design, can be found in the Selected Bibliography.

The 3-dB bandwidth of the filter, the shape factor, and the termination impedance are determined by the shunt capacitors to ground and the coupling capacitors. These determine the bandwidth of our filter with a minor effect from the values of L_s and C_s. The entire process of design is complex and beyond the scope of this book, which is intended only to make a designer aware of what is involved in the process. However, for those interested, a full explanation of the equations and procedures used in our design can be found in [4].

All elements are interactive and dependent on the Q of the quartz crystals. The final bandwidth and interactions are determined from the design equations in the reference mentioned above. With these values, we can now approximate a filter design, which will be very close to the final result with a 90% to 95% confidence factor. The electrical model of our eight poles Chebyshev 0.1 ripple quartz ladder filter is shown in Figure 4.34.

It can be seen from looking closely at Figure 4.34, that the series resonance model from Figure 4.31 is present in every quartz node or element in our filter. As mentioned before, in designing quartz ladder filters, only series resonance values should be used for modeling.

We will next use a program called 5Spice to analyze our network and produce a response plot of our new design. The values and nodes from Figure 4.34 are then input exactly as shown in the figure into this program. 5Spice is a proven analysis tool that offers professional capabilities to experienced circuit designers while remaining easy to use. This is a big step up from the many student oriented programs available today.

The final plot of the filter is shown in Figure 4.35. We can now proceed with confidence to the brass boarding of our new filter. Although there will still be a certain amount of tweaking involved to a final product, the new filter will be as close as possible to a final design. Our calculated insertion loss is 1.5 dB and our ripple is 0.1 dB. The actual performance of our new filter will depend mainly on the quality of the quartz used as well as on the rest of the parts. Much remains to be done to maintain port-to-port leakage at a minimum for the best ultimate rejection performance and to be able to reproduce the filter consistently in manufacturing.

In conclusion, we have learned about the behavior of a single individual crystal at resonance being a high Q series resonant circuit, as shown in Figure 4.30. We have then learned about modeling a single quartz crystal element as shown in Figure 4.31.

Figure 4.32 showed the behavior of a single quartz resonator at the fundamental series resonance frequency, accompanied by a parallel resonant component, and

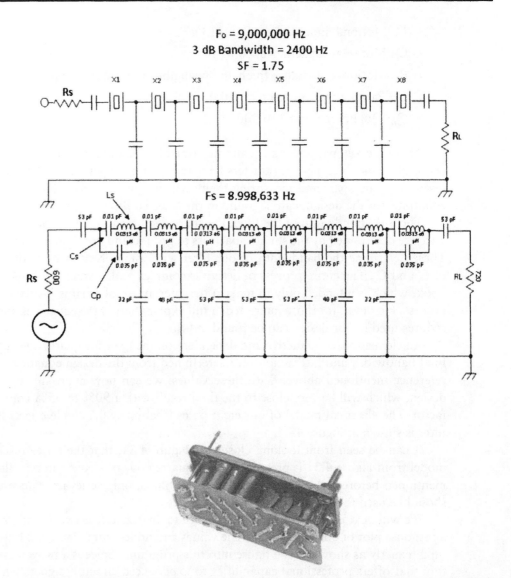

Figure 4.34 Design model and actual implementation of the 9-MHz Chebyshev 0.1 ripple quartz ladder IF filter as input to the 5Spice analysis program [11]. All values have been either calculated or measured from individual quartz crystals before entering the model in order to ensure a 90% confidence factor. Input and output impedance transformation will have to be performed using transformers or split capacitor matching networks in order to provide a 50-ohm impedance.

spaced closely together. We have discussed temperature impacts on quartz elements as shown in Figure 4.33 and finally designed an 8-pole quartz ladder IF filter as shown in Figures 4.34 and 4.35.

Other types of quartz filters can be designed. The schematic diagram of a two-section crystal lattice filter is shown in Figure 4.36. In this example, the shunt inductances for the crystals are part of the input and output transformers. The crystals Y_1 and Y_2 are slightly different in frequency from crystals Y_3 and Y_4. In reality the series resonant crystals (Y_1 and Y_2) are tuned to the parallel resonance frequency of the other two crystals (Y_3 and Y_4) providing for a bandpass character-

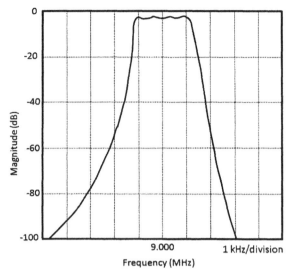

Figure 4.35   Frequency response of the eight poles Chebyshev 0.1 ripple quartz ladder filter design. The simulation was performed using the 5Spice program tool.

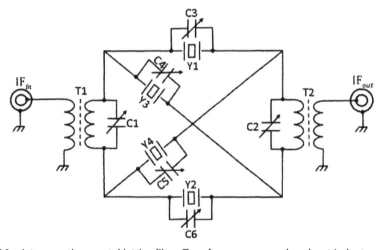

Figure 4.36   A two-section crystal lattice filter. Transformers are used as shunt inductances.

istic as shown in Figure 4.37. In this design, transformers have to be identical and matched for best results.

A more complex crystal-lattice filter is shown in Figure 4.38.

4.35   Monolithic Crystal Filters (MCF)

A simpler way of achieving a quartz filter is the monolithic approach. In a monolithic filter, there are no discreet adjustments such as inductances and capacitors. Two electrode pairs are deposited on a quartz wafer (electrochemical methods of depositing aluminum, silver, or gold are used in conjunction with AT cuts of quartz crystals) as shown in Figure 4.39.

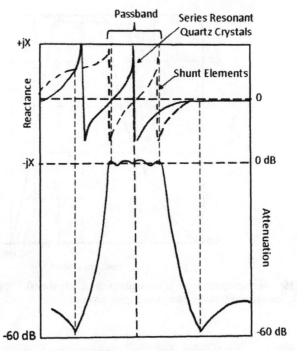

Figure 4.37 Response characteristic of the two-section lattice filter.

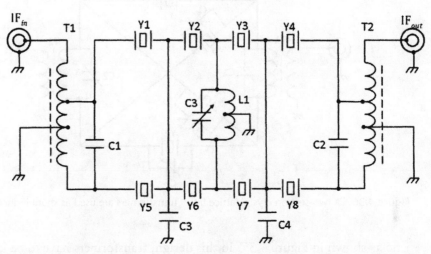

Figure 4.38 Internal configuration of a commercial 8-pole crystal lattice filter (XF-9B) from KVG (Kristallverarbeitun Neckarbischofsheim, Germany). Each stage is made of tuned half-lattice sections with one crystal in each branch.

From Figure 4.39, a sound wave generated at electrode A travels through the crystal wafer to electrode B, at the two resonant regions as shown in our example. The thickness of the wafer at the two points determines the bandpass characteristic of the monolithic filter. This "folded" approach to acoustically coupled resonators was not fully understood until 1965 when R. A. Sykes at Bell Telephone Laboratories and M. Onoe in Japan simultaneously discovered the mathematics governing this mechanism.

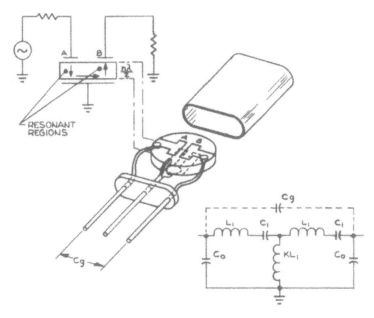

Figure 4.39 A two resonator monolithic filter and its electrical equivalent. The thickness of the crystal wafer equals $n\lambda/2$, where λ is the operating wavelength and n is an odd integer. C_g is the gap capacitance between leads, a limiting factor for the stop band rejection of the monolithic filter.

This event led to the development of multiresonator monolithic filters such as the one shown in Figure 4.40. However, the inability to tune out unwanted elements, such as in a conventional quartz filter, limited the practical number of resonators in a monolithic to four pairs.

While in a conventional quartz filter, the interaction of unwanted parameters could be minimized by adjusting variable elements such as capacitors and inductances. In a monolithic filter, this interaction is a fixed function of the mechanical

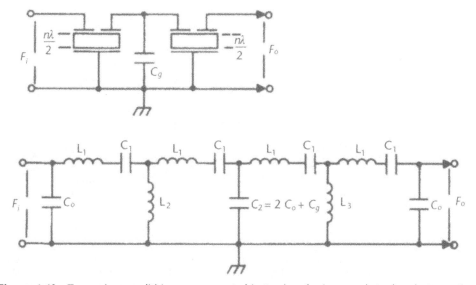

Figure 4.40 Two-pole monolithics are connected in tandem for increased stopband attenuation and ease of manufacturing.

arrangement, and is further aggravated by indirect sound paths of unpredicted nature between the input and output, producing spurious responses and degrading the stopband attenuation of the filter. Controlling factors such as the quality of etching or the polishing of the quartz also play an important role in canceling these effects, making it hard for multiple-pole quartz monolithic filters to be produced economically.

New technological breakthroughs, such as ion etching combined with sophisticated polishing techniques, as well as the use of other than quartz materials such as lithium niobate, lithium tantalate, and bismuth germanium oxide, can allow for increasingly complex monolithic filters to be built on the same substrate.

4.36   The Tandem Monolithic

One way of achieving greater selectivity without the calibration problems of the multiresonator monolithic on a single substrate is to connect two or more two resonator monolithics in a tandem fashion as shown in Figure 4.40.

In this arrangement, control over the electrical parameters of each monolithic is guaranteed by the less demanding manufacturing process of a two-pole device. The tandem design provides for increased stopband attenuation by minimizing the effect of unwanted modes. Filters with as many as 10 resonators have been manufactured by this technique. Figure 4.41 shows an actual implementation of a four-pole tandem monolithic filter with a center frequency of 75 MHz. Two 4171F monolithic filters are used in this example. Figure 4.42 shows the frequency response of the filter.

4.37   Ceramic Filters

Where size and cost are important in a receiver design, the ceramic piezoelectric filters are used in the IF. Such filters usually operate at 455 kHz, although much

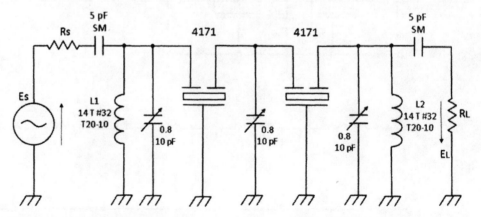

Figure 4.41 A practical implementation of a four-pole tandem monolithic filter, centered at 75 MHz (−6-dB bandwidth is ±13 kHz). Such a filter could be used in the first IF of a double conversion upconvert superheterodyne receiver or transceiver. An amplifier can also be used between the filters in order to provide better isolation.

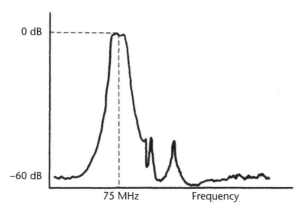

Figure 4.42   Frequency response of the 75-MHz tandem monolithic filter.

higher frequencies have been achieved with this technology. The electrical performance of the ceramic filter resembles that of the quartz crystal filter. However, piezoelectric ceramics do not exist in natural form such as with quartz crystals.

The complicated manufacturing process starts with ceramic compositions such as lead zirconate-titanate and lead metaniobate. These powders are mixed with water in order to produce a paste, which is further formed into disks of a certain size. The disks are then dried out, before being baked at high temperatures much like bricks in a ceramic factory. It is in the baking process that a high voltage "poling" field is applied in order to give them the piezoelectric property.

This shock distorts the physical shape and electron properties of the ceramic structure creating a permanent piezoelectric element, like a powerful magnet would magnetize a ferric material. Silver electrodes are then deposited on each side of the disks, and a ladder arrangement can be accomplished by stacking as many as 15 or more elements in a tubular package. Figure 4.43 shows the electrical equivalent of a ceramic ladder filter.

Other configurations such as monolithics are also achievable with this technology.

Ceramic filters are easy to apply and match with circuits as shown in Figure 4.44. Their biggest drawback is their relative frequency instability with temperature, which could be an important requirement in certain applications.

4.38   Surface Acoustic Wave (SAW) and Bulk Acoustic Wave (BAW) Filters

The SAW filter takes advantage of the acoustic wave propagation in piezoelectric materials such as lithium niobate ($LiNbO_3$), lithium tantalate ($LiTaO_3$), bismuth germanium oxide (BGO), bismuth silicone oxide (BSO) or quartz (SiO_2), but unlike the quartz monolithic filter, which uses transversal propagation properties, the SAW uses mechanical waves that propagate at the surface of the solid substrate via actual motion of the particles, much like a water wave in an ocean. This is shown in Figure 4.45. A SAW filter usually consists of three sets of transducers, which are electrochemically deposited on the surface of a piezoelectric crystal as shown in Fig-

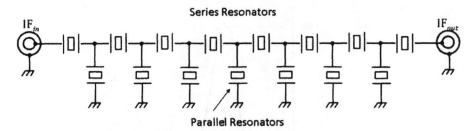

Figure 4.43   Electrical configuration of a 15-disk ceramic ladder filter.

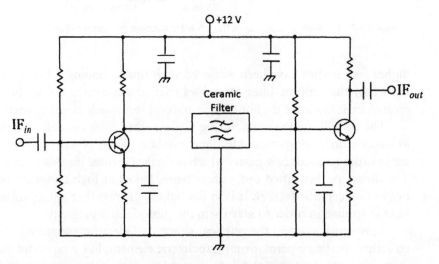

Figure 4.44   Typical circuit implementation of a ceramic filter in an IF.

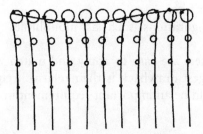

Figure 4.45   Propagation of a surface acoustic wave across a quartz crystal substrate showing actual motion of particles. Secondary vertical propagation is also shown by the smaller circles spaced at about a tenth wavelength from each other in the vertical plane.

ure 4.46. These thin aluminum electrodes (chosen for minimum mechanical mass) are spaced 1/4 or 1/2 wavelengths from each other.

If transducer A is excited at the frequency of interest, a surface wave will be generated and will travel simultaneously toward the intercepting transducers B and C, providing the same output at both locations. If only one receiving transducer was used, half the energy would be lost, and thus a higher insertion loss would result. The two outputs are electrically in phase in order to minimize the insertion

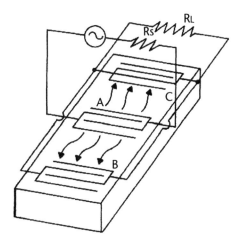

Figure 4.46   Typical implementation of a SAW filter. Transducer A emits surface waves in both directions to be intercepted by transducers B and C, which are electrically coupled together for minimum loss.

loss characteristic of the filter. Other methods of focusing the acoustic energy are also used.

The main advantage of the SAW filters is the relatively slow propagation rate of these types of waves compared with electromagnetic waves. This velocity varies by design but is generally 3×10^3 m/s, or about 100,000 times slower than the speed of light, allowing technological and physical feasibility for filters at very high frequencies. SAW filters for IFs have been manufactured between 30 MHz to 700 MHz, while SAW RF filters have been manufactured typically from 20 MHz to 3 GHz.

SAW filters are relatively lossy devices exhibiting anywhere from a 2-dB to as much as a 35-dB insertion loss. Their ultimate rejection is generally limited to around −35 to −40 dB. Percentage bandwidths of anywhere from 0.02% to 60% are possible. SAW filters are implemented using optical or electron beam lithography.

New 5.2-GHz SAW resonators and bandpass filters have been recently implemented for mobile communications using atomically flat surface aluminum nitride sapphire technology. A percentage bandwidth of 6% was realized along with an insertion loss of only 3 dB. The stopband was −25 dB.

As we go up in frequency, SAW filters designs become even more of a lithographic challenge. 10-GHz ladder design SAW filters have been recently implemented. They are fabricated on lithium niobate ($LiNbO_3$) using an electron beam direct writing and "lift-off" process. The aluminium electrodes had a width of 95 nm and a thickness of 30 nm. The experimental results showed an insertion loss of 3.7 dB. For more information on this type of filter, the reader is directed to the ample materials listed in the Selected Bibliography. Practical SAW filters have been realized at frequencies from 20 to 2,400 MHz. They are inherently small, rugged, and reliable, and require no adjustments. However, they suffer from the same problems as the monolithics, such as inferior ultimate rejection due to interaction of unpredicted modes of wave propagation. SAW filters are ideally suited for mass production because of the lithographic manufacturing techniques that can be used to produce this type of filter.

BAW filters take advantage of thin film resonators, which are very similar to quartz crystals. They are scaled down in size to adapt to the higher frequencies that they usually operate at. In a BAW filter, a piezoelectric film is sandwiched between two metal films to resemble a ladder or lattice approach implemented usually in Butterworth or Chebyshev approximations. The most common BAW filter implementation is a ladder approach resembling a quartz or ceramic ladder filter consisting of series resonators combined with parallel elements, which are usually tuned to a lower frequency to achieve the bandpass function. As with all high Q filters, the maximum rejection is determined by the number of elements at the cost of increased insertion loss.

BAW filters are typically designed from 0.5 to 16 GHz. However, VHF BAW filters have also been fabricated. They can be smaller than SAW filters and are relatively inexpensive to manufacture, making them useful in wireless applications such as WiMAX and WiLAN. They usually exhibit typical Qs of around 1,000 and insertion losses of about 0.5 dB per pole. Shape factor and stopband are relatively inferior. In general, BAW resonators and filters look very promising when used for high-volume mobile communication systems requiring a small size and low cost.

4.39 Technological Trade-Offs in Intermediate Frequency (IF) Filters

We discussed several methods of achieving selectivity in the IF portion of a superheterodyne receiver. New ways of IF filtering have been developed, and among them are the optical, baseband, and bandpass digital signal processing (DSP) filters, which will be discussed in separate chapters. These new filters provide certain advantages over the previous technologies and introduce their own problems as well. Table 4.3 shows a list of technological trade-offs between the various methods discussed, and provides us with a better understanding of their merits and limitations as we apply them to receivers in the following chapters.

Table 4.3   Advantages and Disadvantages of Several Technologies Used in IF Filter Design

Technology	Advantages	Disadvantages
L/C	Inexpensive	Low Q.
Mechanical	Outstanding shape factor. Stable with temperature.	Limited in frequency to below 1 MHz. Relatively expensive.
Quartz ladder, lattice, monolithic	Very high Q. Stable with temperature.	Limited in frequency to about 150 MHz. Narrow bandwidths cannot be achieved at higher frequencies.
Ceramic, ladder	Smaller than mechanical. Inexpensive.	Limited in frequency to below 1 MHz.
Surface acoustic wave (SAW), bulk acoustic wave (BAW)	Higher in frequency than quartz ladder, lattice, and monolithic, Small, inexpensive.	Poor ultimate rejection. Relatively high insertion loss.
Optical	Higher in frequency than SAW. Very good percentage bandwidth.	Unstable with temperature, expensive
Digital (DSP)—baseband and bandpass	Easily adaptive and versatile.	Limited in frequency below 2 GHz.

References

[1]   *Radio News Superheterodyne Book*, 1926.

[2]   Drentea, C., "Ultra-Wideband Fully Synthesized High-Resolution Receiver and Method," U.S. Patent 7,139,545.

[3]   Drentea, C., "Bragg-Cell Application to High Probability of Intercept Receiver," U.S. Patent 7,324,797.

[4]   Watkins, L. R., "Comprehensive Filter Design," Motorola Inc., 1997, pp. 5–54, available from the author at watkins.lee@ieee.org.

[5]   Zobel, O., "Theory of Electric Wave Filters," *Bell Systems Technical Journal*, January 1923.

[6]   Saal, R., and E. Ulrich, "On the Design of Filters by Synthesis," *IRE Transactions on Circuit Theory*, December 1958.

[7]   Geffe, P., *Simplified Modern Filter Design*, New York: John F. Rider, 1963.

[8]   Sverev, A. I., *Handbook of Filter Synthesis*, New York: John Wiley & Sons, 1967.

[9]   Adler, R., "Compact Electromechanical Filter," *Electronics*, 1947.

[10]   Becker, D., "Crystal Oscillators," *Electronics Now Magazine*, January 1993.

[11]   "5Spice Circuit Analysis and Simulation Program," http://www.5spice.com/.

Selected Bibliography

Adler, R., "Compacg Electromechanical Filter," *Electronics*, 1947.

Bouche, G., et al., "System Integration of BAW Filters," *MEMSWAVE 04*, 2004.

Butterworth, S., "On the Theory of Filter Amplifiers," *Wireless Engineer*, Vol. 7, 1930, pp. 536–541.

Cauer, W. "Die Verwirklichung der Wechselstromwiderst ände vorgeschriebener Frequenzabh ängigkeit," *Archiv für Elektrotechnik*, Vol. 17, 1926, pp. 355–388.

Cauer, W., *Siebschaltungen (Filter Circuits)*, (in German) Berlin: VDI-Verlag, 1931.

Cohn, S., "Dissipation Loss in Multiple Coupled Resonators," *Proceedings IRE*, August 1959.

Dilworth, I. J., "Quartz Crystals and Aperiodic Oscillators in RF Systems," *RF Magazine*, August 1988.

Drentea, C., "Local Area Networks," *Scientific Honeyweller*, December 1984.

Hashimoto, K., et al., "Optimum Leakey-SAW Cut of LiTaO$_3$ for Minimized Insertion Loss Devices," *IEEE Ultrasonic Symposium Proceedings*, 1997.

Ikata, O., et al., "Development of 800 MHz Band SAW Filters Using Weighting for the Number of Finger Pairs," *IEEE Ultrasonic Symposium Proceedings*, 1990.

Kawachi, O., et al., "Optimum Cut of LiTaO$_3$ for High Performance Leaky Surface Acoustic Wave Filters," *Proceedings IEEE Ultrasonic Symposium*, 1996.

Lakin, K. M., et al, "High Frequency Stacked Crystal Filters for GPS and Wide Bandwidth Applications," *IEEE 2001 Ultrasonics Symposium*, 2001.

Lakin, K. M., "Thin Film Resonator Technology," *IEEE 2003 FCS-EFTE*, 2003.

Mahon, S., and R. Aigner, "Bulk Acoustic Wave Devices—Why, How, and Where They Are Going," *CS MANTECH Conference*, Austin, TX, May 14–17, 2007.

Morita, T., et al., "SAW Resonator Filters for Mobile Communications," *Proceedings International Symposium SAW Devices for Mobile Communications*, 1992.

Nakamura, K., M. Kazumi, and H. Simizu, "SH-Type and Rayleigh-Type Surface Wave on Rotated Y-Cut LiTaO$_3$," *Proceedings IEEE Ultrasonic Symposium*, 1977.

Saal, R., *The Design of Filters Using the Catalog of Normalized Lowpass Filters*, Germany: Telefunken, 1966.

Sverev, A. I., and H. Blinchikoff, *Filtering in the Time and Frequency Domains*, New York: McGraw-Hill, 1981.

Sverev, A. I., "Introduction to Filters," *Electro-Technology*, June 1964.

Yamanouchi, K., "Generation, Propagation, and Attenuation of 10 GHz-Range SAW in LiNbO$_3$," *Proceedings IEEE Ultrasonic Symposium*, 1998.

Yamanouchi, K., and K. Shibayama, "Propagation and Amplification of Rayleigh Waves and Piezoelectric Leaky Surface Wave in LiNbO$_3$," *J. Appl. Phys.*, Vol. 43, No. 3, 1972.

CHAPTER 5

Implementing Double Conversions

We have seen that the choice of IF frequency has a great impact on a receiver's image rejection and its front-end preselector complexity. If using a low IF in a receiver with a wide RF input coverage, image can actually be in the input RF bandwidth requiring complicated switched front end filters, which can be cumbersome.

This is true not only in LF, HF, and VHF radios, but it can also impact ultrawideband downconvert professional microwave receivers, which can become physically bulky because of complicated front ends.

We have also learned about the rule of 35% upconvert concept, which moves the image frequency out of the RF passband to simplify the front end filters mechanism, minimizing the size of a receiving system and maximizing a receiver's performance.

Over the years, radio receivers (especially communications receivers) have seen a constant increase their IF frequency. Higher-frequency IFs have been implemented at frequencies still falling in the RF passband, but high enough to allow front end filters and switching to be simplified. This has been especially true in single conversion HF receivers, which utilize ultimate bandwidth IFs implemented with quartz filters at 9 MHz and 10.7 MHz.

At VHF/UHF and the higher frequencies, the trend remained to downconvert to lower IFs, thus combating image through complex switching front-end RF filter networks.

It is only relatively recent that receivers have used upconvert technologies to improve this situation. At HF, upconvert radios with the first IFs ranging from about 45 MHz through 115 MHz have been implemented; at microwave frequencies the trend has been similar. A state-of-the-art intelligent microwave receiver, which tunes from 2 GHz to 20 GHz using a first IF of 26 GHz (35% higher than 20 GHz) will be discussed in detail in Chapter 25.

However, the narrowband ultimate information bandwidth has been harder to realize in IF filters at the higher frequencies. The quartz crystal IF filters used in HF radios could not generally be narrow enough at these high IFs to match the information bandwidth because of the percentage bandwidth phenomenon. In addition, end-to-end gains in excess of 130 dB at the same IF frequency made it almost impossible to keep these IFs out of oscillation.

As a result, double, triple, and quadruple conversion receivers have resulted. They minimized the image problem by using high first IFs, thus simplifying the front end design, while balancing the large gains between different IF frequencies and using the best technologies in each of the conversions. This solved a few problems and introduced new ones, especially relating to intermodulation distortion produced by several mixers instead of just one. We will discuss the impact of products on the design of a receiver in Chapters 12 and 13.

An upconvert double conversion receiver scheme converts twice, just as its name implies. First, it uses a fixed or variable local oscillator (a synthesizer) to convert to a relatively higher frequency than our highest frequency to be received by at least 35%. As mentioned before, in an HF receiver, this IF can be anywhere from 45 MHz to 115 MHz. Then, a second conversion is performed to a lower frequency where ultimate bandwidth can be achieved easily using quartz crystal technology or the equivalent.

The choices for the IF center frequencies are not arbitrary, as a complex intermodulation distortion analysis has to be performed to ensure that in-band products do not exist below a seventh order, and out of band products can be rejected with proper IF filters shape factors.

We will use a convenient block diagram of a receiver, which is part of the author's Star-10 transceiver to prove some points. Throughout this book, we will refer to this model to discuss various areas of the design. These areas will be marked in darker gray in figures.

Shown in Figure 5.1 is the block diagram of the Star-10 transceiver. Although this is a fully bilateral HF transceiver (meaning the receiver and a transmitter use common circuits), we will use its receiver chain to discuss various design topics and state-of-the-art technologies.

We will first get acquainted with the system. Star-10 is a fully synthesized double conversion transceiver with a first IF at 75 MHz and the second IF at 9 MHz. Looking at Figure 5.1, the Star-10 transceiver system works as follows: The main antenna (at left) is being switched between the receiver and transmitter by the T/R control. The transceiver is always in the receive mode by default. The received RF signal from 2 to 30 MHz is passed through the automatically switched half-octave bandpass filter bank, which is controlled by the command and control board (DFCB). Only one filter out of eight is selected at any given time, ensuring equal image rejection anywhere within the RF frequency range. From here, the signal is fed to the first upconvert assembly (IF75BC) via a high dynamic range preamplifier and a programmable attenuator (AIPA).

It is in this assembly where the input signals are upconverted to the 75-MHz IF via a high level H-mode mixer using a high-performance synthesized LO, which tunes from 77 to 105 MHz. From here, the 75-MHz IF signal goes through a lowpass filter and a diplexer splitter circuit. Part of the 75-MHz IF information is directed to the 10-kHz-wide, 75-MHz quartz crystal roofing filter bank for further processing. The other half, which is 500 kHz wide, is directed to the IF9BC board for further conversion to the 9-MHz IF and to IF9NB for spectrum analyzer and noise blanker functions.

The 75-MHz roofing filter assembly is followed by the bilateral amplifier (BILAT AMP) assembly. This assembly allows the 75-MHz signals to be amplified with high fidelity by a CA2832 amplifier. This amplifier is a high gain (+36 dB), high intercept, class A type similar to those used in the IF75BC assembly.

The FL75 roofing filter assembly and BILAT AMP assembly are followed by the 9-MHz bilateral IF, IF9BC assembly where the second conversion to 9 MHz takes place. This assembly is equipped with the second AGC loop, called BIPA,

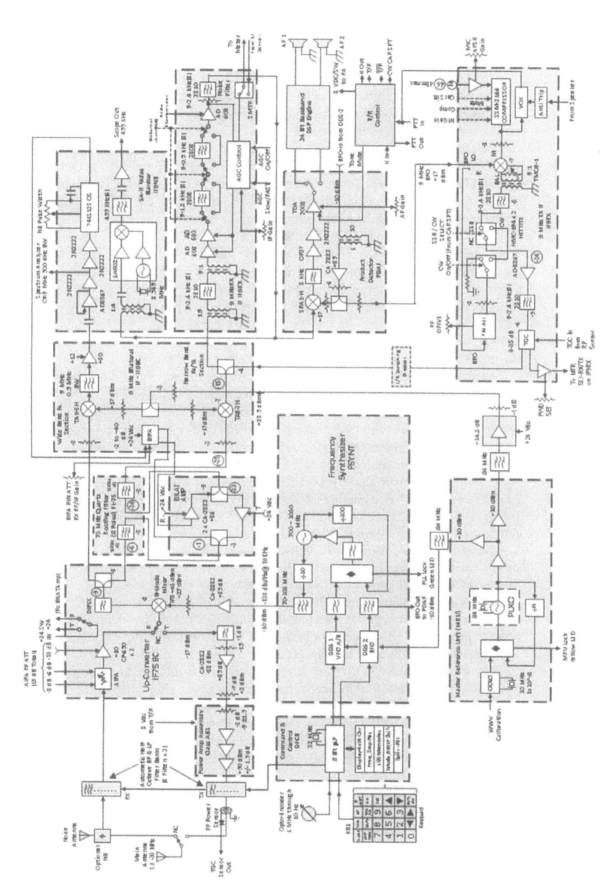

Figure 5.1 Block diagram of the Star-10 transceiver showing the double conversion receiver path.

which allows for 30 dB of adjustable gain action from the front panel RF/IF gain control.

The 75-MHz IF signals coming from the BILAT AMP assembly are converted here to the 9-MHz IF for the IF9NB and for the main 9-MHz receive IF assembly called IF9RX. This IF achieves the ultimate receiver bandwidth selection and amplification as commanded by the command and control assembly DFCB.

The IF9RX board provides 100 dB of gain (80-dB AGC) using three high dynamic range (high IP3) logarithmic/linear IF blocks. These are the Analog Devices AD-603 logarithmic linear amplifiers and provide the third amplifier used to compensate for variations in narrowband filter insertion loss and equal AGC/S meter indications regardless of the filter combination.

The last 2.4-kHz quartz filter is used to clean up noise from previous amplifiers. The IF bandwidth selection is provided by four 8-pole 9-MHz quartz crystal filters. Instead of selecting individual filters like in conventional IF designs, these filters are combined in a novel cascaded AND function (rather than an OR function) for a total of 32 poles (plus the 8 pole roofing filter) of total selectivity. The first and last 2.4-kHz filters set the maximum IF bandwidth of 2.4 kHz, while the 1.8-kHz and 0.5-kHz filters set narrow selections for different modes depending on the mode selected from the command and control.

As shown in Figure 5.1, the two 8- pole 9-MHz quartz crystal filters with a bandwidth of 2.4 kHz are always used at the beginning and the end of the 9-MHz IF chain for good noise management. Then, additional 8 pole crystal filters of narrower bandwidths are inserted or removed from the circuit using miniature RF relays between the gain stages (for a maximum of 32 poles) depending on the mode selection and as commanded by the DFCB.

Finally, the receiver chain is completed by feeding the IF9RX output to the product detector PDAF assembly. Here, the 9-MHz coherent BFO signal coming from the coherent DDS-driven synthesizer FSYNT enters the product detector section to be demodulated by the mode commands received from the DFCB assembly. The filtered and AGC-conditioned IF9RX output is finally presented to the PDAF product detector assembly. Here, the 9-MHz signals from the IF are mixed in a high-level mixer (the product detector) with a high-level BFO LO signal coming from the coherent synthesizer FSYNT. The resulting baseband is further filtered using a lowpass filter, which is then amplified and output to the baseband DSP section for processing. This double conversion 75-MHz upconvert, downconvert to a 9-MHz receiver is typical of modern HF receiver design.

The trend of double converting has not been limited only to HF receivers. At microwave frequencies, multiple conversion receivers have eliminated the need for very narrow first IFs by using brute force high dynamic range design elements and considerably wider first IF filters, which, in turn, allowed tuning inside the first IFs with the second conversion. This provided for fast scanning and analysis of large blocks (as large as the first IF bandwidth) of the RF spectrum using high performance brute force synthesizers, which are inherently clean and fast switching.

Such receivers, which can scan the microwave spectrum in blocks of 4 GHz or more at a time, will be discussed later in Chapter 25.

CHAPTER 6

Implementing Multiple Conversions

Triple-conversion receivers and even quadruple-conversion HF receivers appeared in the 1960s. These receivers did not necessarily upconvert but went so far as to use higher first IFs and bring the last conversion down in frequency to as low as 85 kHz or even 50 kHz, in order to achieve good L/C filters selectivity economically.

The Drake 2C was a famous example of such a receiver, and it used a triple-conversion scheme with the first IF between 3.5 and 4 MHz, the second IF of 455 kHz, and a third IF of 50 kHz. For economical reasons, the selectivity was not achieved at 455 kHz but at 50 kHz with the help of L/C filter circuits. It consisted of four tuned circuits of very high Q that were capacitively coupled in cascade as previously shown in Figure 4.20(b).

It was first believed that the more conversions used, the better a receiver was. These receivers featured a lot of gain, but the industry soon learned that more conversions meant noisier receivers because of the intermodulation distortion problem discussed earlier. Today, designers try to stay away from complex multiple conversion receivers; however, DSP filters in HF receivers are sometimes implemented at a low third IF sometimes around 36 kHz or lower despite the fact that bandpass A/Ds and D/As have been available at frequencies of up to 2 GHz.

Implementing Direct Conversions

Over the years, the search was on to find a simpler and more effective way to design superheterodyne receivers. As a result, a new kind of superheterodyne receiver was introduced: the direct-conversion receiver (DCR) type, also known as the zero IF receiver.

At first, the direct-conversion receiver seems attractive from an implementation point of view because of its simplicity. Rather than converting to an intermediate frequency IF, it converts directly to baseband promising imminent savings of using less expensive and fewer low-frequency parts, when compared with the higher costs of IF parts such as quartz crystal IF filters used in conventional upconvert or downconvert superheterodyne receivers.

With the rapid development of wireless communications, direct-conversion transceivers on a chip have experienced a forceful evolution. A very high degree of integration is now possible allowing for the use of low-voltage, low-power circuits with built-in image cancellation circuits right on the chip. These new designs have been used in many modern wireless transceivers, communications receivers, and radar.

Because of this perceived simplicity, the direct-conversion approach has gained popularity in modern applications such as WiMAX, WLAN, GSM, and Wi-Fi systems. Chances are good that you may be using a direct-conversion receiver in your mobile telephone today.

However, the perceived advantages of the direct-conversion receiver have met with new challenges, which have complicated the design of such systems beyond what was anticipated. The first problem is image rejection. Unlike some beliefs, image in a zero IF receiver is not "zero," but is actually located on the other sideband of the baseband signal being received within the baseband's bandwidth.

Image and its mathematics were discussed in detail in Chapter 4. Figure 7.1 shows the mathematical model for the image phenomenon. From a mathematical point of view, image in a superheterodyne receiver is a third-order mixer product (a "product" is not a harmonic but a product of harmonics, e.g., n times LO $+/-$ m times RF) that is received together with the desired signal in the IF). As explained in Chapter 4, image is caused by a signal removed in frequency from the desired signal by two times the IF frequency value. Both signals plus the image noise content (3 dB) are added and heard concurrently with the desired signal within the receiver's IF if precautions are not taken. Consequently, image is a direct function of the IF frequency choice.

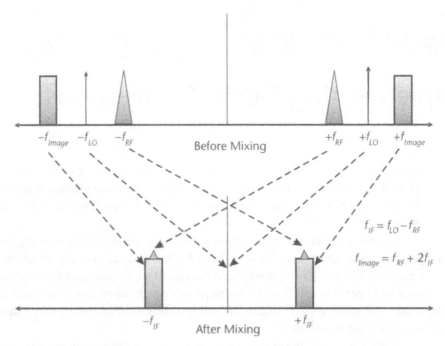

Figure 7.1 Image mathematics.

Once the image signal corrupts the first mixer of a superheterodyne receiver, it cannot be removed unless an image reject preselector filter is used at the input of the receiver. In a typical superheterodyne receiver, this problem is resolved before mixing occurs by using tracking or switched-in preselector filters. This was previously discussed in detail along with the choice of IF frequency.

In a direct-conversion receiver, however, implementing an image reject preselector filter that will separate the image frequency from the desired signal is not possible because of very close-in image frequency, which cannot be eliminated with normal preselector filters (see percentage bandwidth). This is depicted in the example from Figure 7.2, which shows a simple block diagram of a direct-conversion receiver. Its IF is at audio frequencies since the VFO runs practically at the received frequency, offset only by the center of the audio (or baseband) frequency range of interest, in our case 4 kHz. As can be seen in this example, the image is very close in, practically on the other sideband of the received signal. A front-end preselector filter would be impractical or impossible to construct for such a close proximity frequency component. The image and its noise content (3 dB) would actually fold over in the desired part of the signal and appear in the baseband IF disrupting the desired information.

Although the simplicity of a direct-conversion receiver is very attractive, rejecting the image content in this type of superheterodyne became the main challenge for designers of direct conversion architecture. Not only the image content folds over into the desired IF passband, but also the 3-dB noise content at the image frequency from the opposite sideband. This impacts the noise figure and sensitivity of such a system. Much hard work has been done recently to mathematically subtract the image content from the actual information using modern adaptive DSP algorithms in software-defined radios (SDR). We will discuss this later in this chapter.

The solution at hand was to cancel the image right in the mixer; thus, the image reject mixers resulted.

There have been several solutions regarding how to do this job, each with its advantages and disadvantages. Modern implementations of direct-conversion receivers have concentrated on mixer and postmixer image cancellation schemes combined with software and DSP algorithms as we will discuss next.

7.1   Image Reject Mixers

The objective of a direct-conversion image reject mixer is to process and suppress the image without utilizing any or minimum preselector filters. The idea is to process the signal and the image through different paths and cancel the image by its mirrored replica. Distinction between the signal and the image is possible because the two lie on different sides of the LO in the frequency domain (as shown in Figure 7.2), and phase and amplitude transformations can be exploited to separate and cancel the image signal, while allowing the actual IF content to pass. The practical implementation of such techniques have used precision film resistors and capacitors in conjunction with careful layout. Such an image reject mixer and an entire direct-conversion receiver can be fully integrated on a single chip, but physical and temperature instability of the components can impact DC offsets, the matching of the I/Q signals, and can cause crossmodulation between RF and LO signals (see Figures 7.3 and 7.4).

We will now discuss in detail some modern image reject methodologies. There are three well-known image reject architectures. Each has strengths and limitations. They are the Hartley, the Weaver, and the self-calibrating architecture that employs negative feedback.

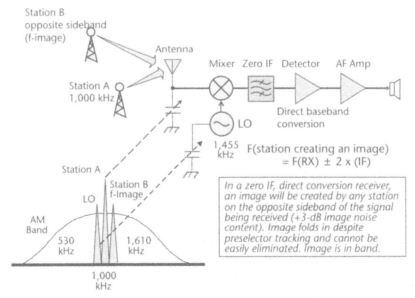

Figure 7.2   Image in a typical direct-conversion superheterodyne receiver is on the opposite sideband of the station being received, making it impractical to use a front-end image reject preselector filter.

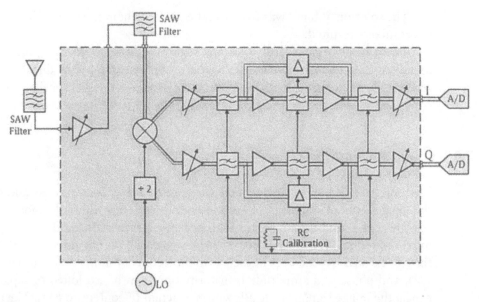

Figure 7.3   Modern direct-conversion receiver on a chip using precision RC calibration techniques for image rejection.

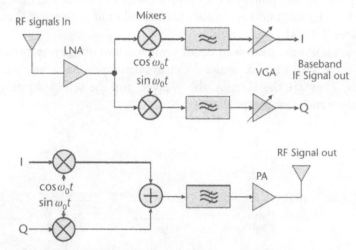

Figure 7.4   Modern implementation of a direct-conversion transceiver.

7.2   Hartley Architecture

The Hartley architecture is shown in Figure 7.5. The incoming RF signal is mixed with quadrature outputs of a local oscillator signal, namely $\sin \omega_{LO}t$ and $\cos \omega_{LO}t$. The mixers outputs are fed through lowpass filters as shown. The resulting signal at node X is out of phase by 90° with respect to the signal at node Y. The RC-CR networks provide fine tuning to achieve a near perfect cancellation of the image component. These parts are precision-trimmed film resistors and capacitors. The signals are then added at A and B resulting in cancellation of the image signal and leaving only the desired signal.

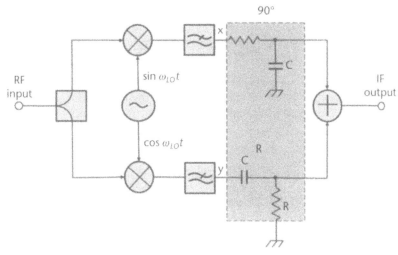

Figure 7.5   Hartley image reject mixer configuration.

The main disadvantage of the Hartley architecture is its sensitivity to mismatch-es. If the gain and phase of the two signal paths are not perfectly matched, image rejection is greatly reduced. Sources of mismatch include quadrature generation er-rors and the inaccuracy of R and C parameters due to the physical implementation process and/or temperature variations.

7.3   Weaver Architecture

To alleviate the mismatch problem associated with the instability of the 90° shift network in the Hartley architecture, the precision RC-CR networks have been re-placed by a second quadrature mixing stage. The resulting topology is the Weaver architecture, which is illustrated in Figure 7.6. It can be seen from this Figure that by adding the new second-stage quadrature mixers, the system produces outputs whose difference cancels the image while maintaining the desired signal.

By avoiding the RC-CR networks of the Hartley approach, the Weaver archi-tecture achieves greater image rejection despite process and temperature variations.

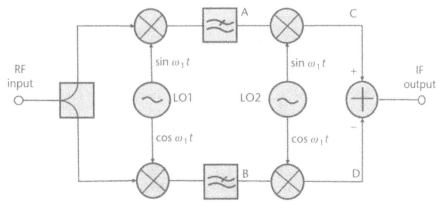

Figure 7.6   Weaver image reject mixer configuration.

However, image rejection remains a critical problem, because the circuit still depends on the precise cancellation of the image tone through resolving gain and phase discrepancies between the signal paths.

The Hartley and Weaver architectures eliminate the need for a preselector filter, but their performance degrades considerably in the presence of gain and phase mismatches. They depend on good symmetry of layout techniques to ensure optimum matching. A typical image rejection of 30 to 35 dB is achieved with these methods. This is below the 60- to > 70-dB image rejection performance of an upconvert superheterodyne scheme, which uses a preselector.

7.4 Self-Calibrating Architecture

An improvement over the Hartley and Weaver techniques has been obtained with the self-calibrating topology. It determines the phase and amplitude errors similarly to the Weaver architecture. However, a better performance is achieved by using the signals obtained to force the correction signal magnitude towards zero by utilizing an additional negative feedback loop. A system block diagram showing the generation of the phase error signal is shown in Figure 7.7. The darker shadowed elements are additions to the previous schemes.

The two additional mixers in the path generate the signal, $\cos(\omega_{IF}t)$, which is mixed again with the output voltage, V_{out}, in order to correct the phase mismatch. The phase error signal obtained is given by:

$$V_\theta = AV_m \sin(\theta/2) \qquad (7.1)$$

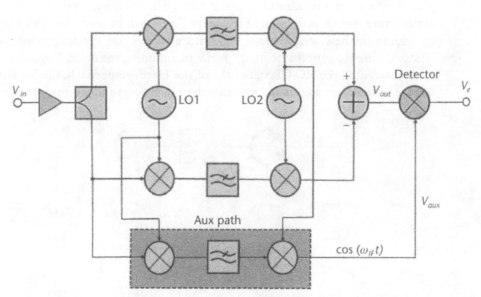

Figure 7.7 Self-calibrating image reject mixer improves on the Hartley and Weaver methods by detecting mismatches caused by temperature and layout changes and introducing corrective negative feedback.

where A is the nominal voltage gain of each quadrature path and V_m is the amplitude of the image signal.

A new variation on this approach is the self calibrating method shown in Figure 7.8. It can be seen from this design that the phase variance information provided by V_θ can be used in a feedback loop utilizing variable delay cells that can automatically correct the signal amplitude and fine tune the local oscillators phase. This process occurs periodically during a semiautomatic calibration process in which LNA_1 is turned off, LNA_2 is turned on at $S2$, and $S1$ is turned on to close the feedback loop. The delay cells (shown at Δ) are then quickly varied differentially using a calibration image signal, which is applied at V_{cal} (it is sometimes swept) in order to tune out the phase and amplitude inconsistencies. After the feedback loop settles, $S1$ opens and LNA_1 is switched back in the circuit at $S2$. The final value of V_θ is stored automatically across capacitor C1 until the next calibration time. This process repeats periodically at an arbitrarily chosen time.

A problem with this image rejection technique is the need to turn off (mute) the receiver, not necessarily a desirable feature in certain applications, and periodically recalibrate the system by updating the value of V_θ stored in capacitor C_1. Furthermore, with the increased complexity, the power consumption increases accordingly, which could be important in certain applications. This method is reported to improve the image rejection ratio from 17 dB to 57 dB, a considerable improvement over previous open-loop systems.

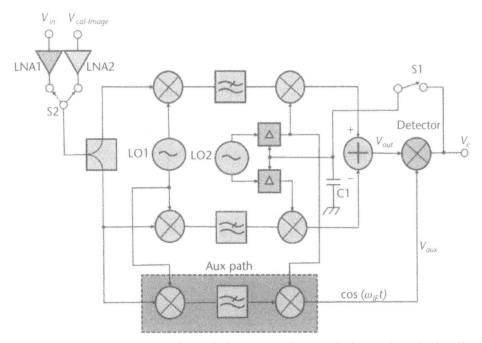

Figure 7.8 Image reject tone sampling technique uses semiautomatic phase and amplitude calibration loop. The process occurs during a calibration process in which LNA_1 is turned off, LNA_2 is turned ON with $S2$ (to allow for a periodic image calibration signal), and $S1$ is turned on to facilitate the feedback process. The brief calibration process is repeated periodically. The entire receiver is muted during the calibration period.

7.5   Image Reject Mixer with Sign-Sign Least Mean Square (SS-LMS) Calibration Method

A more recent image reject mixer using digital tuning is shown in Figure 7.9.

In this design, which is based on the Weaver architecture, the gain and phase differences are being calibrated simultaneously by a sign-sign least mean square (SS-LMS) algorithm.

The SS-LMS adaptive image cancellation system, automatically adjusts the variable gain and phase in the delay cells so as to minimize the differences in the quadrature signal paths. The resultant difference of the two signal paths, $y(t)$, feeds a comparator to generate an error signal, $\varepsilon(t)$. The system also contains two additional mixers, M_x and M_y, which downconvert the signals at points A and B to produce $x_1(t)$ and $x_2(t)$, respectively. Using these inputs, the system updates the control coefficients w_1 and w_2, according to the LMS algorithm.

This complex calibration process is repeated in discrete time intervals until $\varepsilon(t)$ = 0. With this process, the proper values of w_1 and w_2 are found continuously and the calibration procedure is always complete.

Digital SS-LMS image reject systems offer several advantages over the previous methods. First, the phase and amplitude differences control signals are constantly available from the SS-LMS algorithm circuit. This eliminates the need for periodic refreshing. This is important in receivers that cannot afford to mute during calibration time. However, note the increased complexity; this is a considerable departure from the simple idea of the direct-conversion approach.

7.6   Image Reject Mixers Conclusions

Much work remains to be done in image reject mixers. With the demand for complex multimode RFICs in support of the new 3G (and beyond) services, fully

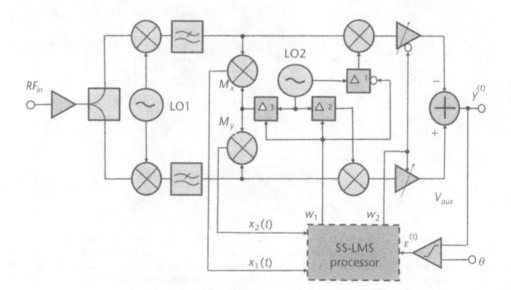

Figure 7.9   Image reject mixer with SS-LMS calibration.

monolithic wireless receivers on a chip continue to fuel the development of new architectures that eliminate the need for upconvert receivers and their front-end tracking or switching preselectors.

There are a multitude of direct-conversion image reject RFICs on the market today. When contemplating a new receiver design, the designer is advised to do a comprehensive technology versus performance and cost trade-off analysis between the envisioned direct-conversion image reject methodology against the classic downconvert or upconvert approaches using the preselector filter techniques.

Most image reject mixer parts achieve image cancellation by tuning out the gain and phase differences to match image rejection requirements at the cost of increased complexity and the need of periodically muting off the receiver for the re calibration process. While certain applications (e.g., TDMA, pulsed radar systems), allow for interlacing calibration times with receiver idle time slots, more work remains to be done in this area.

Direct-conversion receivers are still prone to LO leakage causing unwanted DC offsets. The more complicated Weaver and Hartley approaches avoid this to some degree. The cell phone handset industry have integrated DC receivers in their designs using low power analog IC's; however, they still have amplitude and phase errors to compensate for. Some use digital dividers to remove one source of error from the LO, but the analog errors still need trimming somewhere in the system. There are many IC devices available with integrated polyphase filters built in. Some use simpler implementations like the Havens technique. Low-noise, accurate analog quadrature demodulation still seems to be an intensive area for research, particularly for IC design.

Future trends in image reject mixers include the use of bandpass delta-sigma modulators where the I/Q unbalance can be compensated for in the frequency domain using digital signal processing. Functionally, bandpass delta-sigma modulators have been previously used as digitally-tuned, frequency notch filters, which enhance SNR in DSP IFs, DDS, or PLL synthesizers. Using this technology in image cancellation circuits can provide in excess of 70-dB image rejection. This competes with the upconvert superheterodyne receiver techniques that use front-end preselector filters. However, the complexity and cost of the newer technology has also increased.

A new double-quadrature mixer used in conjunction with a complex delta-sigma modulator filter architecture has been recently reported in the literature. It can achieve a phase accuracy of less than 0.3°. Although this is a powerful way to overcome the sensitivity of the current technology to phase unbalance, it does not yet resolve the problem of sensitivity to amplitude unbalance. More work remains to be done in this area.

Image reject direct-conversion receivers have been used extensively in current wireless as well as other equipment. They appear to provide savings by using lower frequency gain blocks and components instead of the more expensive, higher frequency active components and filters utilized in upconvert receivers. However, they have met with increasing technical challenges. Additional problems with the direct-conversion receiver have been associated with in-band LO leakage back through the antenna. The high linearity needed is not achievable without paying a price to minimize second and third order distortion. As a result, dynamic range can suffer if the offsets are not constantly maintained over temperatures, which presents additional technological challenges.

7.7 Image Recovery Receivers

Image does not necessarily have to be our enemy. It can be employed to our advantage in image recovery receiving systems that can be used for frequency diversity. Frequency diversity receiving systems have been employed to increase the probability of intercept of information by providing concurrent receiving paths either by frequency or space/antenna diversity. The idea is simple: if one of the propagation paths is impeded, communication is still possible through the remaining active paths, increasing the probability of intercept to an ideal 100%. Such systems have been used in voice and data radio communications, but require at least doubling the amount of equipment used.

A unique mathematical opportunity exists to exploit the image in a superheterodyne receiver, which provides frequency diversity by utilizing the image problem.

Using cleverly chosen IFs in receiver design, we can do the job of two receivers with half the parts in a single receiver design.

A case in point is the patent 4,584,716—Automatic Dual-Diversity Receiver [1]. This invention allows for a single radio receiver to receive two preselected frequencies at the same time, by choosing an LO frequency midway between the two frequencies and an IF equal to one half of the difference of the two selected frequencies.

The patent builds on a particular mathematical example of a receiver, which takes advantage of image to hear two identical signals on two different frequencies at once. Such is the case with the WWV stations, which emit time signals at 5 MHz, 10 MHz, 15 MHz, and 20 MHz. In the absence of one signal, a typical radio receiver has to be retuned to a frequency that is not blocked. In critical situations, however, important information could be lost during the retuning process. A single image recovery receiver, which could receive more than one station at the same time would be more advantageous. Figure 7.10 shows the block diagram of the image recovery receiver as applied to the WWV transmissions.

An image recovery receiver such as the one shown in Figure 7.10 is equipped with a single antenna and an IF of 2.5 MHz. It is capable of receiving simulcast WWV time and frequency broadcasts on two separate frequencies, 5 MHz and 10 MHz, both at the same time. When propagation conditions prevent one signal frequency (10 MHz) from being heard, the other signal frequency (5 MHz) could still be heard, increasing the probability of intercept (POI) and vice versa. The image caused by two times the 2.5-MHz IF facilitates the diversity function as explained earlier in Figure 7.10.

Such a receiver has been successfully implemented at the cost of the 3-dB noise addition, which is not that important at HF frequencies. Additional RF microwave receivers based on this concept have been successfully implemented.

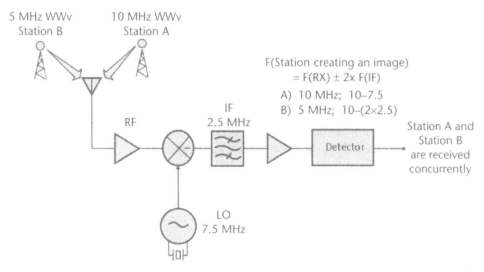

Figure 7.10   Image-recovery receiver concept uses an LO frequency midway between two WWV frequencies (at 5 and 10 MHz) and an IF equal to one half of the difference of the two selected frequencies. This receiver can receive either or both of the WWV signals at the same time providing frequency diversity with only half the amount of parts. See [1].

Reference

[1]   "Automatic Dual-Diversity Receiver," U.S. Patent 4,584,716.

Selected Bibliography

Abdennadher, S., and S. A. Shaikh, "Practices in Mixed-Signal and RFIC Testing," *IEEE Design and Test of Computers*, July–August 2007.

Abidi, A. A., "Direct-Conversion Radio Transceivers for Digital Communications," *IEEE Journal of Solid-State Circuits*, December 1995.

Abidi, A. A., G. J. Pottie, and W. J. Kaiser, "Power-Conscious Design of Wireless Circuits and Systems," *Proceedings of the IEEE*, Vol. 88, No. 10, October 2000.

Abidi, A. A., "Direct-Conversion Radio Transceivers for Digital Communications," *IEEE Journal of Solid-State Circuits*, Vol. 30, No. 12, December 1995.

Akbay, S. S., et al., "Alternate Test of RF Front Ends with IP Constraints: Frequency Domain Test Generation and Validation," *Proceedings of IEEE International Test Conference (ITC)*, October 2006.

Arabi, K., and B. Kaminska, "Testing Analog and Mixed-Signal Integrated Circuits Using Oscillation-Test Method," *IEEE Trans. on Computer-Aided Design of Integrated Circuits and Systems*, Vol. 16, No. 7, July 1997.

Aspin, F. J., "RF System Board Level Integration for Mobile Phones," in *Circuits and Systems for Wireless Communication*, M. Helfenstein and G. S. Moschytz, (eds.), Boston, MA: Kluwer Academic Publishers, 2000.

Bateman, A., and D. M. Haines, "Direct Conversion Transceiver Design for Compact Low Cost Portable Mobile Radio Terminals," *Proceedings IEEE Vehicular Technology Conference.*

Behbahani, F., et al., "CMOS Mixers and Polyphase Filters for Large Image Rejection," *IEEE Journal of Solid-State Circuits*, Vol. 36, No. 6, 2001.

Bopp, M., et al., "A DECT Transceiver Chip Set Using SiGe Technology," *IEEE International Solid State Circuits Conference.*

Cavers, J. K., and M. W. Liao, "Adaptive Compensation for Imbalance and Offset Losses in Direct Conversion Transceivers," *IEEE Trans. on Vehicular Technology*, Vol. 42, November 1993, pp. 581–588.

Chang, Y., and J. Choma, Jr., "A Monolithic RF Image Reject Filter," *Proceedings IEEE Southwest Symposium on Mixed Signal Design*, San Diego, CA.

Chang, Z. Y., and D. Haspeslagh, "A CMOS Differential Buffer Amplifier with Accurate Gain and Clipping Control," *IEEE Journal of Solid-State Circuits*, Vol. 30, No. 7.

Chou, C. -Y., and C. -Y. Wu, "The Design of Wideband and Low Power CMOS Active Polyphase Filter and Its Application in RF Double-Quadrature Receivers," *IEEE Trans. on Circuits and Systems*, Vol. 52, No. 5, 2005.

Chow, J., "ECE 1352F, RF Image Reject Receivers," http://www.eecg.utoronto.ca/~kphang/papers/2002/jchow_imagereject.pdf.

Colebrook, F. M., "Homodyne," *Wireless World and Radio*, Rev. 13, 1924, p. 774.

Crols, J., and M. S. J. Steyaert, "A 1.5 GHz Highly Linear CMOS Downconversion Mixer," *IEEE Journal of Solid-State Circuits*, Vol. 30, No. 7, 1995.

Crols, J., and M. S. J. Steyaert, "A Single-Chip 900MHz CMOS Receiver Front-End with a High Performance Low-Topology," *IEEE Journal of Solid-State Circuits*, Vol. 30, No. 12, 1995.

Croon, J. A., et al., "An Easy to Use Mismatch Model for MOS Transistor," *IEEE Journal of Solid-State Circuits*, Vol. 37, No. 8.

Dabrowski, J., "Fault Modeling for RF Blocks Based on Noise Analysis," *Proceedings of IEEE International Symposium on Circuits and Systems (ISCAS)*, May 2004.

Demmerle, F., "Integrated RF-CMOS Transceivers Challenge RF Test," *Proceedings of IEEE International Test Conference (ITC)*, October 2006.

Der, L., and B. Razavi, "A 2-GHz CMOS Image Reject Receiver with LMS Calibration," *IEEE Journal of Solid-State Circuits*, Vol. 38, No. 2, February 2003.

Der, L., and B. Razavi, "A 2GHz CMOS Image Reject Receiver with Sign-Sign LMS Calibration," *International Solid-State Circuits Conference*, 2001.

Drentea, C., "Automatic Dual Diversity Receiver," U.S. Patent 4,584,716.

Efstathiou, D., and Z. Zvonar, "Enabling Components for Multi-Standard Software Radio Base Stations," *Wireless Personal Communications*, Vol. 13, No. 1, 2000.

Elmala, M. A. I., and S. H. K. Embabi, "Automatic Mismatch Calibration in Hartley Image Reject Receiver," *Proceedings IEEE SOC Conf.*, September 2003.

Elmala, M. A. I., and S. H. K. Embabi, "Calibration of Phase and Gain Mismatches in Weaver Image Reject Receiver," *IEEE Journal of Solid-State Circuits*, Vol. 39, No. 2.

Ferrario, J., et al., "A Low-Cost Test Solution for Wireless Phone RFICs," *IEEE Communications Magazine*, September 2003.

Franke, S. J., "ECE 353—Radio Communication Circuits," Department of Electrical and Computer Engineering, University of Illinois, Urbana, IL, 1994.

Goldfarb, M., et al., "Analog Baseband IC for Use in Direct Conversion W-CDMA Receivers," *Proceedings 2000 IEEE RFIC Symposium*, Boston, MA, July 2000.

Grace, D., and H. Iwatsubo, "RF Filter Technology for Wireless Communications," *Wireless Design & Development*, June 1996.

Gray, P. R., and R. G. Meyer, *Analysis and Design of Analog Integrated Circuits*, 3rd ed., New York: John Wiley & Sons, 1993.

Hairapetian, A., "An 81 MHz receiver in CMOS," *IEEE International Solid State Circuits Conference*.

Hartley, R., "Modulation System," U.S. Patent 1,666,206, April 1928.

Hartley, R., "Single-Sideband Modulator," U.S. Patent 1,666,206, April 1928.

Hassan, "A New Generation of Global Wireless Compatibility," *IEEE Circuits and Devices Magazine*, January 2001.

Hassan, et al., "A Buffer-Based Baseband Analog Front End for CMOS Bluetooth Receivers," *IEEE Trans. on Circuits and Systems-II: Analog and Digital Signal Processing*, August 2002.

Hassan, O. Elwan, and M. Ismail, "CMOS Low Noise Class AB Buffer," *Electronics Letter*, Vol. 35, No. 21, October 1999.

Hsu, C. -C, and J. -T, Wu, "Highly Linear 100 MHz CMOS Programmable Gain Amplifier," *Proceedings IEEE*.

Huelsman, L. P., and P. E. Allen, *Introduction to the Theory and Design of Active Filters*, New York: McGraw-Hill, 1980.

Itoh, K., et al., "2 GHz Band Even Harmonic Type Direct Conversion Receiver with ABB-IC for W-CDMA Mobile Terminal,'" *Proc. 2000 IEEE MTT-S Microwave Symposium*, Boston, MA, July 2000.

Lee, T. H., and S. S. Wong, "CMOS RF Integrated Circuits at 5 GHz and Beyond," *Proceedings of the IEEE*, Vol. 88, No. 10, October 2000.

Long, J. R., and M.C. Maligeorgos, "A 1V 900 MHz Image Reject Downconverter in .5μ CMOS," *Proceedings CICC*.

Ma, H., et al., "Novel Active Differential Phase Splitters in RFIC for Wireless Applications," *IEEE Transactions on Microwave Theory and Techniques*, Vol. 46, No. 12, 1998.

Macedo, J., and M. Copeland, "A 1.9-GHz Silicon Receiver with Monolithic Image Filtering," *IEEE Journal of Solid-State Circuits*, March 1998.

Maligeorgos, J., and J. Long, "A 2.5 V 5.1-5.8 GHz Image Reject Receiver with Wide Dynamic Range," *IEEE International Solid State Circuits Conference*.

Maurer, L., et al., "Influence of Receiver Front end Nonlinearities on W-CDMA Signals," *Proc. 2000 Asia-Pacific Microwave Conference*, Sydney, Australia, December 3–6, 2000.

Maurer, L., et al., "On the Design of a Continous-Time Channel Select Filter for a Zero-UMTS Receiver," *Proceedings 2000 IEEE 51st Vehicular Technology Conference*, Tokyo, Japan, May 2000.

Meyers, B.A., et al., "Design Considerations for Minimal-Power Wireless Spread Spectrum Circuits and Systems," *Proceedings of the IEEE*, Vol. 88, No. 10, October 2000.

Mikkelsen, J. H., T. E. Kolding, and T. Larsen, "RF CMOS Circuits Target IMT-2000 Applications," *Microwaves & RF*, July 1998.

Milor, L. S., "A Tutorial Introduction to Research on Analog and Mixed-Signal Circuit Testing," *IEEE Trans. on Circuits and Systems*, Vol. 45, No. 10, October 1998.

Mitola, J., "Software Radios," *IEEE Communications Magazine*, No. 5.

Mohr, W., and W. Konhäuser, "Access Network Evolution Beyond Third Generation Mobile Communications," *IEEE Communications Magazine*, Vol. 38, No. 12.

Montalvo, A., et al., "A 22mW NDAC Receiver Chip with Integrated Second Channel Filtering," *Digest of Technical Papers, IEEE International Solid-State Circuits Conference.*

Montemayor, R., and B. Razavi, "A Self Calibrating 900 MHz CMOS Image Reject Receiver," *European Solid State Circuits Conference*, September 2000.

Murmann, B., "Digitally Assisted Analog Circuits," *IEEE Micro.*, Vol. 26, No. 2, March–April 2006.

Ogawa, S., N. Kondo, and N. Watanabe, "A Buffer-Based Algorithmic Analog-to-Digital Converter," *Proceedings IEEE ISCAS*, 1989.

Ozev, S., C. Olgaard, and A. Orailoglu, "Multi-Level Testability Analysis and Solutions for Integrated Bluetooth Transceivers," *IEEE Design and Test of Computers*, Vol. 19, No. 5, September–October 2002.

Pärssinen, A., et al., "A 2-GHz Wide-Band Direct Conversion Receiver for WCDMA Applications," *IEEE Journal of Solid-State Circuits*, Vol. 34, No. 12, December 1999.

Pelgrom, M. J. M., A. C. J. Duinmaijer, and A. P. G. Welbers, "Matching Properties of MOS Transistors," *IEEE Journal of Solid-State Circuits*, Vol. 24, No. 5.

Pretl, H., et al., "A SiGe-Bipolar Down-Conversion Mixer for a UMTS Zero- Receiver," *Proceedings IEEE Bipolar/BiCMOS Technology Meeting*, Minneapolis, MN, September 2000.

Pretl, H., et al., "A WCDMA Zero-Front-End for UMTS in a 75 GHz SiGe BiCMOS."

Pun, K. P., J. E. Franca, and C. A. Leme, "A Quadrature Sampling Scheme with Improved Image Rejection for Complex Receiver," *Proceedings IEEE ISCAS.*

Pun, K. P., J. E. Franca, and C. A. Leme, "Basic Principles and New Solutions for Analog Sampled-Data Image Rejection Mixers," *Proceedings IEEE ICECS.*

Pun, K. P., J. E. Franca, and C. A. Leme, *Circuit Design for Wireless Communication Improved Techniques for Image Rejection in Wideband Quadrature Receivers*, Boston, MA: Kluwer Academic Publishers.

Rampmeier, K., et al., "A Versatile Receiver IC Supporting WCDMA, CDMA and AMPS Cellular Handset Applications," *Proceedings 2001 IEEE RFIC Symposium*, Phoenix, AZ, May 2001.

Razavi, B., "Architectures and Circuits for RF CMOS Receivers," *IEEE Custom Integrated Circuits Conference*, 1998.

Razavi, B., "CMOS Technology Characterization for Analog and RF Design," *IEEE Journal of Solid State Circuits*, Vol. 34, No. 3, March 1999.

Razavi, B., "Design Consideration for Direct-Conversion Receivers," *IEEE Trans. on Circuits and Systems-II: Analog and Digital Signal Processing.*

Razavi, B., *Design of Analog CMOS Integrated Circuits*, New York: McGraw-Hill, 2001.

Razavi, B., *RF Microelectronics*, Englewood Cliffs, NJ: Prentice-Hall, 1997.

Rijns, J.J.F., "CMOS Low Distortion High Frequency Variable Gain Amplifier," *IEEE Journal of Solid-State Circuits*, Vol. 31, No. 7.

Rudell, J.C., "Issues in RFIC Design," lecture notes, University of California Berkeley/National Technological University, 1997.

Rudell, J. C., et al., "Recent Developments in High Integration Multi-Standard CMOS Transceivers for Personal Communication Systems," *International Symposium on Low Power Electronics and Design*, 1998.

Rudell, J. C., and J. J. Ou, "A 1.9 GHz Wide-Band Double Conversion CMOS Receiver for Cordless Telephone Application," *IEEE Journal of Solid-State Circuits*, Vol. 32, No. 2.

Ryynänen, J., et al., "A Dual-Band RF Front-End for WCDMA and GSM Applications," *IEEE Journal of Solid-State Circuits*, Vol. 36, No. 8, August 2001.

Sam, B., "Direct Conversion Receiver for Wide-Band CDMA," *Proc. Wireless Symposium*, Spring 2000.

Sampei, S., and K. Fecher, "Adaptive DC-Offset Compensation Algorithm for Burst Mode Operated Direct Conversion Receivers," *Proceedings IEEE Vehicular Technology Conference*.

Schacherbauer, W., et al., "A Flexible Multiband Frontend for Software Radios Using High and Active Interference Cancellation," *Proceedings 2001 IEEE MTT-S International Microwave Symposium*, Phoenix, AZ, May 2001.

Schelmbauer, W., et al., "A Fully Integrated Analog Baseband IC for an UMTS Zero-Receiver," *Proc. Austro Chip 2000*, Graz, Austria, October 2000.

Schwartzel, J., "Filtering and Frequency Control for the Next Generation of Mobile Communication Systems," *Proceedings EFTF*, 1994.

Sedra, A. S., and K. C. Smith, "Filters and Tuned Amplifiers," in *Microelectronic Circuits,* 3rd ed., New York: Saunders College Publishing, 1991.

Skyworks CX74017 Application Note, "On the Direct Conversion Receiver," http://www.ic-online.cn/IOL/viewpdf/CX74017_901462.htm.

Soma, M., "An Experimental Approach to Analog Fault Models," *Proceedings of IEEE Custom Integrated Circuits Conference (CICC)*, May 1991.

Song, B. S., "Low-Spurious ADC Architectures for Software Radio," in *Circuits and Systems for Wireless Communication*, M. Helfenstein and G.S. Moschytz, (eds.), Boston, MA: Kluwer Academic Publishers.

Song, B. S., and J. R. Barner, "A CMOS Double-Heterodyne FM Receiver," *IEEE Journal of Solid-State Circuits*, Vol. 21, No. 6.

Steyaert, M. S. J., et al., "A 2-V CMOS Cellular Transceiver Front-End," *IEEE Journal of Solid-State Circuits*, Vol. 35, No. 12, 2000.

Suma, S., et al. "Surface Mount Type Saw Filter for Hand-Held Telephones," *Proceedings Japan Electric Manufacturing Technology Symposium*, June 1993.

Tabbane, S., *Handbook of Mobile Radio Networks*, Norwood, MA: Artech House, 2000.

Tucker, D. G., "The History of the Homodyne and the Synchrodyne," *Journal of the British Institution of Radio Engineers*, April 1954.

Tucker, D. G., "The Synchrodyne," *Electronic Engineering*, Vol. 19, March 1947, pp. 75–76.

Tuttlebee, W. H. W., "Software-Defined Radio: Facets of a Developing Technology," *IEEE Personal Communications*, Vol. 6, No. 2, April 1999.

Ugajin, M., J. Kodate, T. Tsukahara, "A 1V 12 mW Receiver with 49 dB Image Rejection in CMOS/SIMOX," *2001 IEEE International Solid-State Circuits Conference Digest*, February 2001.

Valdes-Garcia, A., J. Silva-Martinez, and E. Sanchez-Sinencio, "On-Chip Testing Techniques for RF Wireless Transceivers," *IEEE Design and Test of Computers*, July–August 2006.

Vance, I. A. W., "Fully Integrated Radio Paging Receiver," *IEEE Proc.*, Vol. 129, No. 1, 1982, pp. 2–6.

Weaver, D. K., "A Third Method of Generation and Detection of Single Sideband Signals," *Proceedings of the IRE*, Vol. 44, December 1956, pp. 1703–1705.

Weigel, R., et al., "Highly Integrated Si/SiGe RFIC's for 3G Wideband-CDMA Mobile Radio Terminals," *Proc. MIKON 2002*, Gdansk, Poland, May 2002.

Willingham, S. D., et al., "A BiCMOS Low Distortion 8 MHz Low Pass Filter," *IEEE Journal of Solid-State Circuits*, Vol. 28, No. 12.

Wu, S., and B. Razavi, "A 900 MHz 1.8 GHz CMOS Receiver for Dual-Band Applications," *IEEE Journal of Solid-State Circuits*.

Yamaji, T., N. Kanou, and T. Itakura, "A Temperature Stable CMOS Variable Gain Amplifier with 80 dB Linearity Controlled Gain Range, " *IEEE Journal of Solid-State Circuits*.

Yuanjin, Z., "Fully Integrated Self-Tuned Image Rejection Downconversion System," U.S. Patent 6892060.

CHAPTER 8

Special Conversions and Their Implementation

As an offshoot of the direct conversion receiver, some very demanding specialized HF applications have been known to use switchable quartz filters as preselectors at the input of a radio receiver. The idea allows for a very high dynamic range to be achieved at the cost of versatility. Some companies manufactured a single-conversion downconversion receiver with the IF around 16 kHz (see Figure 8.1).

The antenna is directly fed into a multipole crystal lattice or ladder filter with a bandwidth of approximately 2 kHz, allowing final selectivity to be achieved at the input of the radio before any mixing takes place. This filter is relatively lossy and is followed by a low-noise amplifier and mixer, which translate the RF frequency directly to a 16-kHz IF where additional selectivity and amplification is obtained before detection takes place.

It can be seen that the image frequency is 32 kHz away from the received frequency and is greatly reduced by the preselector crystal filter; also, a minimum amount of products are created.

This is truly a superbly performing radio from a signal handling point of view. The receiver is intended for maritime communication where fixed channels are usually allocated to users. However, the percentage bandwidth phenomenon does not allow for a single quartz filter to cover additional channels. This receiver is practically a single channel receiver requiring additional quartz filters for more coverage.

Physically, the receiver measures approximately $1 \times 2 \times 4$ inches with the filter taking most of the space. About 12 separate receivers could be packaged in a box, and frequency change could be accomplished by turning the power on to the desired unit.

If we were to create a general-coverage HF receiver (0.5–30 MHz) based on this principle, we would be using about 14,750 crystal filters, which would tend to cost several million dollars. This would be very impractical, but it has its application. The idea has merit if the application is correct and could be carried into implementations at higher frequencies by changing the preselector technology to SAW or BAW filters.

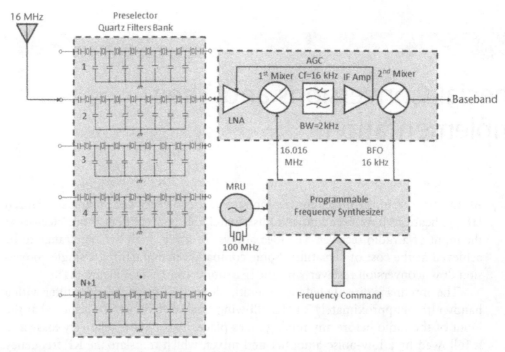

Figure 8.1   Block diagram of the special conversion receiver.

Drift-Canceling Loops and the Barlow-Wadley Receiver

One of the most ingenious receivers ever created is the Barlow-Wadley receiver, invented by Dr. T. L. Wadley of South Africa. This receiver features a drift and phase noise canceling loop technique, which was introduced by Dr. Wadley in the 1970s before frequency synthesis was widely available. The unusual method still remains with us today and can enter the design of many new receivers and synthesizer applications.

The Barlow-Wadley method uses a subtracting local oscillator mixing arrangement, combined with a COMB frequency generator, that actually cancels drift in the receiver without using many high-stability parts.

Figure 9.1 shows an example of such a receiver. Looking at the block diagram, this is a triple-conversion superheterodyne approach with the first IF at 55 MHz, the second IF at 2.5 MHz, and the third IF at 455 kHz. A COMB generator uses a 1-MHz quartz crystal oscillator to provide a multitude of signals at exactly 1-MHz intervals anywhere between 3 MHz to 32 MHz. These 29 harmonics are all fed to one side of the loop mixer, as shown.

A separate free-running variable oscillator, which oscillates between 55.5 MHz to 84.5 MHz (called the MHz oscillator), is manually controlled by the operator and is fed simultaneously to the loop mixer, as well as to the first mixer of the receiver where it combines continuously with the incoming preselected signal, generating the first IF centered at 55 MHz.

This information is further filtered and amplified through the first IF filter. The signal, which now has a bandwidth of 1 MHz (54.5 MHz to 55.5 MHz) and contains all of the RF signals present at the antenna within 1 megahertz of the tuned frequency, is finally fed to one side of the second mixer. The other side of the second mixer is powered from the loop mixer, through a narrow bandpass filter amplifier centered at 52.5 MHz.

It can be seen that the 52.5-MHz signal will only be true at precisely 1-MHz intervals as a result of the selective mixing process, which takes place in the loop mixer, as shown in Table 9.1.

The biggest advantage of this scheme is the drift-canceling mechanism provided for the MHz oscillator by the double-mixing approach. The drift is completely eliminated by the subtraction process in the second mixer, providing stable conversion for the second IF.

Table 9.1 Example of Drift-Canceling Selective Mixing Process in the Loop Mixer of a Typical Barlow-Wadley Receiver

MHz Oscillator Frequency (MHz)	Harmonic Generator Frequency Selected (MHz)	Local Oscillator Input to Loop Mixer (MHz)
55.5	3	52.5
56.5	4	52.5
57.5	5	52.5

The second IF is 1 MHz wide. It allows for the second IF to track inside the 3-MHz to 2-MHz IF (1 MHz) via a variable capacitor, which tracks together with the fine-tuning VFO, operating from 3.455 MHz to 2.455 MHz. This provides the fine resolution conversion inside the second IF to the third IF, which is centered at 455 kHz within any of the 1-MHz bands previously selected. The 455-kHz IF is further processed through narrow filters and amplifiers, and is finally detected using the BFO and a product detector and fed to the speaker via the audio amplifier.

A slight disadvantage of the Barlow-Wadley scheme is that the signals produced by the harmonic generator are usually heard at the beginning and the ends of each 1-MHz band as a result of system leakage, despite the fact that the COMB generator is usually shielded and placed far away from the critical areas. This problem, however, can be considered an advantage as it provides a means of dial calibration for the receiver.

Many variations on the Barlow-Wadley receiver have been implemented over the years, from the original Barlow-Wadley XCR-30 to Racal's RA-17 communications receiver, to the National HRO-500, and the FRG-7. The Barlow-Wadley scheme remains as a widely used scheme and a symbol of technical ingenuity to this day.

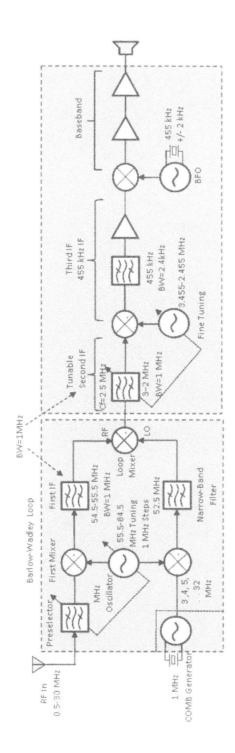

Figure 9.1 Block diagram of a typical Barlow-Wadley receiver.

CHAPTER 10

High Probability of Intercept (HPOI) and the Ideal Receiver

So far we have discussed some of the key receiver concepts and technologies with an accent on the superheterodyne as used for communications. However, radio receivers address many other applications ranging from guidance, to detecting low probability of intercept (LPI) radars, to radio surveillance, radio astronomy, SETI, and more.

Extensive efforts have been made to design and develop new kinds of receivers capable of performing in an increasingly complex RF environment while maintaining the high fidelity of the data. The purpose of these receivers is to capture as much of the RF spectrum as possible in a minimum amount of time and with the highest frequency resolution and dynamic range possible. These are the high probability of intercept (HPOI) receivers. They perform such tasks as gathering information from short pulse emitters like radars or pulsars, detecting faint reflections from flying objects, mapping the surface of the planet Venus through its sulfuric acid clouds, and capturing various manmade modulation formats, including spread spectrum.

Probability of intercept is simply defined as a measure of how likely it is that a transmission event will be detected. It depends on many things and is expressed in percentage. For instance, in the case of a known continuous AM transmission station, the probability of intercept (POI) is 100% even if using an inexpensive receiver. However, in other applications, such as surveillance or SETI, it depends on many other factors like knowing the exact frequency or frequency range over which the signal can be detected, its bandwidth, its transmission time, its repetition rate, and its antenna directivity factors. Equally important are such design elements as dynamic range with all its technical implications including the quality of local oscillators and synthesizers, the noise figure, and compression points of devices used in a receiver design. All these and more contribute to the POI.

To dramatize how receiver technology and design impacts POI, Figure 10.1 shows the difference between an inexpensive receiver [Figure 10.1(a)] versus a high dynamic range receiver [Figure 10.1(b)], both attempting to receive a weak CW signal hidden in the sidebands of a large signal close in frequency to it. In this case, the inexpensive receiver uses low dynamic range circuits and a noisy synthesizer, which interacts with the undesired signal obscuring the desired signal in the IF output and rendering its probability of intercept to zero, while the high performance receiver's probability of intercept would be 100%.

Most HPOI receivers today involve the superheterodyne approach. In the case of the scanning superheterodyne, the primary criterion in detection performance

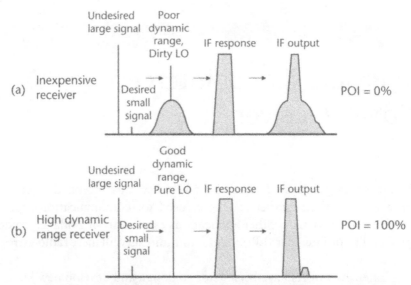

Figure 10.1 Detecting a small desired signal in the sideband of a powerful signal. (a) POI is zero in an inexpensive receiver with poor dynamic range and an inferior LO. (b) Probability of intercept is 100% in a receiver equipped with high dynamic range circuits and a pure LO.

uses the probability that an emitter will be detected on a single scan of a receiver almost instantaneously. An alternative criterion uses the cumulative probability of detection defined as the probability that an emitter has been detected at least once within a given close-in frequency range.

POI is the probability that a receiver system will detect, process, and identify an emitter within a specified time. It depends on many factors such as the probability of the emitter beam width being in the receiver's beamwidth (instantaneous field-of-view), pulse width of emitter, instantaneous bandwidth, receiver sensitivity, receiver resolution, receiver dwell time, receiver scan time, system throughput, stored emitter parameter data validation, reaction time constraints, the number and types of emitters that must be tracked per unit time, and much more.

Most of these properties are contradictory so a lot of trade studies need to be done upon the design of a receiver in order to come up with a compromise solution for a particular receiver design application.

The designs for very high dynamic range are of particular importance, as to avoid the problems associated with images and other spurious products. High performance, high resolution synthesizers, and master reference units (MRU) have been equally important. We will discuss all these design issues in the next chapters.

New, direct sampling digital receivers have recently appeared. They digitize and process the information as close as possible to the antenna.

Ideally, the perfect receiver would be a totally digital receiver defined as a high resolution, high dynamic range analog-to-digital converter (A/D) and a signal processor connected directly to the antenna. It would theoretically translate pulse or modulated RF signals of any frequency, bandwidth, duration, phase, and amplitude into digital words, which can be processed and manipulated using digital signal processing (DSP) techniques. In the purest definition form, this is also called a software-defined radio. In its ideal form, its software would also recognize the type of information received and adapt to it. The receiver could be reconfigured at any time to adapt to any possible application. In this case, the receiver would be called a cognitive radio. Such an ideal receiver is shown in Figure 10.2. Today, it is only realizable within the analog-to-digital data conversion technology limitations. It may one day become a reality.

Although the advances in analog-to-digital converters have been remarkable, they have not followed the geometric predictions of the Moore's law.

Unfortunately, while direct-sampling digital receivers can be implemented at the lower frequencies (mainly HF and VHF), their realization at higher frequencies has been limited due to technological limitations of the analog-to-digital (A/D) technology.

A digital sampling receiver operating at 1.5 GHz has been described in U.S. patent 6,882,310. Beyond 3 GHz, progress in high dynamic range A/Ds and direct sampling receivers has been relatively slow. In addition, with the increase in frequency, these devices have been the subject of high power consumption and a progressively reduced number of bits, meaning reduced dynamic range.

Problems remain with aperture jitter and transistor matching, as well as with increased transistor drain-bulk capacitance as the frequency goes up. Despite these problems, slow but constant progress has been achieved. New delta-sigma A/Ds implemented in gallium arsenide, gallium nitride, or indium phosphide, as well as superconducting A/Ds, have evolved. Following these trends, a six-bit A/D converter with a 10-GHz effective resolution bandwidth has been developed. Its power consumption has been predicted at 4 watts.

Furthermore, undersampling has had limited results. Oversampling has been the rule rather than the exception. Digital sampling of RF behaves by the Nyquist criteria, which samples signals via a high-frequency reference of at least twice the highest received frequency. This sampling technique is actually a mixing process, which is mathematically identical to superheterodyning.

It can be seen that as the received frequency goes up, the reference oscillators require higher and higher frequencies with very low jitter. For instance, a 2-GHz direct sampling receiver requires a world-class master reference unit (MRU) with

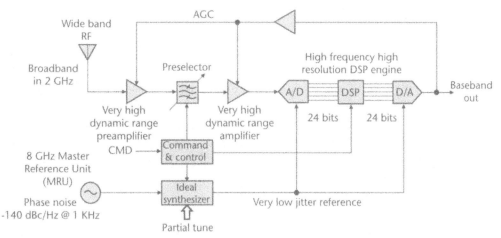

Figure 10.2 The ideal receiver processes RF directly at the antenna. It uses a Nyquist sampling technique utilizing a high-quality, synthesized MRU, which is at least twice the highest frequency to be received, and preferably up to eight times that frequency.

extremely low jitter, which in turn implies a phase noise performance of better than –140 dBc/Hz at 1 kHz at a frequency of 8 GHz. Such MRU can be very expensive. These topics and more will be discussed in much more detail in Chapters 16 and 25.

Beyond 2 GHz, radio receivers today still use various implementations of the scanning superheterodyne approach followed by digital signal processing engines at lower IF frequencies or by using Bragg cell Fourier transformers at IF frequencies in upconvert, downconvert approaches.

Real-time analysis of electromagnetic signatures using sweeping superheterodyne receivers has been somewhat problematic because of their time-consuming operation. Discovering and capturing single nonrepetitive transient RF events has been very demanding for analog and digital receiver designers. A minimum RF event is defined as the narrowest pulse that can be accurately captured by a receiver. It depends not only on the RF technology used, but also on the A/D clock rate in order to compute DFT transforms compatible with the minimum event duration.

In addition, swept-frequency detection can only detect the total power contained within the ultimate bandwidth of a receiving system as it is effectively swept across a wide frequency band of interest at a particular point in time. RF activity, which occurs in a particular range of the band being swept, may not be seen if the time of occurrence for the signal of interest is different than when the receiver is tuned to that frequency. Consequently, as the receiver sweeps, it will provide a low probability of intercept for transient signals of duration less than the entire recurrent sweep time. This makes the scanning superheterodyne receiver best suited for signals that remain stable for one or more complete sweep cycles.

Because of this, new and different types of superheterodyne receivers have evolved culminating with Bragg cell instantaneous processors used in wideband IFs, which allow receivers to scan and analyze wide ranges of RF in a minimum time. We will discuss these and more in Chapter 25.

Selected Bibliography

Drentea, C., "Direct Sampling GPS Receiver for Anti-Interference Operations," U.S. patent 6,882,310.

E2V Data Conversion Products, http://www.e2v.com/products/specialist-semiconductors/broadband-data-converters/.

Eklund, J. E., and R. Arvidsson, "A 10 Bit 120 M/S Multiple Sampling Single Conversion CMOS A/D Converter for I/Q Demodulation," *IEEE International Solid State Circuits Conference*.

Jantzi, S., K. Martin, and A. Sedra, "A Quadrature Bandpass $\Sigma\Delta$ Modulator for Digital Radio," *IEEE International Solid State Circuits Conference*.

Lundberg, K. H., "High-Speed Analog-to-Digital Converter Survey," http://web.mit.edu/klund/www/papers/UNP_flash.pdf.

CHAPTER 11

The Role of the Receiver in a Communications Link

We will now turn to the role of the receiver in a communications link. Receivers are an important part of communications links. The received signal has to be reproduced with high fidelity after the detection process takes place, regardless of what modulation method is being used and what propagation impediments (fading, multipath, and so forth) happen along the path between the transmitter and the receiver.

As a rule, a receiver has to be designed in concert with the communications link application. A communications link is comprised of a transmitter and a receiver accompanied by their respective antennas and a propagation path between them. The classic approach to communications links is shown in Figure 11.1. The detected power present at the receiver is normally addressed through the radar range equation minus the system losses plus the signal processing factor. In a radar case, the path length is double and has an additional element, the target cross-section, which enters the picture.

The propagation path between the two units is a variable not totally predictable, which is impacted by many factors such as free space attenuation, multipath, ducting, fading, and other natural or manmade phenomena. It is because of this variability that the propagation path is sometimes called the "disturbed" path or the "cloud." Measures have to be taken to compensate for this by providing a link margin in the system.

In a communications link, the role of the transmitter is to deliver the modulated RF power to its antenna in order to ensure that enough of a signal is present at the receiver antenna regardless of adverse path conditions. The receiver's role, in turn, is to ensure that sufficient link margin energy exists at the detector to cope with the variable conditions of the path. Thus, the communications link is completed for a specific data reliability expressed by a specific signal-to-noise ratio (SNR) or by a bit error rate (BER).

In order to overcome all possible conditions, a link analysis has to be performed before a receiver can be designed. Within this analysis, the link margin has to be addressed for the entire communications link system at both the transmitter as well as the receiver ends. At the transmitter site, we can increase power and/or provide a better antenna. At the receiver, we can provide more gain, use a better antenna, or use additional protocols and techniques (such as forward error correction) to ensure delivery of data. The link margin can also be increased by choosing more effective forms of modulation.

The link margin is calculated by comparing the expected received signal strength to the receiver sensitivity or threshold, which shows how much margin exists in the link between a positive reception state and a negative reception state for a given modulation. If more reliability is desired, more power and/or more effective modulation schemes can be used.

Although communication is concerned with information transfer, radar is concerned with signal detection in noise. Consequently, one of the most important factors in calculating the communication link margin is the E_b/N_o or signal-to-noise ratio (SNR) depending on the type of modulation being used.

E_b/N_o is expressed in decibels for a given bit error rate (BER) probability. Different types of modulations provide different E_b/N_o values, which can be used to our advantage. Typical E_b/N_o curves for several key RF data modulations are shown in Figure 11.2. They are used in deciding on the receiver (and transmitter) modulation scheme and consequently impact the design. In general, a bit error rate of $1.E^{-05}$ for 99.95% of the time is considered sufficient for most applications.

However, the curves in Figure 11.2 are not always the best solution. There are many other things to be considered in the design of a communications link, and consequently in a receiver design, depending on the physics and sometimes the economics of an application. For instance, in a deep space application at microwave frequencies, phase modulation techniques may not be recommended because of ionospheric scintillation, which can alter the data. Simple, old-fashioned, slow, noncoherent FSK modulation schemes may perform better in such applications despite the fact that the curves for other modulations look better.

Another example is using simplistic amplitude-shift keying (ASK) instead of the more sophisticated phase or coherent modulation schemes despite the fact that E_b/N_o curves are the worst for this form of modulation. ASK [or on/off keying (OOK)] is very economical and forgiving from a receiver frequency stability point of view and economics.

Since it is a form of keyed CW, the receiver can use simple AM diode detectors and it is not sensitive to frequency shifts caused by temperature impacts or vibration. This method is simply an AM modulation form that requires unsophisticated receivers, which can sometimes prove more reliable. Simply put, an ASK communications link can be very robust if using a receiver and demodulator equipped with an automatic threshold detector (see [1]), which can adapt to great variations in amplitude (rapid fading) and increase the probability of intercept by as much as 10 dB. We will discuss a low-cost, high-performance ASK data communications receiver design in Chapter 17.

Although not an AGC, the automatic threshold detector can also be implemented in software or DSP firmware by incorporating it into a digital slicer, which looks at the output of the receiver over a period of time (equivalent to, say, 10 or 20 bits). The slice level is set about one third of the way between the lowest output and the highest output, ensuring a high probability of intercept to decode a multipath or rapid fading data signal in some of the worst propagation conditions. Such an implementation can sometimes rival some of the more exotic forms of modulation, despite its simplicity.

The designer has to be very careful to choose just the right technique and receiver design. The link margin can be found from (11.1).

$$LM = EIRP - L_{path} + G_{Rx} - TH_{Rx} \qquad (11.1)$$

where:

LM is link margin expressed in decibels.

EIRP is the effective isotropically radiated power in dBm (or dBW). It is the total sum and difference of the transmitter power output plus the transmitter's antenna gain, minus miscellaneous losses of cables and connectors.

L_{path} is the total path loss in decibels, which includes all other predicted losses such as fading, rain, antenna pointing, and so forth.

G_{Rx} is the receive gain in decibels.

TH_{Rx} is the desired E_b/N_o expressed in decibels for a given bit error rate performance (see Figure 11.2).

We have seen that the link budget uses the transmitter EIRP power and the sum of all the gains and losses in a communications system to account for propagation losses and reproduce the received power. Then the noise level at the receiver antenna is estimated and compared against the receiver's internal noise. In the classical sense, a receiver has to be designed to ensure that its internally generated noise is less than the external noise present at its antenna for reliable delivery of the data received.

One of the first things a receiver designer has to do in a design is to determine the input frequency or frequency range over which the receiver will operate. This is to understand what natural noise sources and levels of noise are predominant at the frequencies of interest and determine the effective noise figure requirement for

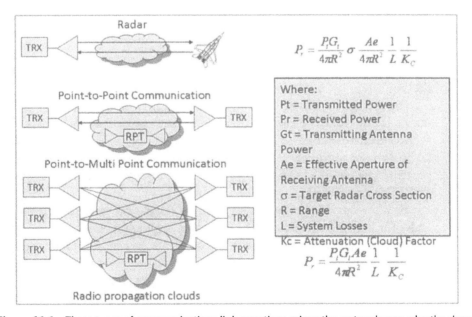

Figure 11.1 Three types of communications link equations minus the system losses, plus the signal processing factors. They are all based on the radar range equation.

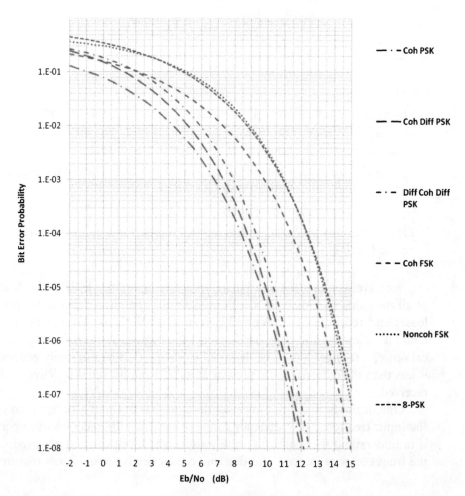

Figure 11.2 E_b/N_o values expressed in decibels for given bit error probabilities using some of the most popular RF data communications modulation schemes. Values are used in link analysis entering a receiver's design parameters.

the receiver. A noise figure lower than what the natural noise contributions are at the frequencies of interest may be unnecessary.

In order to do this, we find the sky noise contributions from Figure 11.3. Looking at Figure 11.3, we can see that the noise content from various noise sources, such as galactic or nonthermal backgrounds at frequencies from 100 MHz to 1,000 GHz, varies considerably. Depending on the frequency of interest, the noise temperature in Kelvin (K) can be anywhere from 10K at 1.4 GHz to about 1,000K at 100 MHz or 150 GHz.

Pick the curve relevant to the pointing (both with respect to galactic latitude and—for higher frequencies—the elevation angle) and read the corresponding sky temperature for the frequency of interest. At the lower frequencies these contributions are composed of a combination of factors including broadband electromagnetic radiation from earthbound sources. The lower the frequency is, the noisier the environment is. At the higher frequencies, the atmosphere acts like a "warm attenuator" in the path and its effect on the propagation has to be calculated on that basis.

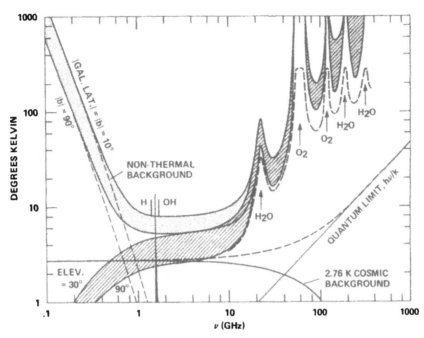

Figure 11.3 Noise contributions in Kelvin (K) seen by an antenna from various terrestrial and extraterrestrial sources from 100 MHz to 1,000 GHz.

Of course, the water hole frequency widow at 1.420 GHz offers the best overall noise performance as shown. This is why it was chosen as a meeting point for extraterrestrial communications. The signal shown represents the hypothetical strength of a transmitter similar to that of the Arecibo Observatory, located some 60 light years away and aimed at us.

Using the information obtained, the designer has to translate noise temperature expressed in Kelvin (K) into decibels (dB) in order to establish the noise figure for the new receiver design. This is found from Figure 11.4 or from the formula shown in this figure.

In addition to the link margin, a high probability of intercept receiver has to be designed to reject strong adjacent signals. This is not necessarily just an IF filter problem, but a complex intermodulation distortion problem. As simple as this may sound, the process of achieving high dynamic range performance is very involved with all aspects of receiver design such as the noise figure, the intermodulation distortion performance of all components, the master reference oscillator, the LO synthesizer phase noise, and more. These topics will be discussed in detail in the remaining chapters of this book.

We have seen that a receiver has to be designed as part of an entire communications link. Armed with the above information, the designer considers the link, including the transmitter power and modulation, its antenna gain or loss, over the path to the receiving antenna and only then, the design of the receiver can begin. Armed with the noise figure requirement and the link margin information, the receiver designer proceeds to find out the gain required of the receiver to reconstruct the transmitted information and provide the additional gain from the link margin to ensure positive data delivery in various path conditions.

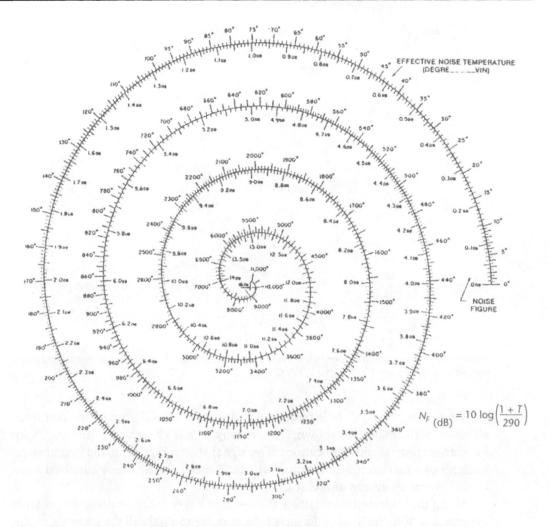

Figure 11.4 Nomogram of temperature translation from Kelvin (K) to decibels (dB). The noise equivalent from Figure 11.2 determines the noise figure of a receiver design.

The free space loss (FSL) of the link plus the link margin have to be mirrored by the receiver's gain dynamically. The receiver can be any of the superheterodyne or direct sampling types discussed, using pertaining technologies in line with the frequencies of operation.

The free space loss (FSL) is calculated using the basic Friis equation from (11.2) or (11.3). For actual calculations in decibels, (11.4) and (11.5) are used. In the calculations, dBs can be directly added or subtracted from dBm using Table 11.1, which translates watts into dBm and dBw. Using this table, one can calculate the EIRP of a transmitter directly in dBm by simply adding or subtracting dB from the dBm readings corresponding to the power in watts. The chart in Table 11.1 accommodates power levels from +60 dBm to −149 dBm.

$$FSL = \left(\frac{\lambda}{4\pi d}\right)^2 = \left(\frac{4\pi df}{c}\right)^2 \qquad (11.2)$$

Table 11.1   Useful Chart in Calculations Involving Power Translation from Watts to dBm

dBm	Power W	Volts	dBm	Power mW	Volts	dBm	Power uW	mVolts	dBm	Power nW	mVolts
60.0	1000.000	223.607	30.0	1000.000	7.071	0.0	1000.000	223.607	-30.0	1000.000	7.071
59.5	891.251	211.098	29.5	891.251	6.676	-0.5	891.251	211.098	-30.5	891.251	6.676
59.0	794.328	199.290	29.0	794.328	6.302	-1.0	794.328	199.290	-31.0	794.328	6.302
58.5	707.946	188.142	28.5	707.946	5.950	-1.5	707.946	188.142	-31.5	707.946	5.950
58.0	630.957	177.617	28.0	630.957	5.617	-2.0	630.957	177.617	-32.0	630.957	5.617
57.5	562.341	167.681	27.5	562.341	5.303	-2.5	562.341	167.681	-32.5	562.341	5.303
57.0	501.187	158.301	27.0	501.187	5.006	-3.0	501.187	158.301	-33.0	501.187	5.006
56.5	446.684	149.446	26.5	446.684	4.726	-3.5	446.684	149.446	-33.5	446.684	4.726
56.0	398.107	141.086	26.0	398.107	4.462	-4.0	398.107	141.086	-34.0	398.107	4.462
55.5	354.813	133.194	25.5	354.813	4.212	-4.5	354.813	133.194	-34.5	354.813	4.212
55.0	316.228	125.743	25.0	316.228	3.976	-5.0	316.228	125.743	-35.0	316.228	3.976
54.5	281.838	118.709	24.5	281.838	3.754	-5.5	281.838	118.709	-35.5	281.838	3.754
54.0	251.189	112.069	24.0	251.189	3.544	-6.0	251.189	112.069	-36.0	251.189	3.544
53.5	223.872	105.800	23.5	223.872	3.346	-6.5	223.872	105.800	-36.5	223.872	3.346
53.0	199.526	99.881	23.0	199.526	3.159	-7.0	199.526	99.881	-37.0	199.526	3.159
52.5	177.828	94.294	22.5	177.828	2.982	-7.5	177.828	94.294	-37.5	177.828	2.982
52.0	158.489	89.019	22.0	158.489	2.815	-8.0	158.489	89.019	-38.0	158.489	2.815
51.5	141.254	84.040	21.5	141.254	2.658	-8.5	141.254	84.040	-38.5	141.254	2.658
51.0	125.893	79.339	21.0	125.893	2.509	-9.0	125.893	79.339	-39.0	125.893	2.509
50.5	112.202	74.901	20.5	112.202	2.369	-9.5	112.202	74.901	-39.5	112.202	2.369
50.0	100.000	70.711	20.0	100.000	2.236	-10.0	100.000	70.711	-40.0	100.000	2.236
49.5	89.125	66.755	19.5	89.125	2.111	-10.5	89.125	66.755	-40.5	89.125	2.111
49.0	79.433	63.021	19.0	79.433	1.993	-11.0	79.433	63.021	-41.0	79.433	1.993
48.5	70.795	59.496	18.5	70.795	1.881	-11.5	70.795	59.496	-41.5	70.795	1.881
48.0	63.096	56.167	18.0	63.096	1.776	-12.0	63.096	56.167	-42.0	63.096	1.776
47.5	56.234	53.026	17.5	56.234	1.677	-12.5	56.234	53.026	-42.5	56.234	1.677
47.0	50.119	50.059	17.0	50.119	1.583	-13.0	50.119	50.059	-43.0	50.119	1.583
46.5	44.668	47.259	16.5	44.668	1.494	-13.5	44.668	47.259	-43.5	44.668	1.494
46.0	39.811	44.615	16.0	39.811	1.411	-14.0	39.811	44.615	-44.0	39.811	1.411
45.5	35.481	42.120	15.5	35.481	1.332	-14.5	35.481	42.120	-44.5	35.481	1.332
45.0	31.623	39.764	15.0	31.623	1.257	-15.0	31.623	39.764	-45.0	31.623	1.257
44.5	28.184	37.539	14.5	28.184	1.187	-15.5	28.184	37.539	-45.5	28.184	1.187
44.0	25.119	35.439	14.0	25.119	1.121	-16.0	25.119	35.439	-46.0	25.119	1.121
43.5	22.387	33.457	13.5	22.387	1.058	-16.5	22.387	33.457	-46.5	22.387	1.058
43.0	19.953	31.585	13.0	19.953	0.999	-17.0	19.953	31.585	-47.0	19.953	0.999
42.5	17.783	29.818	12.5	17.783	0.943	-17.5	17.783	29.818	-47.5	17.783	0.943
42.0	15.849	28.150	12.0	15.849	0.890	-18.0	15.849	28.150	-48.0	15.849	0.890
41.5	14.125	26.576	11.5	14.125	0.840	-18.5	14.125	26.576	-48.5	14.125	0.840
41.0	12.589	25.089	11.0	12.589	0.793	-19.0	12.589	25.089	-49.0	12.589	0.793
40.5	11.220	23.686	10.5	11.220	0.749	-19.5	11.220	23.686	-49.5	11.220	0.749
40.0	10.000	22.361	10.0	10.000	0.707	-20.0	10.000	22.361	-50.0	10.000	0.707
39.5	8.913	21.110	9.5	8.913	0.668	-20.5	8.913	21.110	-50.5	8.913	0.668
39.0	7.943	19.929	9.0	7.943	0.630	-21.0	7.943	19.929	-51.0	7.943	0.630
38.5	7.079	18.814	8.5	7.079	0.595	-21.5	7.079	18.814	-51.5	7.079	0.595
38.0	6.310	17.762	8.0	6.310	0.562	-22.0	6.310	17.762	-52.0	6.310	0.562
37.5	5.623	16.768	7.5	5.623	0.530	-22.5	5.623	16.768	-52.5	5.623	0.530
37.0	5.012	15.830	7.0	5.012	0.501	-23.0	5.012	15.830	-53.0	5.012	0.501
36.5	4.467	14.945	6.5	4.467	0.473	-23.5	4.467	14.945	-53.5	4.467	0.473
36.0	3.981	14.109	6.0	3.981	0.446	-24.0	3.981	14.109	-54.0	3.981	0.446
35.5	3.548	13.319	5.5	3.548	0.421	-24.5	3.548	13.319	-54.5	3.548	0.421
35.0	3.162	12.574	5.0	3.162	0.398	-25.0	3.162	12.574	-55.0	3.162	0.398
34.5	2.818	11.871	4.5	2.818	0.375	-25.5	2.818	11.871	-55.5	2.818	0.375
34.0	2.512	11.207	4.0	2.512	0.354	-26.0	2.512	11.207	-56.0	2.512	0.354
33.5	2.239	10.580	3.5	2.239	0.335	-26.5	2.239	10.580	-56.5	2.239	0.335
33.0	1.995	9.988	3.0	1.995	0.316	-27.0	1.995	9.988	-57.0	1.995	0.316
32.5	1.778	9.429	2.5	1.778	0.298	-27.5	1.778	9.429	-57.5	1.778	0.298
32.0	1.585	8.902	2.0	1.585	0.282	-28.0	1.585	8.902	-58.0	1.585	0.282
31.5	1.413	8.404	1.5	1.413	0.266	-28.5	1.413	8.404	-58.5	1.413	0.266
31.0	1.259	7.934	1.0	1.259	0.251	-29.0	1.259	7.934	-59.0	1.259	0.251
30.5	1.122	7.490	0.5	1.122	0.237	-29.5	1.122	7.490	-59.5	1.122	0.237

Table 11.1   Continued

dBm	Power pW	uVolts	dBm	Power fW	uVolts	dBm	Power fW	nVolts
-60.0	1000.000	223.607	-90.0	1000.000	7.071	-120.0	1.00000	223.607
-60.5	891.251	211.098	-90.5	891.251	6.676	-120.5	0.89125	211.098
-61.0	794.328	199.290	-91.0	794.328	6.302	-121.0	0.79433	199.290
-61.5	707.946	188.142	-91.5	707.946	5.950	-121.5	0.70795	188.142
-62.0	630.957	177.617	-92.0	630.957	5.617	-122.0	0.63096	177.617
-62.5	562.341	167.681	-92.5	562.341	5.303	-122.5	0.56234	167.681
-63.0	501.187	158.301	-93.0	501.187	5.006	-123.0	0.50119	158.301
-63.5	446.684	149.446	-93.5	446.684	4.726	-123.5	0.44668	149.446
-64.0	398.107	141.086	-94.0	398.107	4.462	-124.0	0.39811	141.086
-64.5	354.813	133.194	-94.5	354.813	4.212	-124.5	0.35481	133.194
-65.0	316.228	125.743	-95.0	316.228	3.976	-125.0	0.31623	125.743
-65.5	281.838	118.709	-95.5	281.838	3.754	-125.5	0.28184	118.709
-66.0	251.189	112.069	-96.0	251.189	3.544	-126.0	0.25119	112.069
-66.5	223.872	105.800	-96.5	223.872	3.346	-126.5	0.22387	105.800
-67.0	199.526	99.881	-97.0	199.526	3.159	-127.0	0.19953	99.881
-67.5	177.828	94.294	-97.5	177.828	2.982	-127.5	0.17783	94.294
-68.0	158.489	89.019	-98.0	158.489	2.815	-128.0	0.15849	89.019
-68.5	141.254	84.040	-98.5	141.254	2.658	-128.5	0.14125	84.040
-69.0	125.893	79.339	-99.0	125.893	2.509	-129.0	0.12589	79.339
-69.5	112.202	74.901	-99.5	112.202	2.369	-129.5	0.11220	74.901
-70.0	100.000	70.711	-100.0	100.000	2.236	-130.0	0.10000	70.711
-70.5	89.125	66.755	-100.5	89.125	2.111	-130.5	0.08913	66.755
-71.0	79.433	63.021	-101.0	79.433	1.993	-131.0	0.07943	63.021
-71.5	70.795	59.496	-101.5	70.795	1.881	-131.5	0.07079	59.496
-72.0	63.096	56.167	-102.0	63.096	1.776	-132.0	0.06310	56.167
-72.5	56.234	53.026	-102.5	56.234	1.677	-132.5	0.05623	53.026
-73.0	50.119	50.059	-103.0	50.119	1.583	-133.0	0.05012	50.059
-73.5	44.668	47.259	-103.5	44.668	1.494	-133.5	0.04467	47.259
-74.0	39.811	44.615	-104.0	39.811	1.411	-134.0	0.03981	44.615
-74.5	35.481	42.120	-104.5	35.481	1.332	-134.5	0.03548	42.120
-75.0	31.623	39.764	-105.0	31.623	1.257	-135.0	0.03162	39.764
-75.5	28.184	37.539	-105.5	28.184	1.187	-135.5	0.02818	37.539
-76.0	25.119	35.439	-106.0	25.119	1.121	-136.0	0.02512	35.439
-76.5	22.387	33.457	-106.5	22.387	1.058	-136.5	0.02239	33.457
-77.0	19.953	31.585	-107.0	19.953	0.999	-137.0	0.01995	31.585
-77.5	17.783	29.818	-107.5	17.783	0.943	-137.5	0.01778	29.818
-78.0	15.849	28.150	-108.0	15.849	0.890	-138.0	0.01585	28.150
-78.5	14.125	26.576	-108.5	14.125	0.840	-138.5	0.01413	26.576
-79.0	12.589	25.089	-109.0	12.589	0.793	-139.0	0.01259	25.089
-79.5	11.220	23.686	-109.5	11.220	0.749	-139.5	0.01122	23.686
-80.0	10.000	22.361	-110.0	10.000	0.707	-140.0	0.01000	22.361
-80.5	8.913	21.110	-110.5	8.913	0.668	-140.5	0.00891	21.110
-81.0	7.943	19.929	-111.0	7.943	0.630	-141.0	0.00794	19.929
-81.5	7.079	18.814	-111.5	7.079	0.595	-141.5	0.00708	18.814
-82.0	6.310	17.762	-112.0	6.310	0.562	-142.0	0.00631	17.762
-82.5	5.623	16.768	-112.5	5.623	0.530	-142.5	0.00562	16.768
-83.0	5.012	15.830	-113.0	5.012	0.501	-143.0	0.00501	15.830
-83.5	4.467	14.945	-113.5	4.467	0.473	-143.5	0.00447	14.945
-84.0	3.981	14.109	-114.0	3.981	0.446	-144.0	0.00398	14.109
-84.5	3.548	13.319	-114.5	3.548	0.421	-144.5	0.00355	13.319
-85.0	3.162	12.574	-115.0	3.162	0.398	-145.0	0.00316	12.574
-85.5	2.818	11.871	-115.5	2.818	0.375	-145.5	0.00282	11.871
-86.0	2.512	11.207	-116.0	2.512	0.354	-146.0	0.00251	11.207
-86.5	2.239	10.580	-116.5	2.239	0.335	-146.5	0.00224	10.580
-87.0	1.995	9.988	-117.0	1.995	0.316	-147.0	0.00200	9.988
-87.5	1.778	9.429	-117.5	1.778	0.298	-147.5	0.00178	9.429
-88.0	1.585	8.902	-118.0	1.585	0.282	-148.0	0.00158	8.902
-88.5	1.413	8.404	-118.5	1.413	0.266	-148.5	0.00141	8.404
-89.0	1.259	7.934	-119.0	1.259	0.251	-149.0	0.00126	7.934
-89.5	1.122	7.490	-119.5	1.122	0.237	-149.5	0.00112	7.490

$$FSL(dB) = 20\log 10\left(\frac{4\pi}{c}df\right) \tag{11.3}$$

$$FSL(dB) = 20\log 10(d) + 20\log 10(f) + 32.4(\text{km})^* \tag{11.4}$$

or

$$FSL(dB) = 20\log 10(d) + 20\log 10(f) + 36.6(\text{miles})^{**} \tag{11.5}$$

where:

λ is the wavelength in meters (or kilometers*);

f is the frequency in hertz (or megahertz*);

d is the path length constant in meters (kilometers* or miles**);

c is the speed of light in vacuum, 2.99792458×10^8 meters per second.

Depending on the application, there are several other losses that a link analysis has to take into consideration. Among them are polarization loss and rain loss. Figure 11.5 and 11.6 show screens of Excel programs showing all inputs and outputs needed for such a program. It is relatively easy to write such an application using all inputs including the noise chart from Figure 11.3 and the E_b/N_o curves for given bit error rates from Figure 11.2, which could be imported separately for the various communications modulations such as ASK, FM NRZ, MSK, FSK, MFSK, AFSK, APK, M-ARY FSK, BPSK, and others.

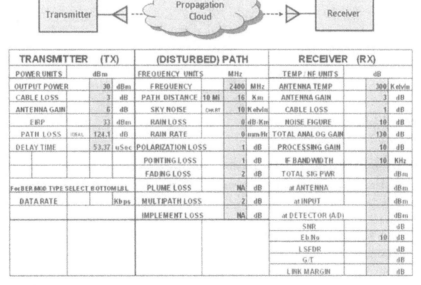

Figure 11.5 Screenshot of an user-friendly Excel path calculator application, which separates the transmitter information from the path and the receiver results.

Link Margin Calculation for Noncoherent FSK Digital Links						
	Enter	Calculated	Units	Link Margin =	2.6	dB
Frequency (MHz)	2400			MDS at RX Input	-103.49	dBm
Wavelength (m)		0.13	m			
Transmit Power (W)	1.0	0.00	dBW	1.00	W	
Transmiter-to-antenna Cable Loss (dB)	3.0	3.00	dB	0.50		
Transmitter Antenna Gain (dBi)	6.0	6.00	dB	3.98		
Transmit EIRP (dBW)		3.00	dBW	2.00	W	
Distance (statute miles)	37	32.13	nm			
Distance (km)		59.53	km			
Power Density (dbm/m²)		-73.49	dBm/m²	4.48E-11	W/m²	
Path Loss (dB)		135.49	dB			
Fade Margin - "Cloud Factor" (dB)	3.0	3.00	dB	0.50		
Received Isotropic Power (dBW)		-135.49	dBW	2.79E-14	W	
Receiver Antenna Gain (dBi)	3.0	3.00	dB	2.00		
Effective Aperture		2.48E-03	m²			
Edge of Coverage Allowance (dB)	0.0	0.00	dB	1.00		
Polarization Loss	0.0	0.00	dB	1.00		
Received Signal Power (dBW)		-132.49	dBW	5.57E-14	W	
Antenna-to-Receiver Cable Loss (dB)	1	75.09	°K			
Signal Input to Receiver (dBm)		-103.49	dBm			
Receiver Noise Figure (dB)	10	2610.00	°K			
Cable/Receiver Temperature (°K)		3360.88	°K			
Receiver Antenna Temperature (°K)	300	300.00	°K			
System Temperature (°K)		3660.88	°K			
System Temperature (dBK)		35.64	dBK			
System G/T		-32.64	dB			
Boltzmann's Constant (dBW/°K-Hz)		-228.60	(dBW/°K-Hz)			
Noise Spectral Density (dBW/Hz)		-192.96	dBW/Hz			
Received Pr/No (dB-Hz)		60.48	dB/Hz			
Data Rate (bits/sec)	2.00E+04	43.01	dB			
Received Eb/No (dB)		17.47	dB			
Bit Error Rate	1.0E-05					
Required Eb/No (dB)		13.35	dB			
Implementation Loss (dB)	1.5	1.50	dB			
Link Margin (dB)		2.6	dB			

Figure 11.6 Another screenshot of a path calculator implemented in Excel.

Table 11.1 is predicated on an impedance of 50 ohms to obtain the values of voltage and extends from 60 dBm to −149 dBm.

Reference

[1]   Thomas, E., "Variable Decision Threshold Computer," U.S. Patent 2,999,925.

Selected Bibliography

Boithias, L., *Radio Wave Propagation*, New York: McGraw-Hill, 1987.

Haykin, S., *Communication Systems*, New York: John Wiley & Sons, 1978.

Inglis, A. F., *Electronic Communications Handbook*, New York: McGraw-Hill, 1988.

Markel, J. D., "Shrinking Intermodulation," *EDN*, August 1967.

Pearl, B., "How to Determine Spur Frequencies," *EDN*, October 1965.

Rousos, W. N., and R. B. Denny, "Threshold Correction System in FSK Transmissions," U.S. Patent 3,947,769.

Seybold, J. S., *Introduction to RF Propagation*, New York: John Wiley & Sons, 2005.

Shores, M. W., "Chart Pinpoints Interference Problems," *EDN*, January 15, 1969.

Simon, M. K., S. M. Hinedi, and W. C. Lindsey, *Digital Communication Techniques*, Upper Saddle River, NJ: Prentice-Hall, 1995.

Sklar, B., *Digital Communications*, Upper Saddle River, NJ: Prentice-Hall, 1988.

CHAPTER 12

System Design Considerations for Modern Receivers

12.1 Introduction

The realization of high-performance, general coverage receivers has brought about the need for a better understanding of their performance. In a radio receiver or any other RF signal processor (e.g., a frequency synthesizer), mixers or sampling A/Ds used to perform conversions are not only frequency-adding or frequency-subtracting mechanisms. They also exhibit a variety of undesired output products in addition to their sum and difference frequencies. Performance can also be impacted by an improper choice of intermediate frequency (IF), center frequency, bandwidth, and/or filter shape factors, as well as incorrect sampling in direct sampling systems. In addition, every element (including amplifiers and filters) in a system can become nonlinear and can behave like a mixer, adding to the problem. These are called spurious intermodulation products and can appear in the passband of IF filters if a system is not correctly designed. This phenomenon becomes more complex with increased RF front-end bandwidth.

If a multimixer situation exists, such as in a multiple conversion receiver, these problems can be further aggravated, as initial unwanted products from the first mixer can carry through and combine with the products from the following mixers, creating a multitude of undesired products at the final IF output. The same can be true in direct digital sampling receivers, which are also mixing mechanisms, if the sampling oscillator is not chosen correctly.

12.2 Understanding Intermodulation Distortion Products

Whether a receiver is a dedicated fixed-frequency type, or a general coverage broadband type, the problem of intermodulation products has to be carefully understood and judged against complex system and economic parameters, such that a minimum number of spurious products will be internally generated and detected within the ultimate IF bandwidth of the receiver.

Products (or spurs) are caused by the nonlinear mixing of various order harmonics of the RF and LO, in up- or downconversion systems as they appear in the bandwidth of a given IF. Contrary to some beliefs, they are not the harmonics, but actually the products of all the harmonics to an infinite order. If the system mathematics are not correctly performed, these products can be right in the passband of these IF filters.

More often, because IF filters do not exhibit a perfect square response, they appear in the sidelobes of these filters as we will see later in an example in this chapter. Products are described by:

$$F_{IF} = \pm mF_{RF} \pm nF_{LO} \tag{12.1}$$

$$F_{RF} = \pm mF_{IF} \pm nF_{LO} \tag{12.2}$$

It can be seen from (12.1) and (12.2) that in a superheterodyne or a direct sampling processor, the products are caused by the sum or difference of the harmonics of the RF and LO, where m and n are integers of 1 to ∞. Products are characterized by their numerical order, indicated by the sum of $m + n$ regardless of the plus or minus mixer operation. For example, $2F_{RF} - 3F_{LO}$ = fifth-order product, while $2F_{RF} + 3F_{LO}$ = fifth-order product, as well. The lower the order is, the higher their level can be (with the third and fifth order being the worst). Products below a seventh order that appear in the IF should be avoided in well-designed RF systems.

Anticipating and avoiding low-order products in a system is a complex matter that takes a lot of system design knowledge. Understanding spurious analysis and designing a proper RF system can be a long and tedious task of balancing all elements in the system. This is especially true in coherent schemes where changing one number can inadvertently impact several other parts or sections of the system, which can impede performance or even make a receiver technologically unrealizable.

Although no receiver is perfect (products will always exist in RF systems), much can be done by a receiver system designer to ensure that only higher order products (beyond the seventh order) will be present. Systems should be investigated and modeled before they are implemented.

Products are analyzed using normalized ratios of all the mixed frequencies involved in a system. This is not only true in fixed frequency systems, but also in broadband systems, which can be much more complex to analyze.

As a rule, a good designer will always mark the normalized ratios or ratio ranges on each of the mixers in the system design using the tools provided here to prove that the predicted products are always beyond a seventh order. Ratios should always appear on all block diagrams of RF systems as shown in the simple example in Figure 12.1. In broadband systems, minimum and maximum normalized ratios should be marked as the R_{min} and R_{max} for each of the mixers as we will see later in this chapter.

At the base of the mixer product analysis method is the bandpass sampling concept which is key to many disciplines, including the superheterodyne RF signal processing. In analyzing spurious conditions for bandpass sampling it is also important to distinguish between the baseband Nyquist rate (twice the highest frequency) and the bandpass Nyquist rate (twice the signal bandwidth).

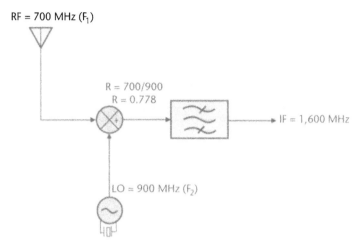

Figure 12.1 Normalized ratio of the RF and LO should always appear on every high-level system block diagram. In this simple example, a fixed set of frequencies is used. If a range of frequencies are involved, an R_{min} and R_{max} should be used for every mixer.

12.3   Predicting Receiver System Spurious Performance: Design Tools for Predicting Intermodulation Distortion

12.3.1   Product Charts and Their Use—The Intermodulation Distortion Web Analysis Tool

Over the years, smart tools have evolved in support of (12.1) and (12.2). They can be used to analyze and predict in-band IF intermodulation distortion in receivers and other RF processors before systems are designed and built by accident. In creating such tools, mathematicians have used a process called the "geometry of aliasing" because aliasing can be expressed graphically with products overlapping slanted lines as shown in Figure 12.2.

Let's now look at some of the analytical tools the RF system designer uses to determine these products. Shown in Figure 12.2 is the normalized intermodulation Web analysis chart. This is a good visual method to express a complex mathematical process especially in a wideband system situation.

This tool has the normalized ratios already built in. The top part from the center line is dedicated to mixers in adding mode, while the bottom part from the center line is dedicated to mixers in a minus mode. The two heavy lines represent the respective desired products: $F_2 + F_1$ for the top and $F_2 - F_1$ for the bottom.

We will now use this chart in Figure 12.2 to understand how it works. Let's assume that we are going to design a simple fixed-frequency receiver for 700 MHz such as the one shown in Figure 12.3.

We chose a somewhat imperfect scheme on purpose in order to support our explanation. With a local oscillator of 900 MHz, the upconvert receiver will have a first IF of 1,600 MHz (with an IF filter bandwidth of 100 MHz at −3 dB points). We find the two normalized frequencies ratios:

$F_1 / F_2 = 700/900 = 0.778$

$F_{IF}/F_2 = 1,600/900 = 1.778$

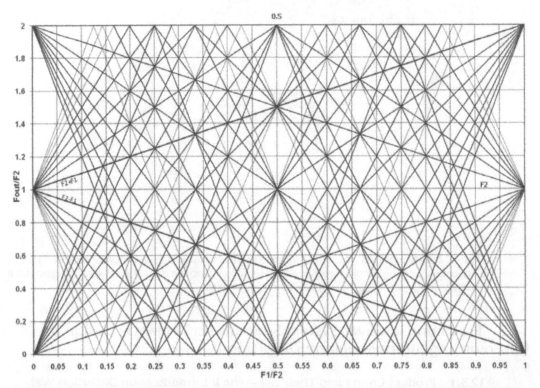

Figure 12.2 Normalized intermodulation product Web chart used for analyzing receivers and other RF processors using mixing processes. This chart was designed using the "geometric bandpass" sampling techniques and can be used for up to a sixteenth order set of products. Generally, good RF systems are designed to be clear of in-band IF spurs below a seventh order.

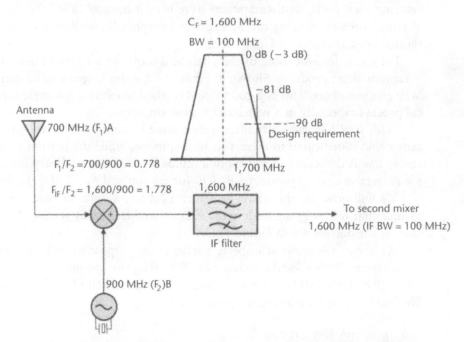

Figure 12.3 Finding in-band intermodulation products dictates the choice of IF filter center frequency, its bandwidth, as well as its shape factor, in order to meet system requirements.

With this information and the mixer product chart from Figure 12.2, we find the locus point for the two ratios, as shown in Figure 12.4. The locus point is then replaced by a rectangle, which represents the IF bandwidth. The purpose of the exercise is to be able to fit the IF bandwidth rectangle between the geometrically oblique lines without touching them, if possible. Although the 3-dB bandwidth of the filter is what we are looking for, the −60-dB bandwidth should also be used because the rectangle will be bigger here and it may be crossed by some of the product lines. This is indeed the case in our example. The more lines cross the IF rectangle, the more products are present in the IF passband.

The chart shows all products produced not only by the fundamental frequencies, but also by multiples of the RF signal and local oscillator (LO) frequencies, which are present in the mixer stage, and correspond to the second, third, fourth, fifth, sixth, seventh, and eighth harmonics of the two mixing signals forming spurious products to sixteenth-order products ($8 + 8 = 16$th order).

Again, the order of the product is determined by the sum of the harmonics order involved. For example, $5F_1 \pm 2F_2$ is a seventh-order product (regardless of the mathematical operation involved) because it involves the fifth harmonic of F_1 combined with the second harmonic of F_2. Higher-order products are also present, but they are usually of a sufficiently low level so as not to cause problems.

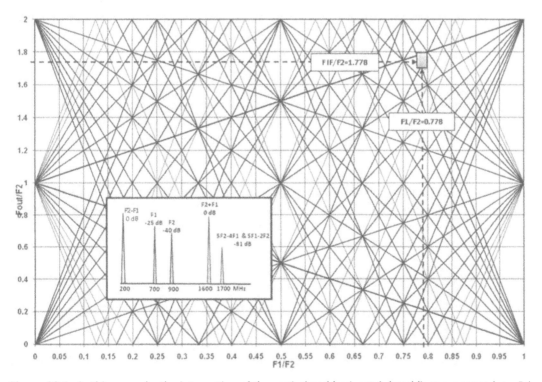

Figure 12.4 In this example, the intersection of the vertical and horizontal doted lines corresponds to F_1/F_2 and $F(\mathrm{IF})/F_2$. This intersection is the locus point of the normalized frequencies. The slanted lines that cross the IF rectangle indicate the in-band intermodulation (spurious) products. The out-of-band spurious outputs, which happen to be in the vicinity of the IF frequency, can also be verified by looking at the products adjacent to the locus point rectangle. The purpose of the intermodulation Web tool is to show that no slanted lines (below a seventh order) intersect the IF rectangle over the IF bandwidth and on its slopes, all the way down to the −60-dB points. This is not the case in our example.

The amplitude of the undesired products identified depends on their particular order number $(m + n)$ with the third being the worst. Most products of the seventh order or higher will be at least 60-dB down from the IF level, and are usually not considered to cause problems. Any slanted line that crosses the locus point IF rectangle corresponds to a product, which is identified by applying the actual numbers in the formulas as shown in Figure 12.4. Values of F_1 and F_2 can then be substituted and the products can be anticipated and avoided by either designing a steeper response IF filter or by changing the frequency plan system design altogether.

If the locus point is examined closely in our example, one can see that there are no in-band products at the −3-dB points of the IF filter, but analyzing the areas adjacent to the locus point (enlarging the IF rectangle) indicates some out-of-band spurs products on the slopes of the IF filter, which will have to be suppressed by the level specified in the requirements. By knowing their order given by the chart, their predicted amplitude can be found. In our case, the seventh- and ninth-order product ($5F_1 - 2F_2$ and $5F_2 - 4F_1$) appear on the upper slope of the IF filter. They are predicted to be 81 dB below the IF level (typical manufacturer prediction). We would like to improve on the situation as this product can interfere with small signals. The IF filter would have to attenuate by about 9 dB at 1,700 MHz to accomplish a system requirement of −90 dB as shown in Figure 12.3.

What this actually means is that we just defined our IF filter order and complexity. The shape factor of the filter will have to be steeper to reject this product to the desired level, which means a more complex filter. Realization of the IF filter (the type, approximation, and technology) should then be determined using the percentage bandwidth concept presented in Chapter 4.

A simpler and possibly easier method of finding in-band products in a system can be achieved by using the normalized charts from Tables 12.1 and 12.2. Table 12.1 is for mixers used in adding mode (B + A) where A and B are the mixing frequencies and B > A. Table 12.2 is for mixers used in a subtracting mode (B − A) with the same conditions applying. It is necessary (especially when entire frequency ranges are analyzed) to use the closest ratio number since the actual ratio may not be in the left column as in our example.

Using the same example from Figure 12.3, and substituting F_1 for A and F_2 for B the same ratios can be obtained. We then use Table 12.1 since the mixer in our example operates in the adding mode, and find the corresponding products as previously indicated in Figure 12.3 (5A − 2B and 5B − 4A). If the numerical values of A and B are inserted in these formulas, the same values can be obtained with the normalized chart method, which is sometimes preferred. This example is shown in Figure 12.5.

If the example receiver was not for a fixed frequency, you can imagine what a job it could be to evaluate all the products generated within a wideband scheme.

Today, computer programs are used successfully to predict these problems. The intermodulation Web analysis tool previously presented can be programmed as an application in Excel, which can output graphically and directly on the IMD Web chart. The following example uses such a program. To get an appreciation of this, we will look at the next example, which uses the broadband receiver in the Star-10 transceiver that was introduced earlier.

Table 12.1   Normalized Intermodulation Product Chart Used for Mixers in an Adding Mode (B+A) Where A and B Are the Mixing Frequencies and B > A (Normalized Ratios Are Shown for Intermodulation Products Up to the Fifteenth Order)

OUTPUT FREQUENCY = B + A where B = RF and A = LO AND B > A

| R = A/B | 1 | 2 | 3 | 4 | 5 | 6 | 7 | 8 | 9 | 10 | 11 | 12 | 13 | 14 | 15 |
| | | | | | | | | | ORDER | | | | | | |
	B	B - A	B +/- 2A	B +/-3A	B +/-4A	B +/-5A	B +/-6A	B +/-7A	B +/-8A	B +/-9A	B +/-10A	B +/-11A	B +/-12A	B +/-13A	B +/-14A
0.000															
0.072															2B-13A & 15A
0.077														2B-12A & 14A	
0.083													2B-11A & 13A		
0.091												2B-10A & 12A			
0.100											2B-9A & 11A				
0.111										2B-8A & 10A					
0.125									2B-7A & 9A						
0.143								2B-6A & 8A							
0.154															3B-12A & 14A-B
0.167							2B-5A & 7A							3B-11A & 13A-B	
0.182													3B-10A & 12A-B		
0.200						2B-4A & 6A						3B-9A & 11A-B			
0.222											3B-8A & 10A-B				
0.250					2B-3A & 5A					3B-7A & 9A-B					4B-11A & 13A-2B
0.273														4B-10A & 12A-2B	
0.286									3B-6A & 8A-B						
0.300													4B-9A & 11A-2B		
0.333				2B-2A & 4A				3B-5A & 7A-B				4B-8A & 10A-2B			
0.364															5B-10A & 12A-3B
0.375											4B-7A & 9A-2B				
0.400							3B-4A & 6A-B							5B-9A & 11A-3B	
0.429										4B-6A & 8A-2B					
0.445													5B-8A & 10A-3B		
0.500			2B-1A & 3A			3B-3A & 5A-B			4B-5A & 7A-2B			5B-7A & 9A-3B			6B-9A & 11A-4B
0.555														6B-8A & 10A-4B	
0.571											5B-6A & 8A-3B				
0.600								4B-4A & 6A-2B							
0.625													6B-7A & 9A-4B		
0.667					3B-2A & 4A-B					5B-5A & 7A-3B					7B-8A & 10A-5B
0.715												6B-6A & 8A-4B			
0.750							4B-3A & 5A-2B							7B-7A & 9A-5B	
0.800									5B-4A & 6A-3B						
0.833											6B-5A & 7A-4B				
0.858													7B-6A & 8A-5B		
0.875															8B-7A & 9A-6B
1.000		2B & 2A		3B-A & 3A-B		4B-2A & 4A-2B		5B-3A & 5A-3B		6B-4A & 6A-4B		7B-5A & 7A-5B		8B-6A & 8A-6B	

Table 12.2 Normalized Intermodulation Product Chart Used for Mixers in a Subtracting Mode (B – A) Where A and B Are the Mixing Frequencies and B > A (Normalized Ratios Are Shown for Intermodulation Products Up to the Fifteenth Order)

OUTPUT FREQUENCY = B - A where B = RF, A = LO and B > A

ORDER

R = A/B	1 B	2 B-A	3 B±2A	4 B±3A	5 B±4A	6 B±5A	7 B±6A	8 B±7A	9 B±8A	10 B±9A	11 B±10A	12 B±11A	13 B±12A	14 B±13A	15 B±14A
0.000	B	B-A													
0.063															15A
0.067														14A	
0.072													13A		
0.077												12A			
0.083											11A				2B-13A
0.091										10A				2B-12A	
0.100									9A				2B-11A		
0.111								8A				2B-10A			
0.125							7A				2B-9A				
0.133															14A-B
0.143						6A				2B-8A				13A-B	
0.154													12A-B		
0.167					5A				2B-7A			11A-B			
0.182											10A-B				3B-12A
0.200				4A				2B-6A		9A-B				3B-11A	
0.214															13A-2B
0.222									8A-B				3B-10A		
0.231														12A-2B	
0.250			3A				2B-5A	7A-B				3B-9A	11A-2B		
0.273												10A-2B			
0.286							6A-B				3B-8A				
0.300											9A-2B				4B-11A
0.308															12A-3B
0.333		2A				2B-4A & 5A-B				3B-7A & 8A-2B				4B-10A & 11A-3B	
0.364													10A-3B		
0.375									7A-2B				4B-9A		
0.400					4A-B				3B-6A			9A-3B			
0.416															11A-4B
0.429								6A-2B				4B-8A			
0.445											8A-3B				5B-10A
0.455														10A-4B	
0.500	A			3A-B	2B-3A		5A-2B	3B-5A		7A-3B	4B-7A		9A-4B	5B-9A	
0.545															10A-5B
0.555												8A-4B			
0.571									6A-3B				5B-8A		
0.600						4A-2B				4B-6A				9A-5B	
0.625											7A-4B				6B-9A
0.667			2A-B				3B-4A	5A-3B				5B-7A	8A-5B		
0.700															9A-6B
0.715										6A-4B				6B-8A	
0.750					3A-2B				4B-5A			7A-5B			
0.778														8A-6B	
0.800							4A-3B				5B-6A				
0.833									5A-4B				6B-7A		
0.858											6A-5B				7B-8A
0.875													7A-6B		
1.000				2B-2A		3B-3A		4B-4A		5B-5A		6B-6A		7B-7A	

OUTPUT FREQUENCY = B + A where B = RF and A = LO AND B > A

ORDER

R=A/B	1 (B)	2 (B−A)	3 (B+/−2A)	4 (B+/−3A)	5 (B+/−4A)	6 (B+/−5A)	7 (B+/−6A)	8 (B+/−7A)	9 (B+/−8A)	10 (B+/−9A)	11 (B+/−10A)	12 (B+/−11A)	13 (B+/−12A)	14 (B+/−13A)	15 (B+/−14A)
0.000	B														
0.072															2B-13A & 15A
0.077														2B-12A & 14A	
0.083													2B-11A & 13A		
0.091												2B-10A & 12A			
0.100											2B-9A & 11A				
0.111										2B-8A & 10A					
0.125									2B-7A & 9A						
0.143								2B-6A & 8A							3B-12A & 14A-B
0.154														3B-11A & 13A-B	
0.167							2B-5A & 7A						3B-10A & 12A-B		
0.182												3B-9A & 11A-B			
0.200						2B-4A & 6A					3B-8A & 10A-B				
0.222										3B-7A & 9A-B					
0.250					2B-3A & 5A				3B-6A & 8A-B						4B-11A & 13A-2B
0.273														4B-10A & 12A-2B	
0.286								3B-5A & 7A-B					4B-9A & 11A-2B		
0.300												4B-8A & 10A-2B			
0.333				2B-2A & 4A			3B-4A & 6A-B				4B-7A & 9A-2B				
0.364															5B-10A & 12A-3B
0.375										4B-6A & 8A-2B					
0.400						3B-3A & 5A-B								5B-9A & 11A-3B	
0.429													5B-8A & 10A-3B		
0.445									4B-5A & 7A-2B			5B-7A & 9A-3B			
0.500			2B-1A & 3A		3B-2A & 4A-B						5B-6A & 8A-3B				6B-9A & 11A-4B
0.555										5B-5A & 7A-3B				6B-8A & 10A-4B	
0.571															
0.600								4B-4A & 6A-2B					6B-7A & 9A-4B		
0.625												6B-6A & 8A-4B			7B-8A & 10A-5B
0.667											6B-5A & 7A-4B				
0.715														7B-7A & 9A-5B	
0.750							4B-3A & 5A-2B						7B-6A & 8A-5B		
0.800									5B-4A & 6A-3B						
0.833															
0.858															
0.875															8B-7A & 9A-6B
1.000		2B & 2A		3B-A & 3A-B		4B-2A & 4A-2B		5B-3A & 5A-3B		6B-4A & 6A-4B		7B-5A & 7A-5B		8B-6A & 8A-6B	

Callout (arrow to row ≈ 0.778): **5A-2B** and **5B-4A**

Figure 12.5 In our fixed frequency receiver example from Figure 12.3, the same results can be obtained as with the normalized intermodulation Web tool by using the normalized intermodulation chart (Table 12.1). If using the ratio A/B = 0.778, the chart points to the same seventh- and ninth-order products (5A − 2B and 5B − 4A).

12.4 System Analysis for a General Coverage Communication Receiver—A Design Case

In the following pages we will consider a system design for a general coverage receiver with a wide input bandwidth of 4 octaves (28 MHz). This is the actual design of the Star-10 transceiver. Unlike the dedicated single-frequency receiver analyzed in the previous pages, this wideband receiver presents an immense product analysis problem, which is due to the many different cases that could be created within the very wide input bandwidth. The problem could be further complicated due to the double-conversion approach used, as products generated in the first IF can multiply in the second IF, which was discussed at the beginning of this chapter.

Many times, the design of a receiver is dictated by economics. This means that inexpensive commercially available IF filters dictate the choice of IF frequencies without considerations for system analysis. Sometimes, a designer does not realize the implications of intermodulation distortion products until it is too late in production and the problem cannot be cured without a total redesign.

A good designer should always do a spurious analysis. Normalized ratios should be indicated on all mixers in the system and the system ratios should be carried through entirely including the synthesizer portion of the design. The synthesizer is part of the receiver and cannot be separated.

The designer should use the analysis tools presented here along with good judgment in the initial choices of frequencies, since no computer or chart can take the place of good engineering procedures, as we will see in the following example.

Assume that a communication receiver is to be designed that will cover the range of 2 to 30 MHz with good image rejection and a minimum of unwanted products generated within its configuration. This receiver is part of a bilateral transceiver system as shown in Figure 12.6.

Looking at Figure 12.6, the system is a wideband double conversion approach with an upconversion first IF compatible with commercially available monolithic crystal filters at 75 MHz. A fully coherent microwave DDS-driven phase-locked synthesizer is to be used as the local oscillator for the first mixer. It will provide an ultimate tuning resolution of 10-Hz steps over the range. The second IF is at 9 MHz because of the availability of good crystal-lattice filters at this frequency. A cascaded quartz filter arrangement is considered in this IF, which will be discussed in Chapter 17.

First, we will attempt to analyze the system with the help of the charts described earlier along with the calculations necessary to support these charts.

Figure 12.7 shows the normalized mathematical ratios for this receiver. An RF signal anywhere from 2 to 30 MHz comes from the antenna into the first mixer where it subtracts from the first local oscillator, which operates so as to always produce a 75-MHz IF. This local oscillator is part of our coherent synthesizer operating from 77 to 105 MHz in 10-Hz steps. The 75-MHz IF frequency dictates the bandwidth of the first IF to be around 10 kHz at minimum, from 74.995 to 75.005 MHz in order for the second 84 MHz fixed local oscillator to provide conversion to the second 9-MHz IF. We will use the subtracting normalized product chart (Table 12.2).

Looking at Table 12.2, if RF = A and LO = B, a minimum and a maximum ratio A/B can be found for the entire range of frequencies from the previous equations.

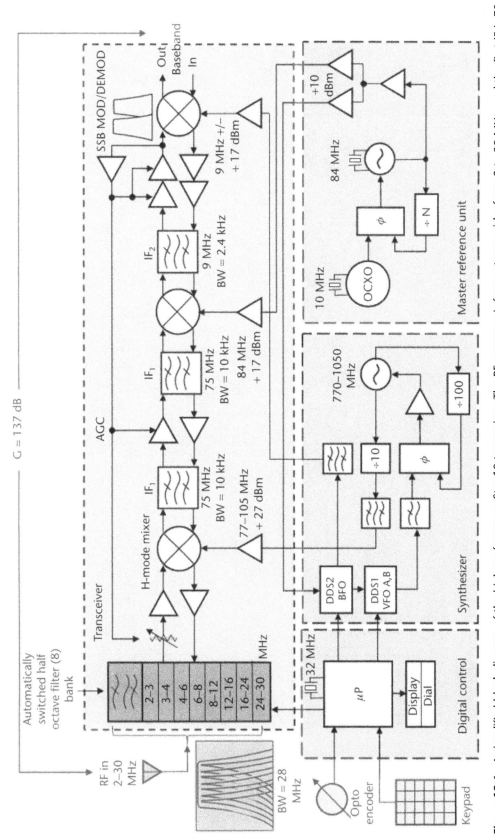

Figure 12.6 A simplified block diagram of the high-performance Star-10 transceiver. The RF coverage is four-octaves wide, from 2 to 30 MHz, and the first IF is 75 MHz, while the second IF is 9 MHz.

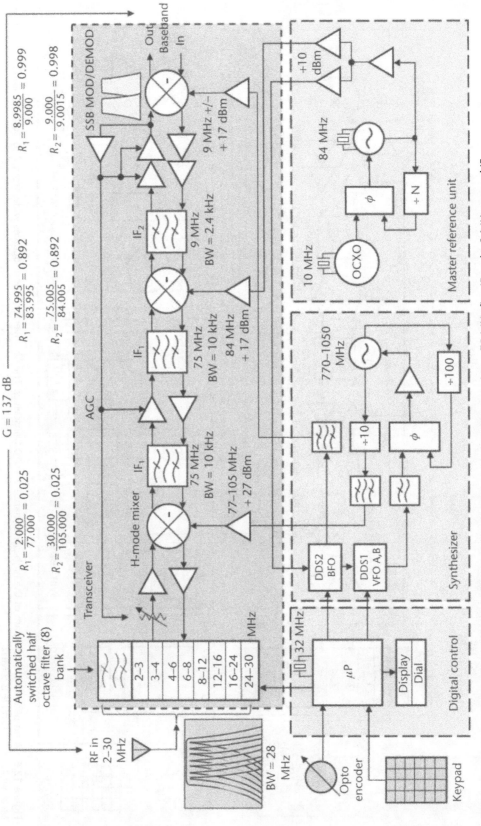

Figure 12.7 Normalized ratios for the double conversion upconvert receiver system using a 75-MHz first IF and a 9-MHz second IF.

Then, the R minimum and R maximum (R_1 and R_2) are defined for the first conversion mixer (M_1) as shown in Figure 12.8. As we can see, this is a very wide normalized range, extending from 0.025 to 0.285.

Looking at this range on the chart from Figure 12.8, we can see a multitude of possible products. Any product indicated within the wideband between R_1 and R_2 could be a potential problem for the corresponding received frequency. A look at our chart indicates a series of possible problems (7A, 6A, 5A, 4A, 3A, 2B – 5A) with the worst one at 3A.

At first glance, we can say that the third harmonic of one of the two mixing signals could be quite powerful and could indeed produce a problem, but a closer look at the system indicates that the offending frequency, A, is actually a received frequency and chances are very small that a distant 25-MHz station has a high level at the third harmonic (75 MHz) appearing at the antenna of our receiver.

The problem is further diminished by our receiver's automatically switched half-octave preselector, which greatly attenuates any 75-MHz signals over the entire 2- to 30-MHz range. This preselector is made of eight steep half-octave bandpass filters, which further clean up our possible problems. If we perform the analysis for each of the half-octave frequency portions, we would not see any problem to at least a seventh-order product. This is a case where engineering judgment is more important than all our tools, which are only used to warn of possible problems.

The same conditions apply to the other products indicated by the chart. They present an even better case since they are even further removed from the received frequencies due to the preselector filtering.

The situation is even better for the M_2 and M_3 mixers since these are narrowband conversions. There are no products to up to a thirteenth order in these cases as shown in Figure 12.8.

In addition to the chart analysis, the same results can be calculated and plotted by using the intermodulation Web chart tool as automated by a computer program. The results for all eight half-octave bandwidths between 2 MHz and 30 MHz using this method are shown in Figure 12.9. The numbers on each rectangle (the 75-MHz IF) correspond to the preselector filter numbers from the filter bank as shown in Figure 12.8. The rectangles are expanded locus points for each one of the half-octave bandwidths as seen in the 75-MHz IF bandwidth of 10 kHz. The dark line is the desired output of a mixer in a subtracting mode ($F_2 - F_1$).

As can be seen, none of the locus points for the filtered frequency ranges touches any of the slanted product lines up to a sixteenth-order product (8 + 8).

Since the first IF bandwidth is only 10 kHz wide, the ratios in the second conversion at M_2 are expressed by the same number (0.892). The IF is centered at 9.000 MHz and its bandwidth is determined by the single-sideband filters.

For simplicity, the –60-dB ultimate bandwidth for these filters was estimated at around 5 kHz for the combined sidebands. Figure 12.8 shows that there is no problem for the second conversion except for a thirteenth-order product, which can be ignored, but since the IF band is so narrow and so close to the 1.000 ratio (which could present problems), the single-sideband filters have to be designed properly (having steep slopes) to ensure proper design. The 9-MHz IF outcome is verified in the Web analysis output shown in Figure 12.9 (lower right).

Let's now proceed to analyze the translation to the baseband ratio from Figure 12.8. The conversion to baseband from 9 MHz at M_3 uses a product detector

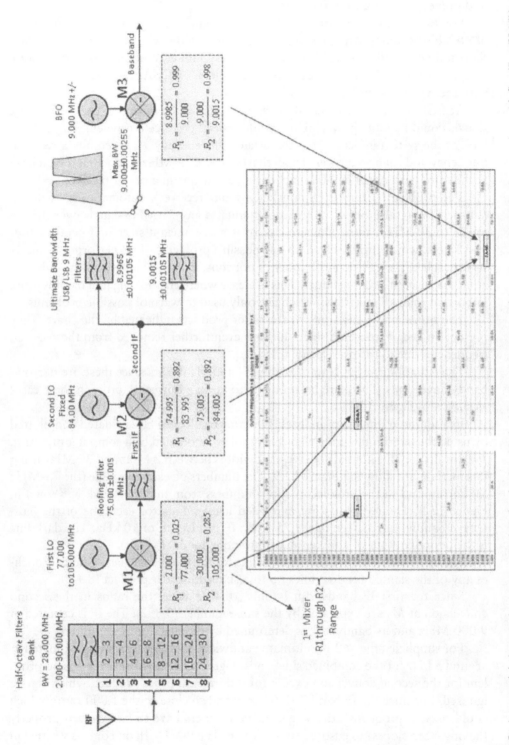

Figure 12.8 Intermodulation spurious system analysis for the Star-10 double conversion, general coverage communications receiver with first IF at 75 MHz and the second IF at 9 MHz.

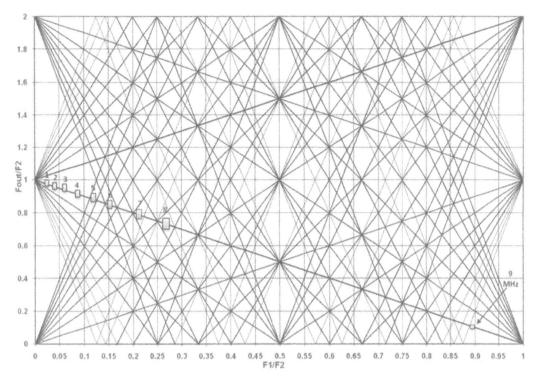

Figure 12.9   Spurious analysis of a 2- to 30-MHz (4 octaves) coverage receiver with an IF of 75 MHz (10-kHz BW) using the spur Web tool. The system uses eight half-octave bandpass filters, which are numbered and correspond to the filters in the block diagram from Figure 12.8. None of the locus points touches any of the slanted product lines up to a sixteenth-order product (8 + 8), indicating a very good system as long as the automatic preselector is in the circuit. The second IF at 9 MHz shown at bottom right is also properly chosen.

and the BFO local oscillator. A normalized ratio of 0.998 to 0.999 is found here. Provided that the IF filters have good shape factors, this, again, puts our products into a thirteenth-order region, which is deemed very good as we are usually not concerned with any products beyond a seventh order.

It can be seen from this analysis that the charts can only be used to the guideline extent; however, they are in sufficient detail to get a good idea of a system's intermodulation performance.

We have determined that our wideband Star-10 receiver system design provides a clean scheme. Some systems may require good performance beyond the seventh-order products. If so, the problem of creating a good system becomes progressively difficult as there is no RF system without products. In some cases, up to a twelfth-order performance has been reported.

This concludes the spurious analysis for the Star-10 double-conversion general coverage receiver. In performing this tedious task, we have scientifically proved that we have a thoroughly analyzed system with a minimum of intermodulation problems. We can now proceed with confidence to the dynamic range analysis of our receiver.

In conclusion, we can safely say that all frequency schemes used in receivers (or other RF processors) have unwanted products within their outputs, and the fact remains that any configuration chosen is simply the best compromise in the opinion of the designer.

Wideband receivers are much harder to analyze than fixed-frequency receivers. It is not unusual to spend a great amount of time to determine "a best of the worse system." A good system should be mathematically designed such as to be free of any possible spurious problems up to a seventh order in its IFs. Tools have been presented on how to perform spurious analysis of fixed frequency as well as broadband RF systems.

Selected Bibliography

Brown, T. T., "Mixer Harmonic Chart," *Electronics Magazine*, April 1951.

Drentea, C., "Designing a Modern Receiver," *Ham Radio*, November 1983.

Drentea, C., *Radio Communications Receivers*, New York: McGraw-Hill, 1982.

Drentea, C., "The Star-10 Transceiver," Part 1, *QEX—ARRL*, November/December 2007.

Drentea, C., "The Star-10 Transceiver," Part 2, *QEX—ARRL*, March/April 2008.

Drentea, C., "The Star-10 Transceiver," Part 3, *QEX— ARRL*, May/June 2008.

Manassewitsch, V., *Frequency Synthesizers, Theory and Design*, New York: John Wiley & Sons, 1976.

"Nonlinear System Modeling and Analysis with Applications to Communications Receivers," Signatron, Inc., June 1973, Rome Air Development Center.

CHAPTER 13

Dynamic Range

13.1 Definitions: The Five Types of Dynamic Range

In today's crowded spectrum, high probability of intercept (HPOI) receivers require high sensitivity, along with minimizing in-band and out-of-band interference of all kinds. These requirements do not go hand-in-hand, as sensitive receivers are usually more prone to intermodulation distortion because of their inherently nonlinear devices. Single tone, multiple tone, as well as wideband phase-noise also come into the picture. In addition to their speed of acquisition, HPOI receivers require the ability to receive small signals in the presence of strong adjacent signals, which may be greater in amplitude by more than 100 dB. This concept also includes a high degree of rejection to spurious products produced by nonlinear interaction of many powerful signals, sometimes far removed from the receiving frequency when mixed with the phase noise of the receiver's local oscillators. This is true not only in superheterodyne receivers, but also in direct sampling digital receivers, which also use local oscillators as sampling references.

Most of the receiver design efforts of the past 50 years have been directed toward replacing already proven vacuum tube technologies with newly developed semiconductors. This was initially considered an easy task, but the results have been surprisingly poor. While such characteristics as high sensitivity and good selectivity were maintained, the new solid-state designs lacked dynamic range. So what is dynamic range?

The ability to handle a wide range of RF input situations with high fidelity is loosely called dynamic range. Much confusion exists today in defining dynamic range. In general, dynamic range is defined as the power range over which a device such as a radio receiver provides useful operation. There are five basic kinds of dynamic range. All are very important, but none are sufficient to fully describe a system's behavior. Each is a limited representation of the dynamic range scenario and it is an indirect measure of a very complex behavior. In order of importance, they are:

13.1.1 Single-Tone Dynamic Range

The following are definitions of single-tone dynamic range.

1. *Linear or instantaneous dynamic range (LDR):* The range over which a receiver can reproduce with fidelity a single-tone signal until the system stops amplifying as designed and compresses. It is measured in decibels from the

minimum discernable signal (MDS), where the gain drops bellow its nominal value by 1 dB (called the -1-dB compression point).

2. *Linear composite dynamic range (LCDR):* The range over which a receiver can reproduce with fidelity a single tone signal beyond the compression points of any single stage using automatic gain control (AGC) applied in reverse order from the back end of the receiver toward the front end, and to all stages. It is measured in dB from the MDS to the composite input level off all elements in the system under the control of AGC.

3. *Linear blocking dynamic range (BDR):* The range over which a receiver can reproduce with fidelity a small signal at the MDS while withstanding compression caused by an adjacent powerful single tone signal, which is somewhat removed in frequency from the listening frequency. It is measured in dB from the MDS, at as close of a delta F to the interferer as possible depending on IF filters' bandwidth. This is sometimes considered the worst-case dynamic range case as it is impacted by factors such as IF filter shape factor, poor phase-noise performance, and nonlinear behavior. The effect is similar to that of linear dynamic range as the receiver tends to compress when reaching a limit.

13.1.2 Two-Tone Dynamic Range

The following are definitions of two-tone dynamic range.

1. *Spurious-free third-order dynamic range (IP3SFDR):* This is sometimes called the *IMD dynamic range.* The range over which a receiver can reproduce with fidelity a single tone signal at the MDS, with the receiver tuned to a third-order product of two combined unmodulated signals under AGC control. Beyond the compression point of the system, the intercept point of the third kind (IP3) is a plotted value on a 3:1 curve and enters the dynamic range picture. The third-order product was chosen because it is the highest amplitude product possible, and therefore the test is the worst possible situation a receiver can encounter from an IMD point of view. The two tones are arbitrarily chosen and could be out-of-band or in-band of the first IF filter (an absolute worst case) in which case, the entire mixer(s), filter(s), preamplifier, and IF amplifier(s) nonlinear behavior enter the picture. In this case, higher IP3s in all elements of the design are a must to withstand the possible distortion. Depending on the frequencies involved, modern receivers can achieve a third-order intercept point (IP3) in excess of +40 dBm (10 watts). However, higher IP3s come at a price of increased power consumption as we will see later on in this book. A state-of-the-art receiver with an IP3 in excess of +45 dBm will also be presented. Depending on the application, a designer is faced with some hard choices when designing for high-intercept points as DC power consumption becomes an issue in LOs and amplifiers, which may not be compatible with particular applications.

2. *Spurious-free second-order dynamic range (IP2SFDR):* The range over which a receiver can reproduce with fidelity a single-tone signal at the

MDS, with the receiver tuned to a second-order product of two combined unmodulated signals under AGC control. Beyond the compression point of the system, the intercept point of the second kind (IP2) is a plotted value on a 2:1 curve and enters the dynamic range picture. Depending on the frequencies and adjacent interference involved, this kind of dynamic range can sometimes be more important than the IP3SFDR.

There are some other *specialized dynamic range* methods of measurement. Among them are multiple-tone signals mixed with noise and measured as bit error rate or false alarm readings over a defined period of time. These methods are numerous and can be tailored to specific situations.

All dynamic range methods share a common element, the *minimum discernable signal* (MDS), which represents a receiver's best case sensitivity within the noise figure considerations discussed in Chapter 11. The MDS is expressed in dBm and it means that the receiver can detect a signal out of its noise floor. It can be at the system's noise floor or noise figure (NF) threshold, above it (within the E_b/N_o values discussed in Chapter 11), or even below it depending on the modulation and signal processing used by the receiver.

The MDS is considered as the lower limit (*PL*) of the dynamic range, and is sometimes defined as a signal 3-dB greater than the equivalent noise level for a specified IF bandwidth. Then, the MDS can be found by using (13.1).

$$PL\,(\text{dBm}) = MDS\,(\text{dBm}) = -171\ \text{dBm} + NF\,(\text{dB}) + 10\log IF\ BW\,(\text{Hz}) \qquad (13.1)$$

where:

MDS is the lower-power limit of dynamic range in dBm;

NF is the system noise figure in decibels;

IF BW is the IF bandwidth in hertz;

-171 dBm = KTB + 3 dB = -174 + 3 = -171 dBm;

KTB is the thermal noise power in a 1-Hz bandwidth at room temperature (-174 dBm).

Note: *KTB* is a product of three values: K is Boltzmann's constant (1.38×10^{-23} Joules/K), T is the absolute temperature (assume 300K room temperature), and B is the noise bandwidth (Hz).

The MDS is specified for a given IF bandwidth and is best with the narrowest bandwidth available. The highest bandwidth should be used for a worst case analysis.

The MDS is determined by the noise figure (*NF*), which in turn is a measure of the noise added by the receiver circuitry over the established external noise (including the noise temperature of the antenna) as calculated by the link analysis and within the link margin specified for the frequency range of interest. This was discussed in Chapter 11.

13.2   Determining Noise Figure Requirements

In addition to the classic plot previously shown in Figure 11.2, additional radio noise models have been created to account for all possible sources of noise at a receiver's antenna. One of these models is presented in the widely used reference report 322 by the International Radio Consultative Committee (CCIR). This report uses omnidirectional antenna measurements made at sixteen locations throughout the world, which can be used for initial receiver design despite some minor discrepancies found with some other sources.

The graph in Figure 13.1 shows the overall external noise power contribution (dB above KT) [1] for the frequency range of 10 kHz to 10 GHz. This information is based on measurements made with an omnidirectional vertical antenna over a perfectly conducting ground [2].

The sky noise data and the solar noise data were obtained with a directional antenna pointed at the sources. Maximum and minimum readings were provided in the case of atmospheric and manmade noise. Equation (13.2) was used to find the available noise power *Fa*.

$$Fa = Pn/KToB = Ta/T \tag{13.2}$$

where:

F_a is the effective antenna noise factor;

Pn is noise power available from an equivalent loss-free antenna (WATTS);

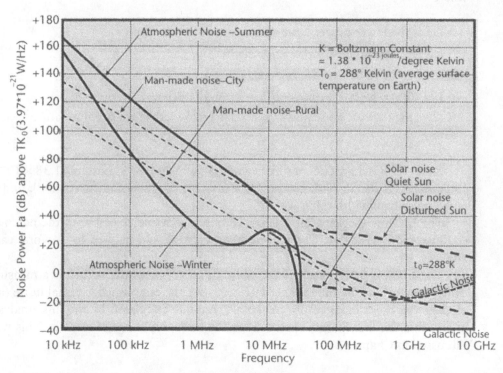

Figure 13.1   Median radio noise power spectral density from various sources determines initial noise figure requirements for HPOI receivers. Data was compiled from [2].

K is the Boltzmann constant = 1.38×10^{-23} joules per kelvin (K);

To is the reference temperature, taken as 288K;

B is the effective receiver noise BW (Hz);

Ta is the effective antenna temperature in the presence of external noise (K).

In high frequency work, environments are quite noisy as seen in Figure 13.1. Looking at the overall picture between galactic, atmospheric, and manmade noise, the external contribution to a receiver is such that the worst noise level expected at HF is about 18 dB. This is the determining factor in choosing the noise figure for such receiver. An 8- to 15-dB noise figure is typical in such a case, and anything quieter is not necessary in such an application. If a VHF/UHF or microwave receiver is contemplated, lower-noise figures are consequently used. The limiting factors for such receivers would be the galactic and solar noise, as shown.

Designing for a given noise figure in an HPOI receiver involves all circuits, from mixers to amplifiers and local oscillators, with mixers being the predominant contributors. The noise figure of a receiver is expressed by the total noise added by this circuitry. An expression for defining this parameter in terms of signal-to-noise ratios is given by

$$NF = 10\log\big[(PSi/PNi)/(PSo/PNo)\big] \tag{13.3}$$

where *NF* is the noise figure in decibels, *PSi* is the signal power at the input, *PNi* is the noise power at the input, *PSo* is the signal power at the output, and *PNo* is the noise power at the output.

The noise figure of a receiver is subject to all the noise figures of all the devices/components and circuits in its makeup. For practical reasons, NF usually equals the MDS of the receiver. In addition to being calculated or measured with special noise figure equipment, the noise figure can also be measured using a simple process. This is known as the poor man's noise figure method.

The noise figure of a a receiver (or an RF device) is a measurement of the noise power generated internally. This power is traditionally referenced to the input of the device and compared with the power generated by a 50Ω termination at 290K. The effective noise temperature is defined as the temperature at which the 50Ω termination generates noise power equivalent to that generated by the device when referenced to the input. Noise figure (in decibels) is related to the effective temperature (T_e) by using:

$$F_{(dB)} = 10\log 10\big(1 + T_e/290\big) \tag{13.4}$$

The following method is used to find a receiver's absolute noise temperature. This method works well up into the VHF/UHF range.

- Make a short coax input line and terminate it with a noninductive resistor (a carbon film resistor) equal to the input impedance of the receiver.
- Cover the resistor with a small plastic bag such as used in cooking, or build up a small surrounding epoxy case using electrically nonconductive high-dielectric epoxy.

- Immerse the terminating resistor for several minutes in ice water (0°C).
- Adjust a programmable attenuator (calibrated in decibels) located at the output of the receiver into an RMS meter such that the output meter reads mid-scale. Note the reading.
- Next, immerse the 50Ω termination resistor in boiling water (100°C) for several minutes.
- Adjust the decibel attenuator such that the RMS meter reads again mid-scale (same value as before).
- Then take an accurate difference of the attenuator in decibels for the two conditions.
- Convert the decibel difference to noise power by the relation:

$n = 10^{x/10}$ where x = dB differential

then the noise temperature of the entire receiver is found with the following equation:

$$T_e(\text{K}) = (373 - 273n)/(n - 1) \tag{13.5}$$

13.3 Sensitivity

Although MDS is usually sufficient to determine the requirement for the noise floor of a receiver, the sensitivity terminology is sometimes used. This term is sometimes confused with the MDS. In reality, it is related to the MDS, but it introduces some new parameters such as $S + N/N$ (Ksn) and the modulation characteristics factors (Km). In communications receivers, these parameters depend on the E_b/N_o link margin for a given bit rate predictions as we discussed earlier in Chapter 11. In radar and surveillance receivers, $S + N/N$ and compounded probability of intercept are some of the ways to specify sensitivity.

Sensitivity is a way of expressing the ability of a radio receiver to positively detect a signal of a certain level with a predicted probability of intercept. There are many ways of defining how sensitivity is calculated. Included are tangential sensitivity, signal-to-noise ratio, output signal-to-noise ratio, accumulated false-alarm rate over a given time period, bit-error rate, probability of detection, and so forth. The type of method chosen usually depends on application factors, such as type and degree of modulation, IF bandwidth employed and type of detected output, as well as the receiver's noise figure. We will deal with two simple ways of expressing receiver sensitivity. Equation (13.6) shows one method of determining sensitivity.

$$S = -174 \text{ dBm} + NF + 10\log IF\, BW + Ksn + Km \tag{13.6}$$

where:

S is the sensitivity in dBm;

−174 dBm = KTB = Thermal noise power in a 1-Hz bandwidth at room temperature;

NF is the noise figure in decibels;

IF BW is the predetection IF bandwidth in hertz;

Ksn is the desired *S* + *N/N* in decibels of the detected signal;

Km is a variable expressed in decibels, which is a function of the modulation used.

As can be seen from this equation, sensitivity improves (becomes a smaller number) with decreased NF and IF bandwidth.

Another method of measuring sensitivity of a receiver is expressed by the signal input, expressed in microvolts, necessary to produce a baseband output, which is 10-dB greater than the noise figure of the receiver. For example, if the specification reads: 10 μV for 10-dB S/N, it means 10 microvolts of signal at the antenna (30% amplitude modulated) will be heard at the output, 10 dB over the internal noise of the receiver. This method is usually used in specifying commercial receivers. In professional data communications receivers, the link margin E_b/N_o is normally used.

In conclusion, sensitivity can take many quantitative forms. One fact that must be kept in mind is that sensitivity varies throughout the frequency coverage of a receiver because of gain variations in the front end. Good designers will usually publish sensitivities at several points in the frequency coverage.

13.4   Design Considerations for the Front End—Composite Noise Figure

In order to keep intermodulation products at a low level, good receivers use a minimum of preamplification, usually just enough to compensate for the loss in the preselector filters and the first mixer. Figure 13.2 shows a block diagram for the front end of such a receiver. This arrangement can provide a third-order intercept point of +20 dBm and a typical noise figure of 12 dB, as shown. From this example, it can be seen that in order to achieve these specifications, a high-level mixer was chosen requiring +27 dBm of local oscillator drive. Such a mixer can provide a third-order intercept point of greater than +28 dBm (typical doubly balanced mixer LO drive requirements is +7 dBm).

A better approach and a more economical one is shown in Figure 13.3. By eliminating the preamplifier, and by reconsidering each stage for noise figure, the signal-handling capability of the entire front end is maintained with less local oscillator drive.

13.5   Understanding the Third-Order Intercept Point Spurious-Free Dynamic Range (IP3SFDR)

As we previously discussed, one of the five dynamic range methods is the third-order intercept point spurious free dynamic range (IP3SFDR) also known as the

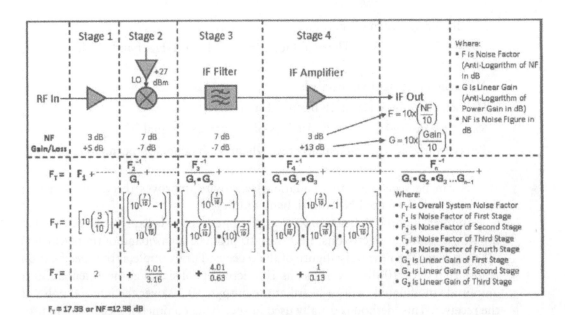

Figure 13.2 System composite noise figure. The receiver front end has a low noise figure due to careful choice of devices. Note that losses in filters or mixers equal noise figures for the particular parts.

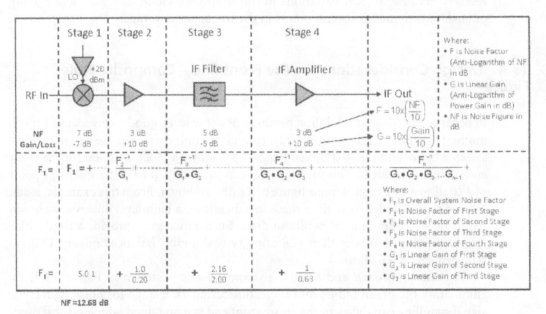

Figure 13.3 Eliminating the preamplifier and using a more judicious design and specific parts yield the same noise figure and good signal handling capability.

intermodulation dynamic range. This method is important because it uses a worst case third-order product methodology. The lower level of this dynamic range is the MDS as previously discussed.

The upper limit of this dynamic range (*Pu*) is limited by the level of two equal input signals creating a third-order intermodulation product at the output of the receiver, which is equal in amplitude to the minimum discernable signal (MDS) level.

The upper limit of the dynamic range can then be expressed by using (13.7).

$$Pu(dBm) = 1/3(MDS + 2IP)$$
$$= 1/3(-171(dBm) + NF(dB) + 10\log IF BW(Hz) + 2/3IP(dBm)) \tag{13.7}$$

where Pu is the upper power limit of the dynamic range in dBm and IP is the receiver's third-order input intercept point in dBm.

By combining the two equations, we can find (13.8) for the total spurious free dynamic range:

$$IP2SFDR(dBm) = Pu(dBm) - PL(dBm)$$
$$= 1/3(MDS + 2IP) - MDS = 2/3(IP - MDS) \tag{13.8}$$
$$= 2/3(IP(dBm) - NF(dB) - 10\log IF BW(Hz) + 171(dBm))$$

where $IP3SFDR$ is the third-order spurious-free dynamic range.

It can be seen from this equation that this kind of dynamic range is directly proportional to the intercept point (IP) and inversely proportional to the noise figure (NF), and IF bandwidth (IF BW). We can then say that the IP3SFDR improves with the lowest noise figures, narrower IF bandwidths, and higher intercept points.

The following example shows a practical application for the IP3SFDR equation. Assume a typical high-performance receiver with a noise figure of 8 dB, an IF bandwidth of 2.1 kHz, and an input intercept point of +20 dBm. Substituting these quantities in (13.8) yields:

$$IP3SFDR = 2/3(+20\,dBm - 8\,dB - 10\log 2,100\,Hz + 171\,dBm) = 99.85\,dB$$
$$\therefore IP3SFDR \cong 100\,dB$$

The total distribution of this number can be best understood by examining the graph in Figure 13.4. We know that the total spurious-free dynamic range (SFDR) for our receiver is approximately 100 dB, but what is not known is where this range fits in the total picture of the receiver's sensitivity. Once this is found, what does this range mean from a practical performance point of view? We had previously determined that the lower limit of a dynamic range is given by the minimum detectable signal (MDS). If using (13.1) for our example, we find the lower limit of the receiver's IP3SFDR range to be approximately −130 dBm.

$$MDS = -171 + 8 + 10\log 2,100 = -129.77\,dBm \cong -130\,dBm$$

We can then say that the system's noise figure for an IF bandwidth of 2.1 kHz is 3 dB below this number, or −133 dBm (MDS is sometimes defined as a signal 3-dB greater than the equivalent noise level for a specified IF bandwidth).

Knowing the MDS, the IP (+20 dBm), and with the help of (13.1) we can determine the upper limit of our 100-dB dynamic range.

$$Pu = 1/3(-129.77 + 40) = -29.92 \, \text{dBm} \cong -30 \, \text{dBm} \, \}$$

The same result would be obtained if we added the total dynamic range of 99.85 dB to the MDS.

$$Pu = 99.85 + (-129.77) = -29.92 \, \text{dBm}$$

This last procedure could be used to verify the validity of (13.1). If these numbers are plotted as shown in Figure 13.4, we can conclude that the receiver in our example will perform undisturbed for all input signals varying from approximately −30 dBm to −130 dBm, with the receiver tuned to a third-order intermodulation product produced by two strong signals equal in amplitude and different in frequency from each other. The amplitude of these signals as well as their difference frequency (ΔF) were represented in our example by the +20-dBm input intercept point. In practice, this quantity is a function of the output intercept of all nonlinear elements, such as mixers, amplifiers, and so forth involved in the design of the receiver, as we will see next.

Figure 13.5 shows the intercept method, used as an evaluation method for the strong signal handling capability of a radio receiver. In practice, the IP3SFDR is usually considered as the main method to judge the dynamic range of a receiver. It is measured with the setup shown in Figure 13.6. First, the minimum discernable signal (MDS) is found by disconnecting RF_1 via switch $S1$ as shown.

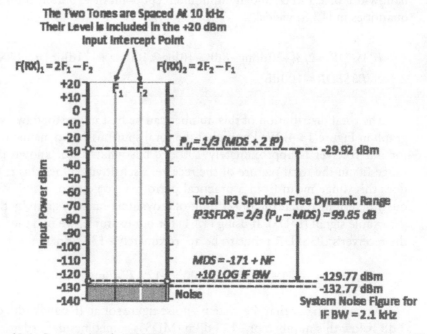

Figure 13.4 Determining the IP3SFDR range for a receiver with a noise figure of 8 dB, an IF bandwidth of 2.1 kHz, and an input intercept point of +20 dBm.

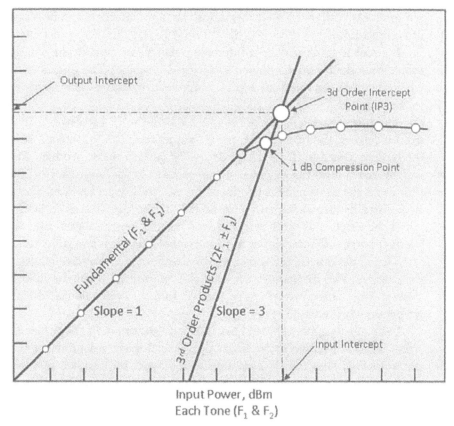

Figure 13.5   Determining output and input intercept points for a radio receiver or other nonlinear device (e.g., a mixer).

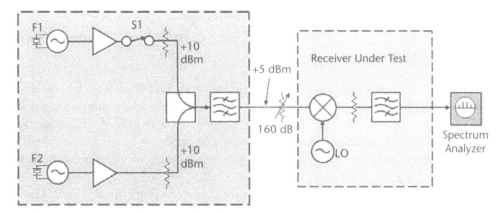

Figure 13.6   Measurement of receiver MDS (*S1* open) and output intercept point (*S1* closed). Input intercept can then be plotted and used in the dynamic range formula. This setup also applies to testing mixers.

The MDS is measured as the power necessary for the generator RF_2 expressed in dBm, to produce a 3-dB increase in baseband output over the noise level of the receiver as indicated by an RMS voltmeter or a spectrum analyzer.

Knowing the MDS, the setup uses the two generators to actually find the output intercept, and with this information the input intercept can be plotted as shown in Figure 13.5.

In order to find the output intercept point the outputs of the two signal generators (RF_1 and RF_2) are combined in a hybrid combiner. The output of the combiner, which contains the two-tone signal, is applied through a calibrated step attenuator to the receiver.

The tone spacing is arbitrary and follows the IF bandwidth. In communications receivers, the two generators are sometimes 20, 10, 5, or 2 kHz apart with the receiver tuned to $2F_2 - F_1$ or $2F_1 - F_2$, a third-order product. The attenuator is then varied until the response of the receiver at the frequency of the third-order product is the same as that produced by the MDS found earlier. The performance is specified by measuring and plotting the output intercept as shown.

If the receiver is well designed, the desired output signal and the distortion product curve will intersect as high as possible, as shown in our example.

This is the output intercept, which describes the intermodulation response of the receiver. The input intercept can also be plotted from the intercept point as shown. This number can then be used to find the spurious-free dynamic range as we previously discussed.

In conclusion, the receiver processes a weak signal in the presence of adjacent strong signals. Because of nonlinear deficiencies in the design of the first mixer and the front end, especially if a preamplifier is used, the receiver may not be able to copy the weak signal and it may be completely blocked when tuned to a third-order frequency of two interfering signals. The receiver's ability to perform under such brutal conditions is expressed by the IP3 spurious-free dynamic range. Recognizing this impediment came from military applications, which required a receiver to perform properly in the presence of a number of nearby high-power transmitters broadcasting on many adjacent frequencies, all at the same time.

13.6   Simulating and Measuring Composite Linear Dynamic Range for an HPOI Receiver

We will next look at a real case analysis of an HPOI receiver. As previously mentioned, the Star-10 transceiver covers seamlessly and continuously the entire HF frequency range of 2 to 30 MHz in a single band with a 10-Hz frequency resolution, and with an ultimate receiver composite linear dynamic range of 150 dB or better. A block diagram of the Star-10 transceiver showing all receiver sections (in darker gray) is shown in Figure 13.7.

The transceiver is a bilateral double conversion upconvert/downconvert design that features automatically switched half-octave filter banks in the front end and a high first IF of 75 MHz for superior image rejection over its four octaves RF frequency coverage.

As can be seen in the figure, the receiver system employs as many as three AGC loops (the main 9-MHz AGC, BIPA, and AIPA) to achieve the extended composite linear dynamic range indicated. We will next perform a composite linear dynamic range analysis on this receiver. There are many public domain programs, which can be used to model a receiver's linear dynamic range behavior.

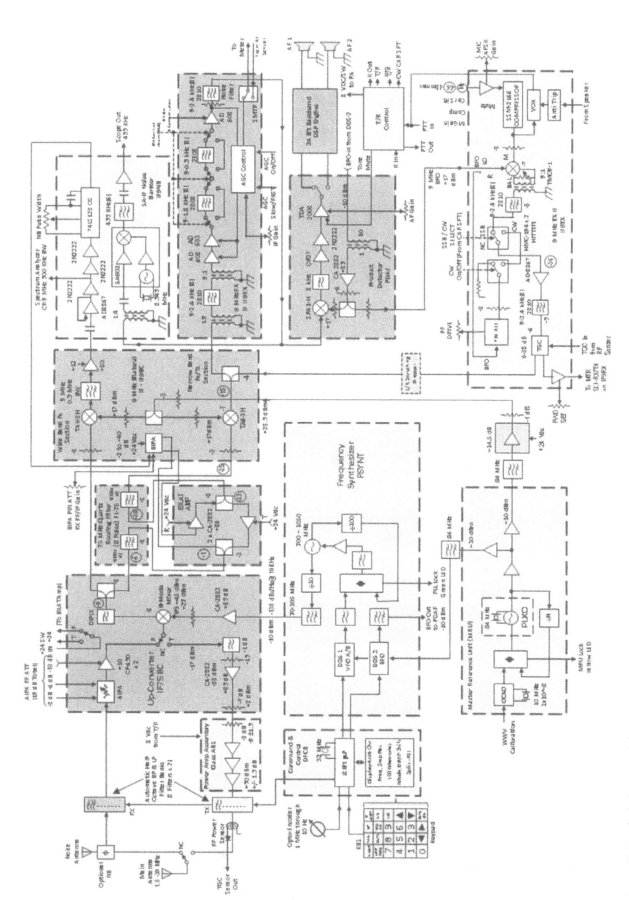

Figure 13.7 Block diagram of the receiver sections (darker gray) of the Star-10 transceiver.

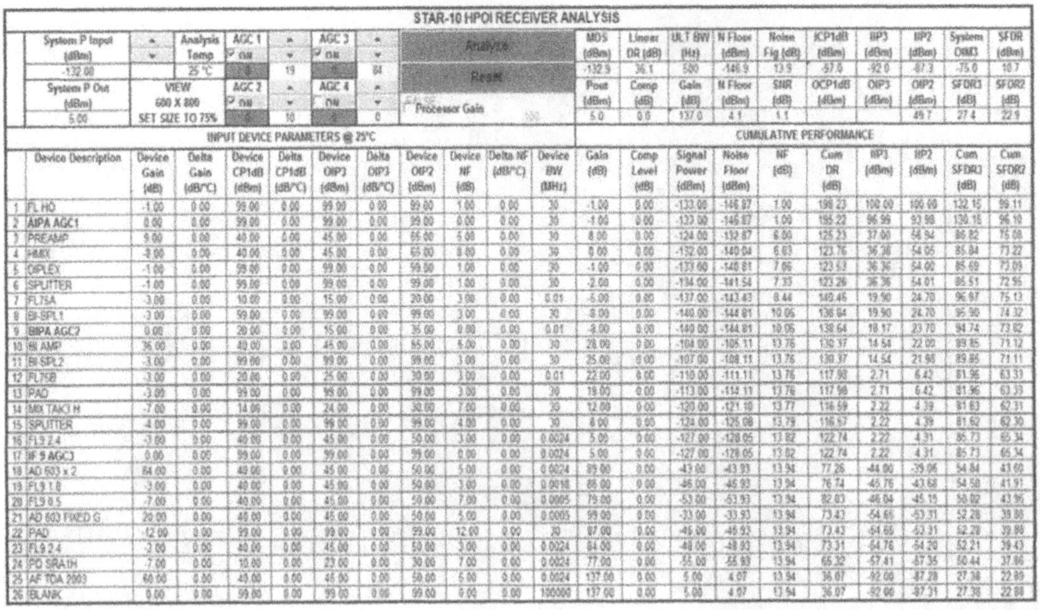

STAR-10 HPOI RECEIVER ANALYSIS

System P Input (dBm)	Analysis Temp	AGC 1		AGC 3			MDS (dBm)	Linear DR (dB)	ULT BW (Hz)	N Floor (dBm)	Noise Fig (dB)	ICP1dB (dBm)	IIP3 (dBm)	IIP2 (dBm)	System OIM3	SFDR (dBm)
-132.00	25 °C	ON	19	ON	84	Analyze / Reset	-132.9	36.1	500	-146.9	13.9	-97.0	-92.0	-87.3	-75.0	10.7
System P Out (dBm)	**VIEW 600 X 800**	**AGC 2**		**AGC 4**			**Pout (dBm)**	**Comp (dB)**	**Gain (dB)**	**N Floor (dBm)**	**SNR (dB)**	**OCP1dB (dBm)**	**OIP3 (dBm)**	**OIP2 (dBm)**	**SFDR3 (dB)**	**SFDR2 (dB)**
5.00	SET SIZE TO 75%	ON	10	ON	0	Processor Gain	5.0	0.0	137.0	4.1	1.1			49.7	27.4	22.9

#	Device Description	Device Gain (dB)	Delta Gain (dB/°C)	Device CP1dB (dBm)	Delta CP1dB (dB/°C)	Device OIP3 (dBm)	Delta OIP3 (dB/°C)	Device OIP2 (dBm)	Device NF (dB)	Delta NF (dB/°C)	Device BW (MHz)	Gain (dB)	Comp Level (dB)	Signal Power (dBm)	Noise Floor (dBm)	NF (dB)	Cum DR (dB)	IIP3 (dBm)	IIP2 (dBm)	Cum SFDR3 (dB)	Cum SFDR2 (dB)
		INPUT DEVICE PARAMETERS @ 25°C										CUMULATIVE PERFORMANCE									
1	FL HO	-1.00	0.00	99.00	0.00	99.00	0.00	99.00	1.00	0.00	30	-1.00	0.00	-133.00	-146.87	1.00	198.23	100.00	100.00	132.15	99.11
2	AIPA AGC1	0.00	0.00	99.00	0.00	99.00	0.00	99.00	0.00	0.00	30	-1.00	0.00	-133.00	-146.87	1.00	195.22	96.99	93.98	130.15	96.10
3	PREAMP	9.00	0.00	40.00	0.00	45.00	0.00	65.00	5.00	0.00	30	8.00	0.00	-124.00	-132.87	6.00	125.23	37.00	56.94	86.82	75.08
4	HMX	-8.00	0.00	40.00	0.00	45.00	0.00	65.00	8.00	0.00	30	0.00	0.00	-132.00	-140.04	6.83	123.76	36.38	54.05	85.84	73.22
5	DIPLEX	-1.00	0.00	99.00	0.00	99.00	0.00	99.00	1.00	0.00	30	-1.00	0.00	-133.00	-140.81	7.06	123.53	36.36	54.00	85.69	73.09
6	SPLITTER	-1.00	0.00	99.00	0.00	99.00	0.00	99.00	1.00	0.00	30	-2.00	0.00	-134.00	-141.54	7.33	123.26	36.36	54.01	85.51	72.95
7	FL75A	-3.00	0.00	10.00	0.00	15.00	0.00	20.00	3.00	0.00	0.01	-5.00	0.00	-137.00	-143.43	8.44	140.46	19.90	24.70	96.97	75.13
8	BI-SPL1	-3.00	0.00	99.00	0.00	99.00	0.00	99.00	3.00	0.00	30	-8.00	0.00	-140.00	-144.81	10.06	136.84	19.90	24.70	95.90	74.32
9	BIPA AGC2	0.00	0.00	20.00	0.00	15.00	0.00	35.00	0.00	0.00	0.01	-8.00	0.00	-140.00	-144.81	10.06	138.64	18.17	23.70	94.74	73.82
10	BI AMP	36.00	0.00	40.00	0.00	45.00	0.00	65.00	5.00	0.00	30	28.00	0.00	-104.00	-105.11	13.76	130.37	14.54	22.00	89.85	71.12
11	BI-SPL2	-3.00	0.00	99.00	0.00	99.00	0.00	99.00	3.00	0.00	30	25.00	0.00	-107.00	-108.11	13.76	130.37	14.54	21.98	89.85	71.11
12	FL75B	-3.00	0.00	20.00	0.00	25.00	0.00	30.00	3.00	0.00	0.01	22.00	0.00	-110.00	-111.11	13.76	117.98	2.71	6.42	81.96	63.33
13	PAD	-3.00	0.00	99.00	0.00	99.00	0.00	99.00	3.00	0.00	30	19.00	0.00	-113.00	-114.11	13.76	117.98	2.71	6.42	81.96	63.33
14	MIX TAK3 H	-7.00	0.00	14.00	0.00	24.00	0.00	30.00	7.00	0.00	30	12.00	0.00	-120.00	-121.10	13.77	116.59	2.22	4.39	81.63	62.31
15	SPLITTER	-4.00	0.00	99.00	0.00	99.00	0.00	99.00	4.00	0.00	30	8.00	0.00	-124.00	-125.08	13.79	116.57	2.22	4.39	81.62	62.30
16	FL9 2.4	-3.00	0.00	40.00	0.00	45.00	0.00	50.00	3.00	0.00	0.0024	5.00	0.00	-127.00	-128.05	13.82	122.74	2.22	4.31	85.73	65.34
17	IF 9 AGC3	0.00	0.00	99.00	0.00	99.00	0.00	99.00	0.00	0.00	0.0024	5.00	0.00	-127.00	-128.05	13.82	122.74	2.22	4.31	85.73	65.34
18	AD 603 x 2	84.00	0.00	40.00	0.00	45.00	0.00	50.00	5.00	0.00	0.0024	89.00	0.00	-43.00	-43.93	13.94	77.26	-44.00	-39.96	54.84	43.60
19	FL9 1.8	-3.00	0.00	40.00	0.00	45.00	0.00	50.00	3.00	0.00	0.0018	86.00	0.00	-46.00	-45.93	13.94	76.74	-45.76	-43.68	54.50	41.91
20	FL9 0.5	-7.00	0.00	40.00	0.00	45.00	0.00	50.00	7.00	0.00	0.0005	79.00	0.00	-53.00	-53.93	13.94	82.03	-46.04	-45.15	58.02	43.96
21	AD 603 FIXED G	20.00	0.00	40.00	0.00	45.00	0.00	50.00	5.00	0.00	0.0005	99.00	0.00	-33.00	-33.93	13.94	73.42	-64.65	-63.31	52.28	39.88
22	PAD	-12.00	0.00	99.00	0.00	99.00	0.00	99.00	12.00	0.00	30	87.00	0.00	-45.00	-45.93	13.94	73.42	-64.65	-63.31	52.28	39.88
23	FL9 2.4	-3.00	0.00	40.00	0.00	45.00	0.00	50.00	3.00	0.00	0.0024	84.00	0.00	-48.00	-48.93	13.94	73.31	-64.76	-64.20	52.21	39.43
24	PO SRA1H	-7.00	0.00	10.00	0.00	23.00	0.00	30.00	7.00	0.00	0.0024	77.00	0.00	-55.00	-55.93	13.94	65.32	-57.41	-67.35	50.44	37.86
25	AF TDA 2003	60.00	0.00	40.00	0.00	45.00	0.00	50.00	5.00	0.00	0.0024	137.00	0.00	5.00	4.07	13.94	36.07	-92.00	-87.28	27.38	22.89
26	BLANK	0.00	0.00	99.00	0.00	99.00	0.00	99.00	0.00	0.00	100000	137.00	0.00	5.00	4.07	13.94	36.07	-92.00	-87.31	27.38	22.88

(a)

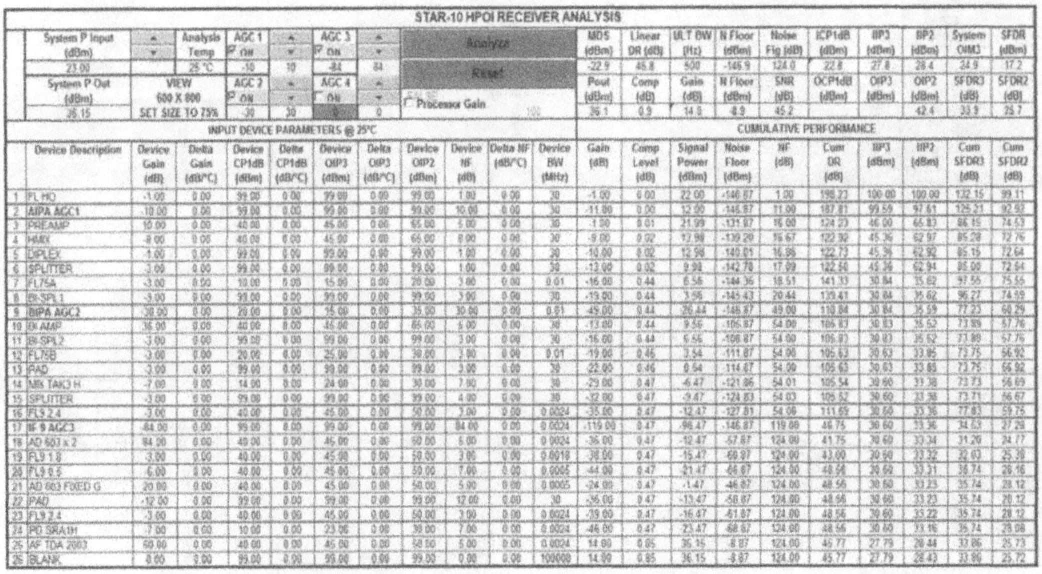

STAR-10 HPOI RECEIVER ANALYSIS

System P Input (dBm)	Analysis Temp	AGC 1		AGC 3			MDS (dBm)	Linear DR (dB)	ULT BW (Hz)	N Floor (dBm)	Noise Fig (dB)	ICP1dB (dBm)	IIP3 (dBm)	IIP2 (dBm)	System OIM3	SFDR (dBm)
23.00	25 °C	ON -10	10	ON -84	84	Analyze / Reset	-22.9	45.8	500	-146.9	124.0	22.8	27.8	28.4	24.9	17.2
System P Out (dBm)	**VIEW 600 X 800**	**AGC 2**		**AGC 4**			**Pout (dBm)**	**Comp (dB)**	**Gain (dB)**	**N Floor (dBm)**	**SNR (dB)**	**OCP1dB (dBm)**	**OIP3 (dBm)**	**OIP2 (dBm)**	**SFDR3 (dB)**	**SFDR2 (dB)**
35.15	SET SIZE TO 75%	ON -30	30	ON	0	Processor Gain	36.1	0.9	14.0	-8.9	45.2			42.4	33.9	25.7

#	Device Description	Device Gain (dB)	Delta Gain (dB/°C)	Device CP1dB (dBm)	Delta CP1dB (dB/°C)	Device OIP3 (dBm)	Delta OIP3 (dB/°C)	Device OIP2 (dBm)	Device NF (dB)	Delta NF (dB/°C)	Device BW (MHz)	Gain (dB)	Comp Level (dB)	Signal Power (dBm)	Noise Floor (dBm)	NF (dB)	Cum DR (dB)	IIP3 (dBm)	IIP2 (dBm)	Cum SFDR3 (dB)	Cum SFDR2 (dB)
		INPUT DEVICE PARAMETERS @ 25°C										CUMULATIVE PERFORMANCE									
1	FL HO	-1.00	0.00	99.00	0.00	99.00	0.00	99.00	1.00	0.00	30	-1.00	0.00	22.00	-146.87	1.00	198.23	100.00	100.00	132.15	99.11
2	AIPA AGC1	-10.00	0.00	99.00	0.00	99.00	0.00	99.00	10.00	0.00	30	-11.00	0.00	12.00	-146.87	11.00	187.81	99.59	97.61	125.21	92.92
3	PREAMP	10.00	0.00	40.00	0.00	45.00	0.00	65.00	5.00	0.00	30	-1.00	0.01	21.99	-131.87	16.00	124.23	46.00	65.83	86.15	74.53
4	HMX	-8.00	0.00	40.00	0.00	45.00	0.00	66.00	8.00	0.00	30	-9.00	0.02	12.98	-139.29	16.67	122.92	45.36	62.97	85.28	72.76
5	DIPLEX	-1.00	0.00	99.00	0.00	99.00	0.00	99.00	1.00	0.00	30	-10.00	0.02	12.98	-140.01	16.86	122.73	45.36	62.92	85.15	72.64
6	SPLITTER	-3.00	0.00	99.00	0.00	99.00	0.00	99.00	1.00	0.00	30	-13.00	0.02	9.98	-142.78	17.09	122.56	45.36	62.94	85.00	72.64
7	FL75A	-3.00	0.00	10.00	0.00	15.00	0.00	20.00	3.00	0.00	0.01	-16.00	0.44	6.56	-144.36	18.51	141.33	30.84	35.62	97.56	75.55
8	BI-SPL1	-3.00	0.00	99.00	0.00	99.00	0.00	99.00	3.00	0.00	30	-19.00	0.44	3.56	-145.43	20.44	139.41	30.84	35.62	96.27	74.59
9	BIPA AGC2	-30.00	0.00	20.00	0.00	15.00	0.00	35.00	30.00	0.00	0.01	-45.00	0.44	-26.44	-146.87	49.00	110.84	30.84	35.59	77.23	60.29
10	BI AMP	36.00	0.00	40.00	0.00	45.00	0.00	65.00	6.00	0.00	30	-13.00	0.44	9.56	-106.87	54.00	106.83	30.83	35.62	73.89	57.76
11	BI-SPL2	-3.00	0.00	99.00	0.00	99.00	0.00	99.00	3.00	0.00	30	-16.00	0.44	6.56	-108.87	54.00	106.83	30.83	35.62	73.89	57.76
12	FL75B	-3.00	0.00	20.00	0.00	25.00	0.00	30.00	3.00	0.00	0.01	-19.00	0.46	3.54	-111.87	54.00	105.63	30.63	33.85	73.75	56.92
13	PAD	-3.00	0.00	99.00	0.00	99.00	0.00	99.00	3.00	0.00	30	-22.00	0.46	0.54	-114.87	54.00	105.63	30.63	33.85	73.75	56.92
14	MIX TAK3 H	-7.00	0.00	14.00	0.00	24.00	0.00	30.00	7.00	0.00	30	-29.00	0.47	-6.47	-121.86	54.01	105.54	30.60	33.38	73.73	56.69
15	SPLITTER	-3.00	0.00	99.00	0.00	99.00	0.00	99.00	4.00	0.00	30	-32.00	0.47	-9.47	-124.83	54.03	105.52	30.60	33.38	73.71	56.67
16	FL9 2.4	-3.00	0.00	40.00	0.00	45.00	0.00	50.00	3.00	0.00	0.0024	-35.00	0.47	-12.47	-127.81	54.06	111.69	30.60	33.36	77.83	59.75
17	IF 9 AGC3	-84.00	0.00	99.00	0.00	99.00	0.00	99.00	84.00	0.00	0.0024	-119.00	0.47	-96.47	-146.87	119.00	46.75	30.60	33.36	34.63	27.29
18	AD 603 x 2	84.00	0.00	40.00	0.00	45.00	0.00	50.00	6.00	0.00	0.0024	-36.00	0.47	-12.47	-57.87	124.00	41.75	30.60	33.34	31.20	24.77
19	FL9 1.8	-3.00	0.00	40.00	0.00	45.00	0.00	50.00	3.00	0.00	0.0018	-38.00	0.47	-15.47	-60.87	124.00	43.00	30.60	33.32	32.63	25.39
20	FL9 0.5	-6.00	0.00	40.00	0.00	45.00	0.00	50.00	7.00	0.00	0.0005	-44.00	0.47	-21.47	-66.87	124.00	48.56	30.60	33.31	35.74	28.16
21	AD 603 FIXED G	20.00	0.00	40.00	0.00	45.00	0.00	50.00	5.00	0.00	0.0005	-24.00	0.47	-1.47	-46.87	124.00	48.56	30.60	33.23	35.74	28.12
22	PAD	-12.00	0.00	99.00	0.00	99.00	0.00	99.00	12.00	0.00	30	-36.00	0.47	-13.47	-58.87	124.00	48.56	30.60	33.23	35.74	28.12
23	FL9 2.4	-3.00	0.00	40.00	0.00	45.00	0.00	50.00	3.00	0.00	0.0024	-39.00	0.47	-16.47	-61.87	124.00	48.56	30.60	33.22	35.74	28.12
24	PO SRA1H	-7.00	0.00	10.00	0.00	23.00	0.00	30.00	7.00	0.00	0.0024	-46.00	0.47	-23.47	-68.87	124.00	48.56	30.60	33.16	35.74	28.08
25	AF TDA 2003	60.00	0.00	40.00	0.00	45.00	0.00	50.00	5.00	0.00	0.0024	14.00	0.85	36.15	-8.87	124.00	45.77	27.79	28.44	33.86	25.73
26	BLANK	0.00	0.00	99.00	0.00	99.00	0.00	99.00	0.00	0.00	100000	14.00	0.85	36.15	-8.87	124.00	45.77	27.79	28.43	33.86	25.72

(b)

Figure 13.8 (a) Input and output data for all stages of the Star-10 receiver example. All parts are addressed from several points of view as shown. With this information, an MDS of −132 dBm is calculated. While parts specifications are not always available, the most conservative numbers should be entered. The three AGCs have been turned on and their ranges have been addressed through a best judgment process. They are not active yet since the RF input is at the MDS level. (b) Ramping up the RF input to the receiver activates the AGCs progressively and in reverse order (from the back end of the receiver toward the front) to a maximum uncompressed RF input. The system was ramped over a range of −132 dBm to +23 dBm without compressing. At any level between the MDS and the maximum RF power in, the output is always funneled into the last section (and A/D, usually) for further processing. Consequently, the linear composite dynamic range is in excess of 150 dB.

In using one of these programs, the total linear composite dynamic range behavior is modeled using real-life component gains, losses, AGC ranges, compression points, and ultimate bandwidths utilizing the standard formulas presented earlier. Such program simulates a real-life situation where the RF input at the antenna port can be varied from the MDS up to the entire system's compression point, activating all three AGC stages progressively and in reverse order. We can then obtain a plot of the system's linear behavior before an actual implementation. Such analysis will allow us to tweak components before bread boarding the design, ensuring design confidence.

The output of the analysis tool displays automatically the linear composite dynamic range behavior on a spectrum-analyzer-like display, proving the entire performance of over 150 dB in our case.

The input and output data for the Star-10 receiver at the MDS level is shown in Figure 13.8(a). Figure 13.8(b) shows the results after ramping up the RF input, which progressively activates the three AGCs just as in the actual implementation.

A composite linear dynamic range is then plotted graphically as shown in Figure 13.9.

The plot shows the actual behavior of the receiver's total gain, its noise figure changes under AGC action, where compression is about to occur (for any of the stages), and how AGC ranges take over in reverse sequence in order to extend the input RF range without compressing the system.

It can be seen that without the AGC action, the receiver would have a limited linear dynamic range due to the impact of the compression points. The role of the AGC system is to keep the system just under the compression points and actually extend its linear dynamic range as far up in signal strength as possible. The RF input range and corresponding output vertical bands show how the three AGC actions enter the picture in order to keep the receiver uncompressed over the entire extended range.

The simulation results are presented graphically in Figure 13.9. This plot shows exactly how the system will behave dynamically under full triple AGC action. The purpose of the receiver is to funnel the very large RF input range of over 150 dB (the left-side band) into the limited dynamic range of the final stage or transducer (the upper right side) without compressing it. This output range is intended to match the limited analog-to-digital (A/D) converters input dynamic range used by baseband or passband DSP engines usually following the analog portions of a receiver.

Ideally, any point on the 150-dB RF input dynamic range should always be funneled about 1/3 of the way down from the top of the A/D's limited input dynamic range, as shown in Figure 13.9. Keeping the IF or baseband output level around this point takes care of possible A/D saturation variability, which can be caused by temperature changes.

Figure 13.9 is the actual output of the modeling software used in the analysis. If looking at this figure, one can see the system's noise figure indications at two points in the ramp (only two points are shown for simplicity, NF = 15 dB and NF = 84 dB). This shows how the system's noise figure increases as the RF input is ramped up and the composite AGCs enter the picture. This is normal, as any receiver's noise figure is reduced by the AGC action. When active parts loose gain or attenuators are introduced, their loss in decibels changes the noise figure accordingly,

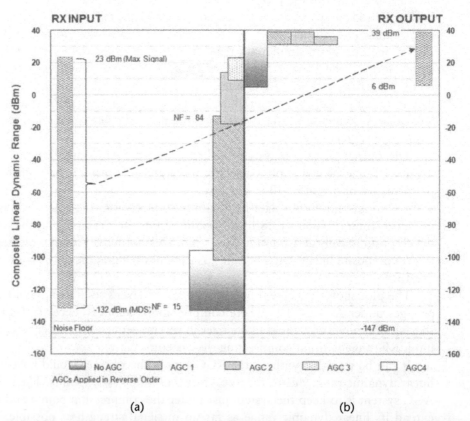

Figure 13.9 (a) RF input and baseband or (b) last IF output results for the Star-10 receiver composite linear dynamic range performance. The bands in the center correspond to the three AGC actions, turned on in progressive sequence over the 150-dB composite linear dynamic range of the RF input. When the first band (bottom) approaches compression, the first AGC turns on and keeps the system out of compression until the RF signal is increased to the point where the second and third AGCs take over. Any signal from the MDS level to the top +23-dBm RF level will be funneled uncompressed to about 1/3 of the way down from the top of the right-side band representing an A/D converter followed by a DSP processor.

contributing to the degradation of the entire system noise figure. However, this is not a problem in our case because the signal level causing the severe AGC action is always higher than the receiver's noise figure at any given point on the dynamic range. What is important is that intercept is possible with increased noise figure because the signal-to-noise level is always maintained higher, as the signal goes up through the uncompressed dynamic range.

In conclusion, the system design and the linear composite dynamic range modeling are usually the most important phases of a receiver design. These tasks, although relatively simple in concept, can be very tedious and take a considerable amount of design time. The software used, no matter how sophisticated, can help the designers but not design a radio for them. This phase of the design is a key part of the design verification methodology to follow after a system is implemented. It sets the system's initial performance goals as close to the final design goal as possible, so a minimum of modifications will be necessary in the circuit design. The system design of a receiver should also include design of the synthesizer portion of the design, which follows a similar process. No RF system should be pursued without doing this important homework.

References

[1]   CCIR Report 322, "Preliminary World Distribution and Characteristics of Atmospheric Radio Noise," *10th Plenary Assembly*, Geneva, 1963.

[2]   CCIR, *World Distribution and Characteristics of Atmospheric Radio Noise*, Report 322, International Radio Consultive Committee, International Telecommunication Union, Geneva, Switzerland, 1964.

Selected Bibliography

Abramowitz, M., and I. A. Stegun, *Handbook of Mathematical Functions, NBS Applied Math*, Series 55, New York: Dover, 1964.

Austin, L. W., "The Present Status of Radio Atmospheric Disturbances," *Proc. IRE*, Vol. 14, February 1926.

Barsis, A. P., et al., "Performance Predictions for Single Tropospheric Communication Links and for Several Links in Tandem," NBS Tech. Note 102, U.S. Department of Commerce, National Bureau of Standards, Boulder, CO, 1961.

BIPM, "Resolution 3 of the 13th CGPM," BIPM brochure on the kelvin, 1967/1968.

Boithias, L., *Radio Wave Propagation*, New York: McGraw-Hill, 1987.

Boute, R., "The Geometry of Sampling"

INTEC, Universiteit Gent, Belgium.

Brown, T. T., "Mixer Harmonic Chart," *Electronics*, April 1951.

Buckner, R. P., and S. M. Doghestani, *Improved Methods for VLF/LF Coverage Prediction*, PSR Report 2380, Pacific Sierra Research Corporation, Santa Monica, California, prepared for Office of Naval Research, Arlington, VA, 1993.

Carson, J. R., "The Reduction of Atmospheric Disturbances," *Proc. IRE*, Vol. 16, July 1928.

Carson, J. R., "Selective Circuits and Static Interference," *Trans. AIEE*, Vol. 43, 1924.

CCIR, *Man-Made Radio Noise*, Report 258-5, International Radio Consultive Committee, International Telecommunication Union, Geneva, Switzerland, 1990.

Cheadle, D. L., "Cascadable Amplifiers," Watkins Johnson Tech-Notes.

Crescenzi, Jr., E. J., and F. A. Marki, "Miniature X-Band Amplifiers and Down Converters," *Watkins Johnson Tech-Notes*.

Dean, S. W., "Correlation of Directional Observation of Atmospherics with Weather Phenomena," *Proc. IRE*, Vol. 17, July 1929.

Dean, S. W., "Long-Distance Transmission of Static Impulses," *Proc. IRE*, Vol. 19, September 1931.

Defense Nuclear Agency, "Combined Threat Effects WABINRES VLF/LF Coverage Prediction," TR90-19, Washington, D.C., 1990.

Defense Nuclear Agency, "TACAMO Pacific Area VLF/LF Communications Effectiveness," TR91-35, Washington, D.C., 1991.

Dexter, C., and R. Glaz, "HF Receiver Design," *Watkins Johnson Tech-Notes*.

Drentea, C., "The Star-10 Transceiver, Part 1," *QEX—ARRL*, November/December 2007.

Drentea, C., "The Star-10 Transceiver, Part 2," *QEX—ARRL*, March/April 2008.

Drentea, C., "The Star-10 Transceiver, Part 3," *QEX—ARRL*, May/June 2008.

Drentea, C., "Beyond Fractional-N, Part 1," *QEX*, March/April 2001.

Drentea, C., "Beyond Fractional-N, Part 2," *QEX*, May/June 2001.

Emery, F. E., "Low-Noise GaAs FET Amplifiers," *Watkins Johnson Tech-Notes*.

Galla, T. J., "Cascaded Amplifiers," *Watkins Johnson Tech-Notes*.

Gandhi, D., and C. Lyons, "Mixer Spur Analysis with Concurrently Swept LO, RF, and IF: Tools and Techniques," *Microwave Journal*, May 2003.

Glaz, R.D., "Signals Typical to HF Spectrum," Watkins Johnson Tech-Notes.

Hagn, G. H., "Man-Made Radio Noise and Interference," *Proc. of AGARD Conf.*, No. 420, Lisbon, Portugal, October 26–30, 1987.

Hagn, G. H., *Selected Radio Noise Topics*, SRI International Final Report, Project 45002, Contract No. NT83RA6-36001, 1984.

Harper, A. E., "Some Measurements on the Directional Distribution of Static," *Proc. IRE*, Vol. 17, July 1929.

Henderson, B. C., "Reliably Predict Mixer IM Suppresion," *Microwaves & RF*, November 1983.

International Telecommunication Union, *Characteristics and Applications of Atmospheric Radio Noise Data*, CCIR Report 322-3, Comité Consultatif International Des Radiocommunications, Geneva, Switzerland, 1968.

Hey-Shipton, G. L., and A. W. Denning, "Millimeter-Wave Block Converters," *Watkins Johnson Tech-Notes*.

International Telecommunication Union, *World Distribution and Characteristics of Atmospheric Radio Noise*, CCIR Report 322, Documents of X Plenary Assembly, Comité Consultatif International Des Radiocommunications, Geneva, Switzerland, 1963.

International Telecommunication Union, "Revision of Atmospheric Radio Noise Data," CCIR Report 65 (revised), Comité Consultatif International Des Radiocommunications, *Documents of IXth Plenary Assembly*, Vol. III, Geneva, Switzerland, 1959, p. 223.

Jansky, K.G., "Directional Studies of Atmospherics at High Frequencies," *Proc. IRE*, Vol. 20, December 1932.

Jansky, K. G., "A Note on the Source of Interstellar Interferences," *Proc. IRE*, Vol. 23, October 1935.

Jeruchim, M. C., P. Balaban, and K. S. Shanmugan, *Simulation of Communication Systems*, New York: Plenum Press, 1992.

Markel, J. D., "Shrinking Intermodulation," *EDN*, August 1967.

Middleton, D., "Statistical-Physical Models of Man-Made and Natural Radio-Noise Environments, Part I: First-Order Probability Models of the Instantaneous Amplitude," Office of Telecommunications Report OT 74-36, April 1975.

Middleton, D., *Statistical-Physical Models of Man-Made and Natural Radio-Noise Environments, Part II: First-Order Probability Models of the Envelope and Phase*, Office of Telecommunications Report OT 76-86, April 1976.

Middleton, D., *Statistical-Physical Models of Man-Made and Natural Radio-Noise Environments, Part III: First-Order Probability Models of the Instantaneous Amplitude of Class B Interference*, NTIA Contractor Report 78-1, June 1978.

Middleton, D., *Statistical-Physical Models of Man-Made and Natural Radio-Noise Environments, Part IV: Determination of the First-Order Parameters of Class A and Class B Interference*, NTIA Contractor Report 78-2, September 1978.

Morgan, H. K., "Rain Static," *Proc. IRE*, Vol. 24, July 1936, p. 959.

Murthy S. N., and G. Krishnamraju, "Interference to Low Earth Orbit Satellite (LEOS) Services in VHF Band from Ground Based Emissions," *IETE Technical Review*, 1995.

Pearl, B., "How to Determine Spur Frequencies," *EDN*, October 1965.

Physical Research, Inc. (for the Office of Naval Research), "Summary Report of the Fifth ONR Workshop on ELF/VLF Radio Noise," Naval Ocean Systems Center (NOSC), San Diego, CA, April 23–24, 1990.

Potter, R. K., "An Estimate of the Frequency Distribution of Atmospheric Noise," *Proc IRE*, Vol. 20, September 1932.

Rice, P. L., et al., "Phase Interference Fading and Service Probability," in *Transmission Loss Predictions for Tropospheric Communication Circuits, Volume 2*, NBS Technical Note 101, Annex V, 1966.

Riley, N. G., and K. Docherty, "Modeling and Measurement of Man-Made Radio Noise in the VHF-UHF Band," *Proceedings of the 9th International Conference on Ant. Prop.*, Vol. 2.

Sashoff, S. P., and J. Weil, "Static Emanating from Six Tropical Storms and Its Use in Locating the Position of the Disturbance," *Proc IRE*, Vol. 27, November 1939.

Shores, M. W., "Chart Pinpoints Receiver Interference Problems," *EDN*, January 15, 1969.

Spaulding, A. D., "The Natural and Man-Made Noise Environment in Personal Communications Services Bands," NTIA Report 96-330, May 1996.

Spaulding, A. D., "The Roadway Natural and Man-Made Noise Environment," *IVHS Journal*, 1995.

Spaulding, A. D., and R. T. Disney, *Man-Made Radio Noise, Part 1: Estimates for Business, Residential, and Rural Areas*, Office of Telecommunications Report OT 74-38, June 1974.

Spaulding, A. D., and F. G. Stewart, *An Updated Noise Model for Use in IONCAP*, NTIA Report 87-212, January 1987.

Spaulding, A. D., and J. S. Washburn, *Atmospheric Radio Noise: Worldwide Levels and Other Characteristics*, NTIA Report 85-173, U.S. Department of Commerce, National Telecommunications and Information Administration, Boulder, CO, 1985.

Stimson, G. W., *Airborne Radar,* Hughes Aircraft, SciTech, and IEEE Press.

Taylor, B. N., *Guide for the Use of the International System of Units*, Washington, D.C.: Government Printing Office, 1995.

"Understanding IP2 and IP3 Issues in Direct Conversion Receivers for WCDMA Wide Area Basestations," *Linear Technology Magazine*, June 2008.

Watt, A. D., et al., "Performance of Some Radio Systems in the Presence of Thermal and Atmospheric Noise," *Proc. IRE*, Vol. 46, December 1958.

CHAPTER 14

High-Performance Receiver Front-End Design Example

14.1 Designing a Front End for an HF Receiver/Transceiver

We will discuss the front end of a high-performance HPOI receiver using the Star-10 receiver example previously shown in Figure 13.7. Although this is an HF example, the same ideas apply to any other frequency range implemented with different parts and technologies. As can be seen from our previous discussions on dynamic range in Chapter 13, the noise figure of a receiving system is set mainly by the front end configuration. This was evident from Figures 13.2 and 13.3, and the equations associated with them.

Looking at the block diagram from Figure 13.7, the Star-10 transceiver is a perfect example of what it takes to obtain very high dynamic range while keeping a low noise figure. Shown in Figure 14.1 is the schematic diagram of the upconvert front end of the Star-10 receiver (IF75BC assembly). As can be seen, the system is bilateral and can also be used in the transmit mode, but we will limit our discussion only to the receiver part.

Looking at Figures 13.7 and 14.1, the RF signal coming from the antenna through the half-octave filter bank enters the system at J2. It is then presented to a set of attenuators called AIPA (or advanced intercept point attenuator) and a preamplifier, which can be switched in or out using miniature RF relays in order to maintain a high intercept point. The preamplifier switching, combined with the various switched attenuator pad combinations, allow for 3 dB through 19 dB of attenuation in several steps if required.

The preamplifier uses a push-pull Norton design equipped with two high-dynamic range CP-650 FETs (Q1 and Q2), and an H-mode, high-level mixer using the SD-5000 (U3), a very high intercept device. The CP650 is an epitaxial n-channel field effect transistor array, which is a good example of a high dynamic range device. It consists of 15 FETs in parallel on a single chip, which allows for a transconductance (gm) in the order of 150,000 μmhos.

Preamplifiers used in receivers at up to 400 MHz usually use JFETs. Beyond this point, bipolar transistors can yield low-noise figures at up to 2 GHz. Above this frequency, gallium arsenide (GaAs) FETs and indium phosphide (InP) transistors are used to achieve a low-noise figure performance. Special care was taken in the layout of the board to equally balance the elements of the push-pull preamplifier.

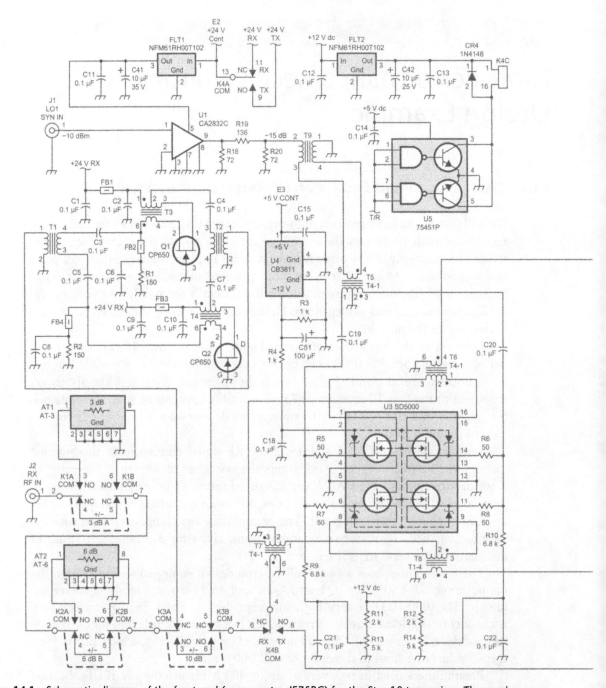

Figure 14.1 Schematic diagram of the front end (upconverter, IF75BC) for the Star-10 transceiver. The receiver uses a push-pull Norton amplifier, a high-level H-mode mixer, and class A amplifiers. An MDS of −132 dBm was calculated (−136 dBm actual was measured) and an IP3 of +45 dBm was predicted and found. Power consumption is over 30 watts DC. (*From:* [1–3]. Figure courtesy of ARRL.)

The Star-10 receiver design also uses high performance monolithic CA2832, class A push-pull amplifiers to provide high LO drive levels (+27 dBm) to the H-mode mixer.

The Star-10 front end dissipates approximately 30 watts of DC power (at 24V DC and 12V DC) to ensure a +45 dBm IP3 and a superior dynamic range for the

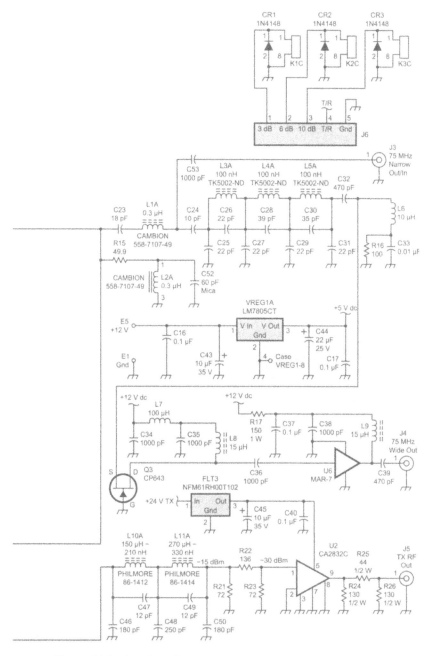

Figure 14.1 (continued)

receiver. The front-end assembly doubles as a bilateral IF in transmit, providing RF drive signals for the follow-up power linear amplifier (Figure 14.2).

The monolithic class A amplifiers are mounted under the board and heat is extracted through the assembly walls using heat sinks and special brushless quiet fans mounted on the sides of the aluminum assembly, and is further exhausted through the top of the transceiver cover. The IF75BC assembly implementation is shown in Figure 14.3.

(a)

(b)

Figure 14.2 (a) Front-end assembly of the HPOI receiver in the Star-10 transceiver. The push-pull Norton amplifier using two CP650 JFETs is visible in the center (the two round heat sinks). The SD-5000 H-mode mixer is also visible just to the right of the transistors. Class A CA2832 amplifiers are mounted on the back of the PC board and heat is extracted using heat sinks and computer grade brushless fans. The AIPA AGC is located on the left side along with the miniature RF relays used to switch the networks in and out under operator control. See the schematic diagram in Figure 14.1. (b) Class A amplifiers are mounted on the back of the front end converter board. Heat is extracted through the assembly walls and is further exhausted through the top of the transceiver cover. The IF75 DC assembly uses 30 watts of DC power to obtain a +45 dBm IP3 for the receiver.

Looking at Figure 14.1, the 75-MHz IF signal is output at J3. From here, the IF signals are fed into a set of specially designed eight-pole quartz roofing filters

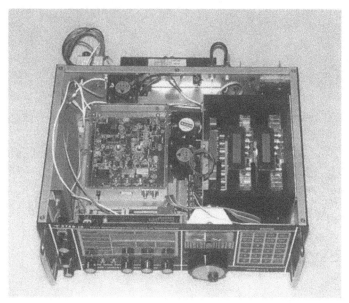

Figure 14.3   Location of the front-end assembly (IF75BC) in the Star-10 transceiver. The board uses 30 watts of DC power to provide a high receiver intercept point of +45 dBm. Brushless fans extract heat from the assembly, which is exhausted through the top of the radio.

with a bandwidth of 10 kHz and a set of high dynamic range bilateral amplifiers, again using class A, CA-2832 monolithic amplifiers as shown. The location of the bilateral amplifier assembly in the total system is shown in Figure 13.7, while its schematic diagram is shown in Figure 14.4.

The 75-MHz IF filters are designed to be able to withstand up to +10 dBm signal levels without being damaged. The top section of the bilateral amplifier at the top of Figure 14.4 is the receiver section. It gets activated by turning its +24V DC power on from the front-end assembly. With the power applied to the top section, the unpowered bottom section (used only in transmit) remains passive as the splitter/combiner arrangement naturally rejects the active side signals. The process reverses in transmit. The assembly also uses an output and a return path from the BIPA (bilateral intercept point attenuator) AGC, which is shown in Figure 14.5. This attenuator uses four PIN diodes to achieve its AGC functionality. It can provide 30 dB of distortion-free automatic or manual attenuation for the 75-MHz signal and enters the linear composite dynamic range analysis previously discussed. The physical implementation of the bilateral amplifier is shown in Figure 14.6.

Looking at Figure 13.7, the 75-MHz quartz filter assembly and the bilateral amplifier are followed by a second converter IF9BC. The schematic diagram of this converter is shown in Figure 14.7. Again, this is a bilateral unit, which uses a pair of high level mixers. A synthesized coherent LO at 84 MHz is utilized to convert from the 75-MHz IF to the 9-MHz IF.

To conclude, we have seen what it takes from a system and a circuit point of view to design a high dynamic range–high probability of intercept (HPOI) receiver. The front end of such a receiver has to be designed to accomplish a high intercept point while keeping a low MDS simultaneously. Such a goal is not trivial and re-quires the utilization of high-level class A push-pull amplifiers with very low noise figures, high-level mixers, and effective quartz IF filters, which can withstand high

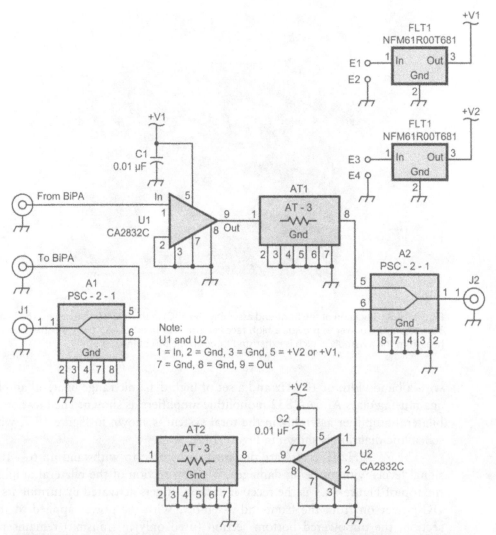

Figure 14.4 The bilateral amplifier schematic diagram. The variable PIN attenuator (BIPA) is inserted into the receive path of this bilateral amplifier. The 75-MHz signal is input at J1 and follows the top section of the schematic, which is powered up by the front end assembly (+24V DC). The power to the bottom section is off during receive. The section is isolated from the top circuits via natural isolation provided by the splitter/combiner arrangement. The process is reversed in transmit. (*From:* [1–3]. Figure courtesy of ARRL.)

level IF signals. Sometimes, the weak link in an upconvert receiver design is the IF quartz filter, which cannot withstand high levels of RF without being destroyed. In such cases, wider bandwidth lumped elements filters or cavity filters are used in the first IF and importance is put on even higher dynamic range front end stages to make up for the wider IF bandwidths. The process is increasingly DC power hungry as the dynamic range is increased.

Sometimes, radios use front-end programmable attenuators to claim higher IP3s at the cost of the MDS. It is useless to insert attenuators at the input of a receiver, claiming higher intercept points, while moving the MDS level up at the same time. Claiming a high intercept point by itself means nothing unless it is used together with a low MDS and consequently enters the dynamic range equations.

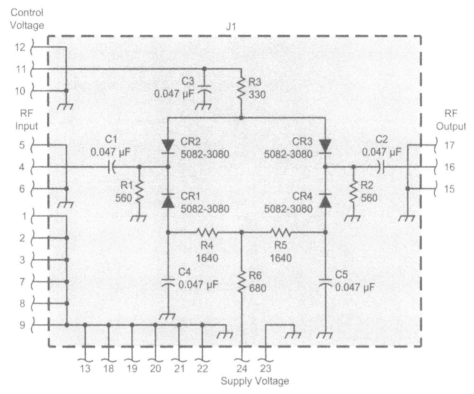

Figure 14.5 The quadruple PIN diode attenuator (BIPA) is part of the AGC system or can be manually adjusted by changing the voltage at pin 11. About 30 dB of IMD-free attenuation can be obtained with this arrangement. (*From:* [1–3]. Figure courtesy of ARRL.)

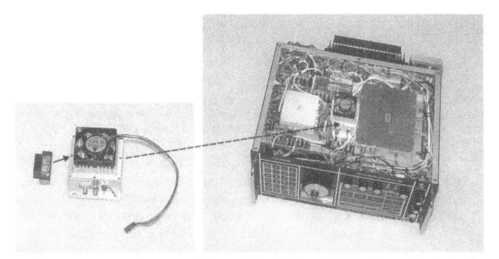

Figure 14.6 A CA-2832 class A monolithic amplifier (left), its physical implementation in the bilateral amplifier section (two are used), and the assembly location on the bottom of Star-10 transceiver. This unit can draw 15 watts of DC power to ensure proper dynamic range.

Good sensitivity coupled with high compression points and IP3s as combined with wide-range AGC systems are a must if performance is the primary goal in a receiver design. Of course, all this comes at a cost of design sophistication and power consumption as there is no magic about getting good results with cheap parts.

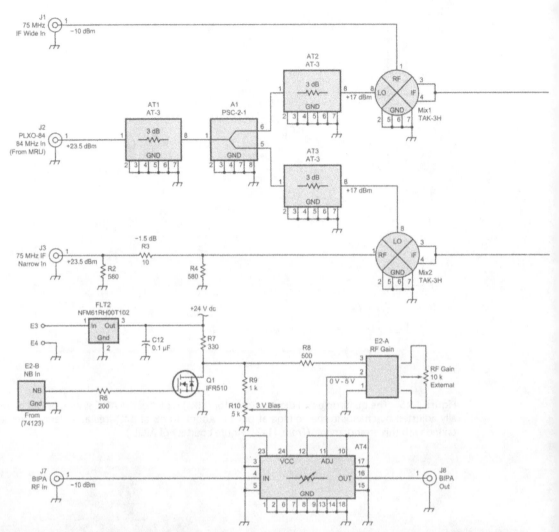

Figure 14.7 The second converter IF9BC converts the 75-MHz IF to the 9-MHz IF where the ultimate bandwidth is achieved. Again, this is a bilateral circuit. It uses an 84-MHz coherent synthesized LO in conjunction with high-level mixers. AGC and noise blanking attenuation is achieved through the PIN diode in the BIPA circuit previously shown in Figure 14.5. The 75-MHz IF input is at J3 and the 9 MHz output is at J6. The transmit functions are not discussed here. (*From:* [1–3]. Figure courtesy of ARRL.)

Depending on frequencies and application, a designer must always balance the power consumption against the dynamic range requirements and decide what can be done in a particular application, which may not necessarily allow for the high power consumption. The Star-10 receiver was designed to compete in a very intense HF spectrum where high-level signals located just a few kilohertz from a desired small signal could obliterate a normal receiver. This performance is similar to that of high performance airborne radar, which has to deal with weak Doppler modified signals in the presence of high-level clutter signals caused by mountains or water.

The system design analysis is probably the most important part of an HPOI receiver design. Several other factors such as good phase noise of local oscillators

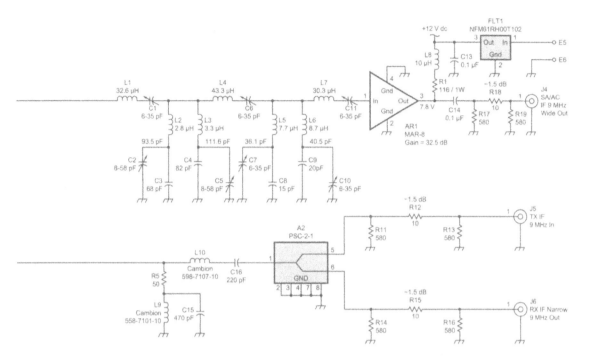

Figure 14.7   (continued)

and synthesizers are also extremely important in the design of HPOI receivers. We will discuss mixers and synthesizer design in Chapter 16.

We will now turn to a key ingredient of an HPOI receiver, the preselector. While single-frequency receivers present a relatively simple preselector filter design problem, broadband systems such as Star-10 require a complex bank of automatically switchable filters as a preselector. We will next discuss this important part of the design.

14.2   Practical Preselector Design: Automatically Switched Half-Octave Filter Banks—A Design Case

Modern broadband HPOI receivers today depart considerably from the traditional preselectors described earlier in Chapter 4 by using banks of automatically switched bandpass filters. The switching is usually accomplished with the use of low-leakage silicon or PIN diodes or with miniature RF relays, which guarantee high intercept points. Signal diodes such as the IN914 or the IN4148 are usually avoided. In addition, back-to-back diode limiters as sometimes seen in the front end of receivers should be avoided if at all possible because they are prone to becoming mixers, which in turn generate intermodulation distortion and spoil the dynamic range of a receiver. Another factor that can contribute to the degradation of the dynamic range

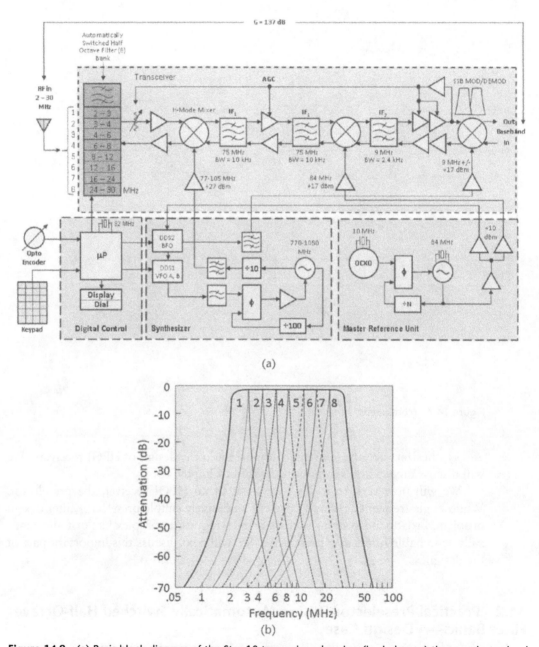

(a)

(b)

Figure 14.8 (a) Basic block diagram of the Star-10 transceiver showing (in dark gray) the preselector bank of automatically switched half-octave filters. (b) Composite frequency response of all eight half-octave filters. Only one filter is selected at any given time.

in some HF receivers is the size of the iron-powder toroid cores sometimes used in the construction of continuously tunable preselector filters, a new trend of an old method of preselection. It has been found that larger cores are superior to small cores from an intermodulation distortion point of view.

Preselector filters in modern upconvert receivers are usually of the half-octave, Bessel design type, in order to keep second- or higher-order products out of the passband of the receiver's front end as well as to provide steep attenuation beyond the cutoff points.

The block diagram in Figure 14.8(a) shows (in gray) how such filters are incorporated in the design of the Star-10 HPOI receiver. Figure 14.8(b) shows the total composite frequency response characteristic for all the half-octave filters from 2 to 30 MHz. Only one filter is switched in automatically at any given time.

These examples show that the old manual "switch and peak" HF method of preselecting has been surpassed by this "hands-off" method. While until recently this method applied only to expensive communication receivers, new lower-cost versions of such preselectors have been introduced by receiver designers. The method is also used at microwave frequencies as explained in [4]. This ultrawideband fully synthesized high-resolution receiver method is shown in Figure 14.9.

This wideband HPOI microwave receiver tunes from 2 to 20 GHz and upconverts to an IF of 26 GHz. The intelligent preselector utilizes half-octave filters, which are automatically switched in by the synthesizer commands with the help of MEMS (microelectromechanical systems). These are tiny mechanical switches, usually electrostatically activated, that are built onto semiconductor chips and are measured in micrometers. We will discuss this microwave receiver, in combination with Bragg cell IFs, in Chapter 25.

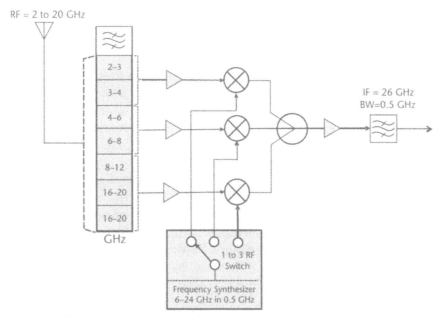

Figure 14.9 Half-octave filters bank as used in an upconvert (first IF = 26 GHz) ultrawideband microwave receiver that tunes from 2 to 20 GHz. See [4].

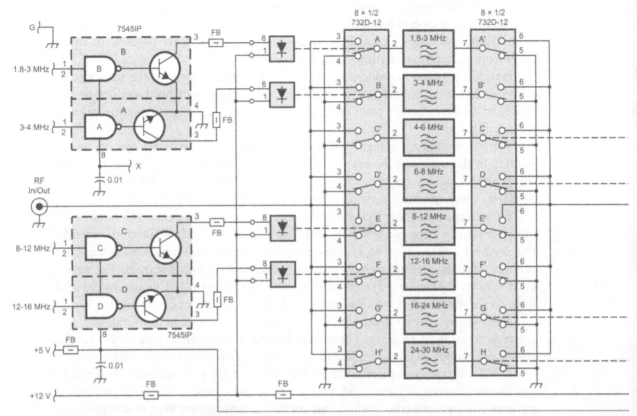

Figure 14.10 Block diagram of the Star-10 receiver half-octave automatic preselector filter bank. The design uses digital line drivers to switch in filters using miniature RF relays for best dynamic range. (*From:* [1–3, 5, 6]. Figure courtesy of ARRL.)

14.3　Switching Mechanisms of Front-End Filters for Best Dynamic Range Performance

The Star-10 transceiver features separate automatically switched half-octave filter banks for receive and transmit. In the receive mode, RF signals from 2 to 30 MHz are automatically selected by the command and control mechanism of the transceiver in concert with the transceiver's tuning frequency information. Automatic frequency selection is achieved anywhere in the frequency range, providing equal image and spurious rejection in receive, as well as equal harmonic and spurious rejection in transmit anywhere in the frequency coverage.

The 75-MHz first IF puts the receiver image away by 150 MHz (2×75 MHz) at any frequency between 2 to 30 MHz. With the proper half-octave filter selected, the amount of image rejection provided over each one and all filters is guaranteed to be uniform throughout the entire RF coverage. Conversely, the proper half-octave lowpass filter selected in the transmit chain ensures equal spurious and harmonic rejection throughout the frequency range. Both banks use the same electrical design with the difference that the transmitter banks are built with high-power inductors (heavier cores) to withstand the high-power RF of the transmitter. The block diagram of the receiver filter assembly with its switching mechanism is shown in Figure 14.10.

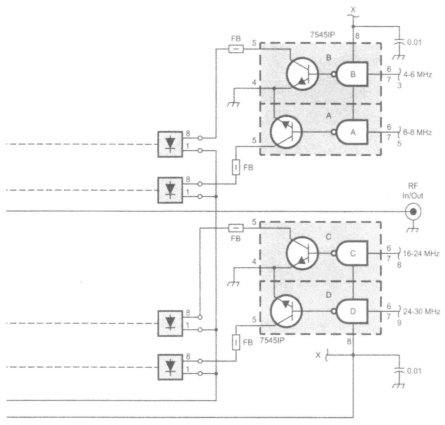

Figure 14.10 (continued)

14.4 Automatically Switched Half-Octave Filters Design

Shown in Figure 14.11 are the half-octave filter banks' physical implementation. As can be seen, they are plug-in shielded assemblies. They insert into the PC boards that contain the control circuits utilizing 75451 digital line drivers, which in-turn switch corresponding RF relays using commands from the command and control assembly (DFCB) to select the proper receive and/or transmit combination. The half-octave filter assemblies insert into a motherboard as shown.

The schematic diagram for the basic lowpass filters banks is shown in Figure 14.12. The receiver filters are designed similarly using the same corner frequencies. There are eight filters in the receiver assembly and eight high-power lowpass filters in the transmit section as shown.

Figure 14.13(a) shows a top view of the Star-10 transceiver with the location of the front-end assembly and the half-octave filter banks assembly (right side). The miniature RF relays are visible on the receiver filter assembly. Also visible is the synthesizer assembly (left side) and the master reference unit (left side back). Figure 14.13(b) shows the bottom side of the transceiver.

In conclusion, while fixed frequency receivers require simple front-end preselectors, wideband high-performance HPOI receivers require complex banks of half-octave filters, which are automatically selected by the receiver synthesizer

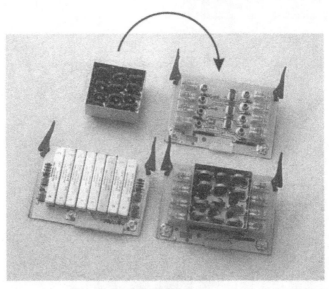

(a)

(b)

Figure 14.11 (a) The automatically switched half-octave filter banks for the Star-10 transceiver. In the receiver bank (bottom left), the switching miniature relays and drivers are located on the back side of the PC board. In the high-power transmitting banks, the filters are located inside the metal units, which in turn plug into the main board as shown. This is to prevent RF from getting into the sensitive driver circuits. (b) All half-octave filter banks are plug-in boards into a motherboard called the cage. Signals from the command and control section as well as RF cables are input through the motherboard located at the bottom of the assembly.

commands in order to ensure equal rejection of image throughout an entire range of RF tuning. The actual switching is performed using PIN diodes, high-performance RF relays, or MEMS depending on the application. These automatic preselectors can be an important part of an HPOI receiver design.

#1

f_C = 3.2 MHz Band = 1.8 – 3 MHz
Reject 6 MHz by 60 dB
f_C : Reject = 1.875:1

Use A C0615C θ = 35° Filter
Minimum Attenuation At 6 MHz = 67.8 dB
Normalized values are:

#2

f_C = 4.5 MHz Band = 3 – 4 MHz
Reject 9 MHz
f_C : Reject ratio = 2:1

Use A C0615C θ = 32° Filter
Minimum Attenuation At 9 MHz = 72.8 dB
Normalized values are:

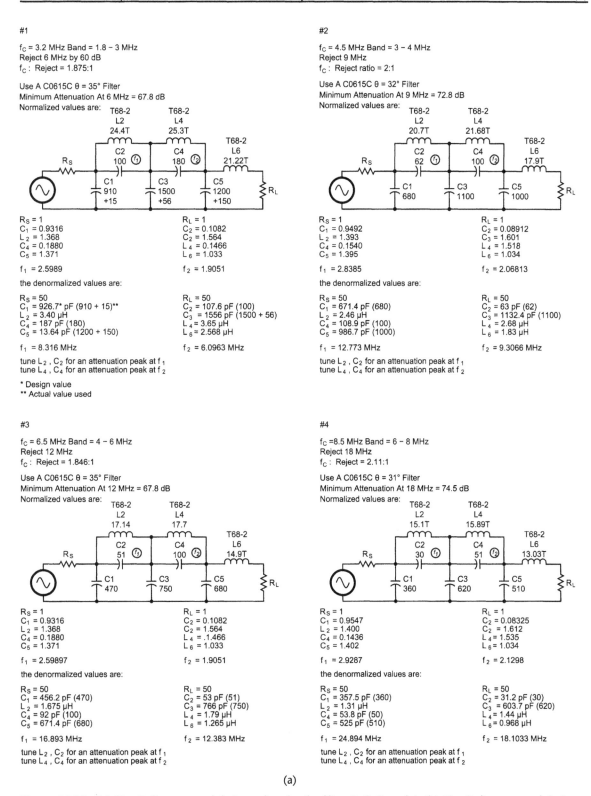

R_S = 1	
C_1 = 0.9316	C_2 = 0.1082
L_2 = 1.368	C_2 = 1.564
C_4 = 0.1880	L_4 = 0.1466
C_5 = 1.371	L_6 = 1.033
f_1 = 2.5989	f_2 = 1.9051

the denormalized values are:

R_S = 50
C_1 = 926.7* pF (910 + 15)**
L_2 = 3.40 μH
C_4 = 187 pF (180)
C_5 = 13.64 pF (1200 + 150)

f_1 = 8.316 MHz

R_L = 50
C_2 = 107.6 pF (100)
C_3 = 1556 pF (1500 + 56)
L_4 = 3.65 μH
L_6 = 2.568 μH

f_2 = 6.0963 MHz

tune L_2 , C_2 for an attenuation peak at f_1
tune L_4 , C_4 for an attenuation peak at f_2

* Design value
** Actual value used

R_S = 1	
C_1 = 0.9492	C_2 = 0.08912
L_2 = 1.393	C_3 = 1.601
C_4 = 0.1540	L_4 = 1.518
C_5 = 1.395	L_6 = 1.034
f_1 = 2.8385	f_2 = 2.06813

the denormalized values are:

R_S = 50
C_1 = 671.4 pF (680)
L_2 = 2.46 μH
C_4 = 108.9 pF (100)
C_5 = 986.7 pF (1000)

f_1 = 12.773 MHz

R_L = 50
C_2 = 63 pF (62)
C_3 = 1132.4 pF (1100)
L_4 = 2.68 μH
L_6 = 1.83 μH

f_2 = 9.3066 MHz

tune L_2 , C_2 for an attenuation peak at f_1
tune L_4 , C_4 for an attenuation peak at f_2

#3

f_C = 6.5 MHz Band = 4 – 6 MHz
Reject 12 MHz
f_C : Reject = 1.846:1

Use A C0615C θ = 35° Filter
Minimum Attenuation At 12 MHz = 67.8 dB
Normalized values are:

#4

f_C =8.5 MHz Band = 6 – 8 MHz
Reject 18 MHz
f_C : Reject = 2.11:1

Use A C0615C θ = 31° Filter
Minimum Attenuation At 18 MHz = 74.5 dB
Normalized values are:

R_S = 1	
C_1 = 0.9316	C_2 = 0.1082
L_2 = 1.368	C_2 = 1.564
C_4 = 0.1880	L_4 = .1.466
C_5 = 1.371	L_6 = 1.033
f_1 = 2.59897	f_2 = 1.9051

the denormalized values are:

R_S = 50
C_1 = 456.2 pF (470)
L_2 = 1.675 μH
C_4 = 92 pF (100)
C_5 = 671.4 pF (680)

f_1 = 16.893 MHz

R_L = 50
C_2 = 53 pF (51)
C_3 = 766 pF (750)
L_4 = 1.79 μH
L_6 = 1.265 μH

f_2 = 12.383 MHz

tune L_2 , C_2 for an attenuation peak at f_1
tune L_4 , C_4 for an attenuation peak at f_2

R_S = 1	
C_1 = 0.9547	C_2 = 0.08325
L_2 = 1.400	C_2 = 1.612
C_4 = 0.1436	L_4 = 1.535
C_5 = 1.402	L_6 = 1.034
f_1 = 2.9287	f_2 = 2.1298

the denormalized values are:

R_S = 50
C_1 = 357.5 pF (360)
L_2 = 1.31 μH
C_4 = 53.8 pF (50)
C_5 = 525 pF (510)

f_1 = 24.894 MHz

R_L = 50
C_2 = 31.2 pF (30)
C_3 = 603.7 pF (620)
L_4 = 1.44 μH
L_6 = 0.968 μH

f_2 = 18.1033 MHz

tune L_2 , C_2 for an attenuation peak at f_1
tune L_4 , C_4 for an attenuation peak at f_2

(a)

Figure 14.12   (a) Circuit diagrams and design values for the filters 1, 2, 3, and 4. (b) Circuit diagrams and design values for the filters 5, 6, 7, and 8. (*From:* [1–3, 5, 6]. Figure courtesy of ARRL.)

#5

f_C = 12.5 MHz Band = 8 – 12 MHz
Reject 24 MHz
f_C : Reject = 1.92:1

Use A C0615C θ = 34° Filter
Minimum Attenuation At 24 MHz = 69.4 dB
Normalized values are:

```
                  T68-6           T68-6
                  L2              L4
                  13.65           14.1T
                                            T68-6
                  C2              C4         L6
                  27   (f1)       47  (f2)   11.8T
        Rs                                        
        ~~~                                   ~~~~
      (~)          C1            C3         C5        > RL
                   T 250         T 390      T 360
```

R_S = 1	R_L = 1
C_1 = 0.9377	C_2 = 0.101
L_2 = 1.376	C_2 = 1.577
C_4 = 0.1761	L_4 = 1.484
C_5 = 1.379	L_6 = 1.034

f_1 = 2.6741 f_2 = 1.9561

the denormalized values are:

R_S = 50	R_L = 50
C_1 = 238.8 pF (250)	C_2 = 25.9 pF (27)
L_2 = 0.876 µH	C_3 = 401.6 pF (390)
C_4 = 44.8 pF (47)	L_4 = 0.945 µH
C_5 = 351.1 pF (360)	L_6 = 0.658 µH

f_1 = 33.426 MHz f_2 = 24.451 MHz

tune L_2 , C_2 for an attenuation peak at f_1
tune L_4 , C_4 for an attenuation peak at f_2

#6

f_C = 16.5 MHz Band = 12 – 16 MHz
Reject 36 MHz
f_C : Reject ratio = 2.18:1

Use A C0615C θ = 30° Filter
Minimum Attenuation At 36 MHz = 76.3 dB
Normalized values are:

```
                  T68-6           T68-6
                  L2              L4
                  12T             12.6T
                                            T68-6
                  C2              C4         L6
                  15   (f1)       27  (f2)   10.3T
        Rs                                        
        ~~~                                   ~~~~
      (~)          C1            C3         C5        > RL
                   T 180         270        T 270
                                 47
```

R_S = 1	R_L = 1
C_1 = 0.96	C_2 = 0.0776
L_2 = 1.408	C_3 = 1.623
C_4 = 0.1337	L_4 = 1.551
C_5 = 1.410	L_6 = 1.034

f_1 = 3.02499 f_2 = 2.19586

the denormalized values are:

R_S = 50	R_L = 50
C_1 = 185.19 pF (180)	C_2 = 14.9 pF (15)
L_2 = 0.679 µH	C_3 = 313.1 pF (270 + 47)
C_4 = 25.8 pF (27)	L_4 = 0.748 µH
C_5 = 272 pF (270)	L_6 = 0.4987 µH

f_1 = 49.9123 MHz f_2 = 36.2317 MHz

tune L_2 , C_2 for an attenuation peak at f_1
tune L_4 , C_4 for an attenuation peak at f_2

#7

f_C = 24.5 MHz Band = 16 – 24 MHz
Reject 48 MHz
f_C : Reject = 1.96:1

Use A C0615C θ = 33° Filter
Minimum Attenuation At 48 MHz = 71.1 dB
Normalized values are:

```
                  T68-6           T68-6
                  L2              L4
                  9.7             10.17
                                            T68-6
                  C2              C4         L6
                  120  (f1)       22  (f2)   8.45
        Rs                                        
        ~~~                                   ~~~~
      (~)          C1            C3         C5        > RL
                   T 120         T 200      T 180
```

R_S = 1	R_L = 1
C_1 = 0.9436	C_2 = 0.09523
L_2 = 1.385	C_2 = 1.589
C_4 = 0.1648	L_4 = 1.501
C_5 = 1.387	L_6 = 1.034

f_1 = 2.75377 f_2 = 2.0103

the denormalized values are:

R_S = 50	R_L = 50
C_1 = 122.6 pF (120)	C_2 = 12.37 pF (12)
L_2 = 0.449 µH	C_3 = 206.4 pF (200)
C_4 = 21.4 pF (22)	L_4 = 0.487 µH
C_5 = 180.2 pF (180)	L_6 = 0.336 µH

f_1 = 67.467 MHz f_2 = 49.292 MHz

tune L_2 , C_2 for an attenuation peak at f_1
tune L_4 , C_4 for an attenuation peak at f_2

#8

f_C = 30.5 MHz Band = 24 – 30 MHz
Reject 72 MHz
f_C : Reject = 2.36:1

Use A C0615C θ = 27° Filter
Minimum Attenuation At 72 MHz = 82 dB
Normalized values are:

```
                  T68-6           T68-6
                  L2              L4
                  8.9             9.4
                                            T68-6
                  C2              C4         L6
                  5    (f1)       12  (f2)   7.57
        Rs                                        
        ~~~                                   ~~~~
      (~)          C1            C3         C5        > RL
                   T 100         T 180      T 150
```

R_S = 1	R_L = 1
C_1 = 0.9747	C_2 = 0.06212
L_2 = 1.429	C_2 = 1.654
C_4 = 0.1066	L_4 = 1.595
C_5 = 1.430	L_6 = 1.035

f_1 = 3.3568 f_2 = 2.4244

the denormalized values are:

R_S = 50	R_L = 50
C_1 = 101.7 pF (100)	C_2 = 6.5 pF (5)
L_2 = 0.373 µH	C_3 = 172.6 pF (180)
C_4 = 11.1 pF (12)	L_4 = 0.416 µH
C_5 = 149.2 pF (150)	L_6 = 0.270 µH

f_1 = 102.38 MHz f_2 = 73.944 MHz

tune L_2 , C_2 for an attenuation peak at f_1
tune L_4 , C_4 for an attenuation peak at f_2

(b)

Figure 14.12   (continued)

(a)

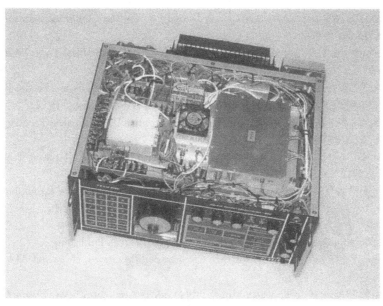

(b)

Figure 14.13   (a) Top view of the Star-10 transceiver showing location of the front end assembly and the half-octave filter banks assembly. The miniature RF relays are visible on the back of the receiver assembly. Also visible is the synthesizer assembly and the master reference unit. (b) The bottom of the Star-10 transceiver and the IF assembly.

References

[1]   Drentea, C., "The Star-10 Transceiver, Part 1," *QEX— ARRL*, November/December 2007.

[2]   Drentea, C., "The Star-10 Transceiver, Part 2," *QEX—ARRL*, March/April 2008.

[3]   Drentea, C., "The Star-10 Transceiver, Part 3," *QEX—ARRL*, May/June 2008.

[4]   Drentea, C., "Ultra-Wideband Fully Synthesized High Resolution Receiver and Method," U.S. Patent 7,139,545.

[5]   Drentea C., and L. R. Watkins, "Automatically Switched Half-Octave Filters, Part 1," *Ham Radio*, February 1988.

[6]   Drentea, C., and L. R. Watkins, "Automatically Switched Half-Octave Filters, Part 2," *Ham Radio*, March 1988.

Selected Bibliography

Analog Devices, "Low Noise Variable Gain Amplifier AD-603," Application Note.

Brigham, O. E., *The Fast Fourier Transform*, Upper Saddle River, NJ: Prentice-Hall.

Drentea, C., "The Art of RF System Design," a three-day comprehensive course.

Drentea, C., "Beyond Fractional-N, Part 1," *QEX*, March/April 2001.

Drentea, C., "Beyond Fractional-N, Part 2," *QEX*, May/June 2001.

Drentea, C., "Designing Frequency Synthesizers," *RF Technology Expo*, Session B-1, Frequency Synthesis, Anaheim, CA, February 10–12, 1988.

Drentea, C., "High Stability Local Oscillators for Microwave and Other Applications," *Ham Radio*, November 1985.

Drentea, C., *Radio Communications Receivers*, New York: McGraw-Hill, 1982.

Larson, L. E., *RF and Microwave Circuit Design for Wireless Communications*, Norwood, MA: Artech House, 1996.

Manassewitsch, V., *Frequency Synthesizers, Theory and Design*, New York: John Wiley & Sons, 1976.

Norton, D., "High Dynamic Range Transistor Amplifiers Using Lossless Feedback," *Proceedings of the 1975 IEEE International Symposium on Circuits and Systems*, 1975.

Norton, D., "High Dynamic Range Transistor Amplifier Using Lossless Feedback," *Microwave Journal*, May 1976.

Norton, D., and A. Podell, "Transistor Amplifier with Impedance Matching Transformer," U.S. Patent 3,891,934, June 1975.

Reisert, J., "VHF/UHF World, High Dynamic Range Receivers," *Ham Radio*, November 1984.

Rohde, U., "Key Components of Modern Receiver Design," *QST*, May, June, July, December 1994.

Tsui, J., *Digital Techniques for Wideband Receivers*, SciTech Publishing.

Vizmuller, P., *RF Design Guide: Systems, Circuits, and Equations*, Norwood, MA: Artech House, 1995.

Waugh, R., "A Low Cost Surface Mount PIN Diode Pi Attenuator," *Microwave Journal*, Vol. 35, No. 5, May 1992.

CHAPTER 15

Mixers

15.1 The Mathematics of Mixers, Laplace, and Fourier Transforms

We have seen that the superheterodyne is a type of receiver that converts incoming RF signals into IF signals for easier signal processing. At the base of the superheterodyne concept is the process of mixing, which was discussed briefly in Chapter 3.

In a superheterodyne receiver, mixers facilitate frequency conversion where the incoming RF signal has the modulation superimposed on it and is mixed with a local oscillator (LO) signal in a manner generating new signal components, which are equal to the sum and the difference of the original frequencies, as previously discussed. One of these products is designated as the intermediate frequency (IF) and is passed by a tuned circuit, which rejects the undesired products as well as the original incoming frequency while still maintaining the information contained in it.

The mixer is a nonlinear component that performs the RF to IF frequency translation. It is a three-port device having two inputs and one output. The input port through which RF is fed to a mixer is usually called the R port. The input port through which the local oscillator (LO) is applied to a mixer is called the L port, and the output port through which the IF is collected for further processing is called the I port.

So, how do mixers work? Mixers use a controlled application of the product generation process of RF and LO harmonics as previously explained in Chapter 12. The creation of harmonics (which is at the basis of product generation) can be explained through the self-mixing mechanism shown in Figure 15.1. Harmonics are generated by mixing a frequency with itself (over and over), taking the sum and throwing away the other result, which is zero, and repeating the process. This is true of both the RF or the LO sources independently of each other, creating RF harmonics and LO harmonics. The product generation process is then completed through the summation and subtraction of the mth order and nth order harmonics to infinity in order to generate products as previously discussed.

In addition, mixing can also be explained through the Fourier-Laplace discrete harmonic decomposition process, which, in turn, takes advantage of the sinusoidal weakly nonlinear behavior of fundamental frequencies in semiconductors or other nonlinear devices.

Weakly nonlinear implies a gradual (not abrupt) departure from the linear state. Products are the sums and differences of harmonically created spurs as previously discussed. In addition to the higher-order products, mixing the RF and LO signals produces a first order set of products (+/−). One of these products is desig-

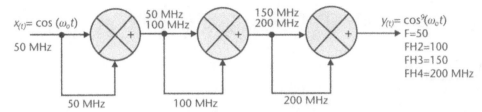

Figure 15.1 The harmonic generation mechanism naturally mixes a frequency source with itself, throws away the subtracted result (zero), and repeats the operation over and over. The process is true of either RF or LO. Then, products result from the summation or subtraction of these harmonics.

nated as the desired IF frequency and the other product is rejected by the IF filter as we previously discussed.

The response of weakly nonlinear systems can also be understood through the classical Volterra power series and the Fourier-Laplace discrete decomposition process as shown in Figure 15.2.

In the Fourier-Laplace transforms, the angular frequency of a signal is assumed to be ω_0. This square wave can be thought to be composed of $a(n)$ (infinite) number of sine wave signals having frequencies ω_0 (fundamental frequency), $2\omega_0$, $3\omega_0$, … (harmonics). Any periodic signal $f(t)$ having a repetition frequency of ω_0 can be written as a sum of its harmonics as shown in

$$f(t) = \sum_{n=-\infty}^{+\infty} F(jn\omega_0)\exp(jn\omega_0 t) \tag{15.1}$$

The Fourier coefficients $F(jn\omega)$ are calculated from

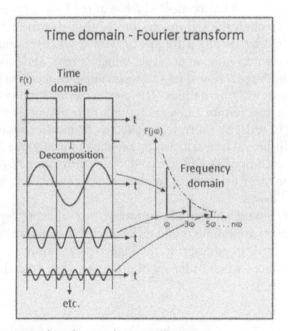

Figure 15.2 The Fourier-Laplace discrete decomposition process.

$$F(jn\omega_0) = \frac{1}{T} \int_{-T/2}^{+T/2} f(t) \exp(jn\omega_0 t)\, dt \tag{15.2}$$

where T is the period of the periodic signal $f(t)$, $T = 2\pi/\omega$, and $F(jn\omega_0)$ is the amplitude of the harmonic component with frequency $n\omega_0$. (We could omit the j operator, but we keep it because any Fourier coefficient F is always a function of $j\omega$ and never of ω alone.) The Fourier coefficients $F(jn\omega_0)$ are complex numbers as shown in

$$F(jn\omega_0) = |F(jn\omega_0)| \exp(j\phi_n) \tag{15.3}$$

where $|F(jn\omega_0)|$ is the amplitude and ϕn is the phase of $F(jn\omega_0)$. When plotting the Fourier transform of a signal $f(t)$, the amplitude $|F(jn\omega_0)|$ and the phase ϕn are plotted against ω; these functions are defined as amplitude and phase spectra. In the case of periodic functions $f(t)$ the Fourier spectra become discrete; that is, the Fourier coefficients $F(jn\omega_0)$ are defined only at the discrete frequencies ω_0, $2\omega_0$, $3\omega_0$, … . Shown is the amplitude spectrum $|F(jn\omega_0)|$ of a symmetrical square wave. For a symmetrical waveform it can be shown that the even harmonics disappear. The Fourier spectrum only shows lines at ω_0, $3\omega_0$, $5\omega_0$, … . In reality, we find many products that are not periodic, such as single pulses. If a signal $f(t)$ is not periodic, we could say that its periodicity approaches infinity, which means its fundamental frequency approaches zero. If the fundamental frequency approaches zero, the spectral lines come closer and closer together, and the sum is replaced by an integral as shown in

$$F(t) = \frac{1}{2\pi} \int_{-\infty}^{=\infty} F(j\omega) e^{j\omega t} dt \tag{15.4}$$

For an aperiodic signal $f(t)$ the Fourier spectrum $F(j\omega)$ becomes continuous and is called the Fourier transform of the signal $f(t)$. The Fourier transform is calculated in the same way as the Fourier coefficients $F(jn\omega)$, where T approaches ∞. We then get

$$F(j\omega) = \int_0^\infty f(t) e^{-j\omega t}\, dt \tag{15.5}$$

15.2   Mixer Topologies

There are many forms of mixers. At the time of this writing, the most popular mixer technology uses Schottky barrier diodes as nonlinear elements because of their abrupt switching ability. The LO signals ride on the steep I/V curves of such diodes causing the spectral decomposition. Silicon Schottky barrier mixers can be operated into the UHF frequency range while GaAs Schottky barrier diodes allow mixers to operate at over 100 GHz. Other mixer designs use silicon transistors or FETs to provide conversion along with gain and inherent noise figure and intercept point depreciation. These mixers require a fourth port for DC biasing. Such is the case of the Gilbert cell active mixers, which will be discussed in Section 15.12.

We have previously seen that mixers produce the sum and difference of RF and LO, and also other undesirable products. These have been discussed in great detail in the previous chapters.

There are several diode mixer topologies available to the RF designer as shown in Figure 15.3. Among them are the single ended (SE), the single balanced (SB), the double balanced (DB), and the quadruple balanced (QB).

15.3   The Single-Balanced Mixer

Figure 15.3 shows an SB mixer, which uses a balanced transformer to feed the LO signal to the two matched mixing diodes, while the RF signal is coupled via a bandpass LC network. The circuit is truly balanced and symmetric and the voltage drop in each diode is the same. Consequently, no LO signal will appear at the RF port and vice-versa. Filtering is generally used in the IF port in order to increase the isolation between the RF and the IF port as shown in this figure.

15.4   The Double-Balanced Mixer and Its Performance Characteristics

From all mixer topologies types, the DB type is probably the most popular kind of mixer used today. This is because it provides better suppression of even harmonics and products than any other mixer type. In addition, it provides better isolation between ports as opposed to other mixer types. This is why the DB mixer is usually chosen in most HPOI receiver designs. The mathematics of a DB mixer are shown in Figure 15.4.

In addition, uneven products can be minimized in DB mixers by increasing the LO drive, which rides harder through the I/V curves of the semiconductors decreasing uneven products.

In a DB mixer, the higher the LO drive level is, the better the mixer is said to be. Based on this fact, the industry has created a de facto standard. According to this convention, there are three classes of DB mixers depending on the LO drive levels. This is shown in Figure 15.5. Class 1 uses LOs from +7 dBm to +9 dBm, class 2 uses LOs from +13 to +17 dBm, and class 3 uses LOs from +23 dBm to +27 dBm.

A DB mixer works in the following manner. Referring to Figure 15.3, it can be seen that the voltage at the secondary of the local oscillator (LO) transformer forms currents that flow through the diode pair D1, D2, or D3, D4 according to polarity, causing alternate conduction. These currents make the ends of the secondary of the RF transformer to appear alternately at ground potential and at the frequency of the signal, which is applied to the LO port.

It can be seen that the signal appearing at the IF output will be influenced by the level and polarity of the signal at the RF transformer's secondary, as well as by which terminal of that secondary is at ground potential at that time (as shown in Figure 15.3).

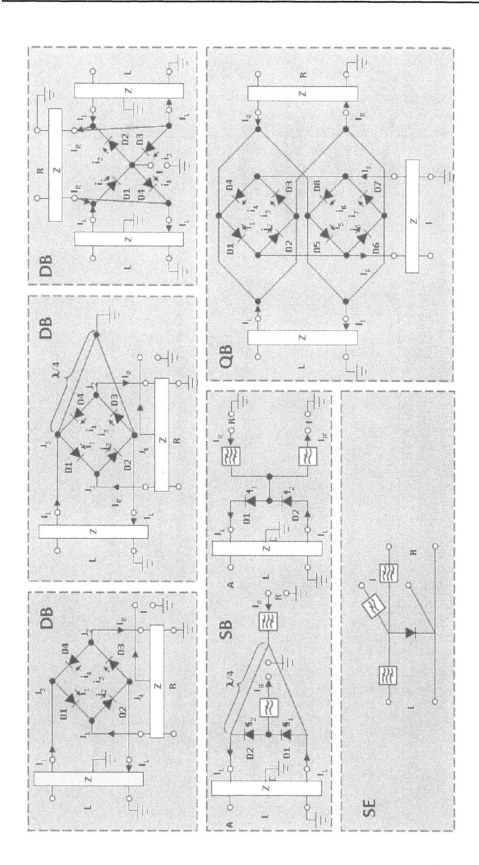

Figure 15.3   Diode mixer topologies. Among them are the single ended (SE), the single balanced (SB), the double balanced (DB), and the quadruple balanced (QB).

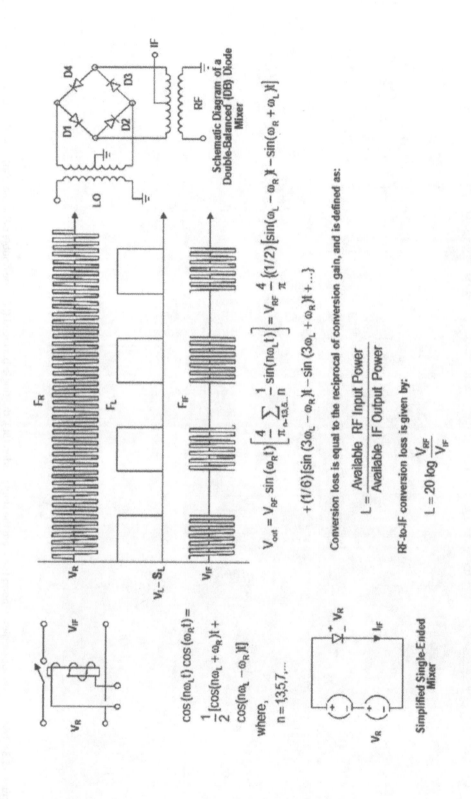

$$V_{out} = V_{RF} \sin(\omega_R t) \left[\frac{4}{\pi} \sum_{n=1,3,5,...} \frac{1}{n} \sin(n\omega_L t) \right] = V_{RF} \frac{4}{\pi} \left\{ (1/2) \left[\sin(\omega_L - \omega_R) t - \sin(\omega_R + \omega_L) t \right] \right.$$

$$\left. + (1/6) \left[\sin(3\omega_L - \omega_R) t - \sin(3\omega_L + \omega_R) t \right] + ... \right\}$$

Conversion loss is equal to the reciprocal of conversion gain, and is defined as:

$$L = \frac{\text{Available RF Input Power}}{\text{Available IF Output Power}}$$

RF-to-IF conversion loss is given by:

$$L = 20 \log \frac{V_{RF}}{V_{IF}}$$

$$\cos(n\omega_L t) \cos(\omega_R t) =$$

$$\frac{1}{2} \left[\cos(n\omega_L + \omega_R) t + \cos(n\omega_L - \omega_R) t \right]$$

where,

$$n = 1, 3, 5, 7, ...$$

Figure 15.4   Electrical operation of a DB mixer.

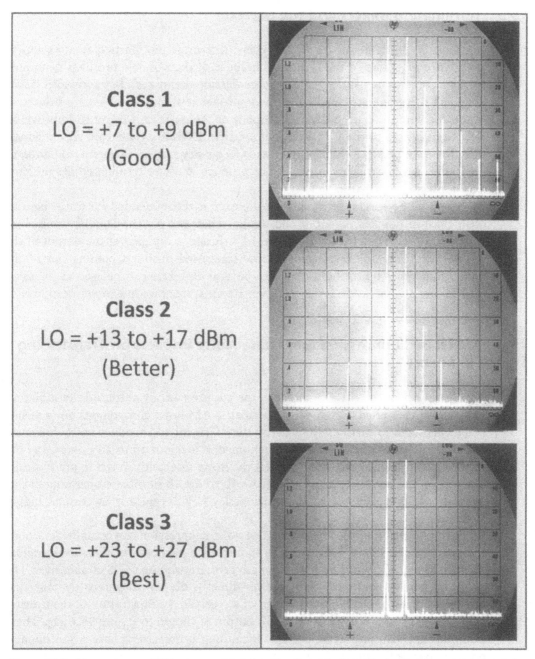

Figure 15.5 Three classes of DB mixers exist defined by their LO drive. Typical two-tone intermodulation performance for the three classes of mixers shows improvement in dynamic range performance with increased LO drive. Class 1 shows that the IMD level is relatively high. With increasing LO drive, classes 2 and 3 provide improved performance as seen in these tests. Class 3 mixers use additional DC power to obtain this performance, so using them may not apply to applications where power consumption is at a premium.

The output at the IF port contains the sum and difference of the frequencies of the signals present at the LO and RF ports, plus all other mixing products created by the harmonic frequencies as discussed earlier in Chapter 12.

15.5   Terminating Mixers and the Diplexer

In using the diode mixer (or other types), one must pay particular attention to impedance matching of the IF output because of the possible products generated by the mixer mechanism, even if they are outside the mixer's IF bandwidth. Some designers use attenuator pads to provide proper termination of such products. It is not uncommon to see 1-dB to 3-dB pads at the ports of a mixer to improve on reflections caused by poor VSWR. Pads used throughout a system are very common remedies to impedance changes over wide frequency ranges. Proper termination of the IF port is of particular interest since products must be terminated despite how well a mixer can perform.

The use of a special filter, called a diplexer, is recommended for achieving this goal. Case in point is the front end design schematic for the Star-10 transceiver, which was previously shown in Figure 14.1. It uses a diplexer at the output of the 75-MHz IF. The diplexer terminates the unwanted products coming out of the mixer, sometimes up to the sixth order. Several diplexers can be used at the same time following a mixer in order to terminate most troublesome mixer products.

15.6   AM Noise Suppression and Phase Noise Impacts on Transferring Signals in Mixers

Another property of the DB mixer is the suppression of amplitude modulation (AM) noise associated with the local oscillator. AM noise components on each side of a local oscillator carrier will mix with the RF signal and produce noise sidebands within the IF passband. While this phenomenon is common to all mixers, the DB mixer tends to cancel this effect so that the noise sidebands in the IF are typically down from the converted carrier level by 40 to 50 dB in mixers operating at frequencies of up to 100 MHz. The spurious levels will typically increase at higher frequencies.

It can be seen that the amount of AM noise suppression can actually determine the mixer's noise figure, which can be spoiled by a particularly noisy local oscillator such as a noisy synthesizer. This can, in turn, impact the MDS of a receiver. The phase noise of a synthesizer can transfer directly, dB-per-dB (minus the insertion loss of the mixer), into the IF passband of a receiver. An illustration of the transfer of phase noise of a noisy LO to the IF output is shown in Figure 15.6 [1]. These examples are also important for understanding synthesizer phase noise impacts on a receiver MDS and dynamic range, since synthesizer phase noise performance impacts receiver performance directly through the receiver converters.

AM noise suppression can determine the maximum local oscillator drive level and the phase noise allowed for a particular mixer to achieve a desired noise figure. In spite of this fact, not all AM noise is canceled in a double-balanced mixer. AM noise that is not canceled is sometimes caused at the IF port by the FM noise (partial phase noise problem) of the local oscillator moving through the slopes of the following IF filter as explained previously. AM noise suppression plays an important part in choosing a good double-balanced mixer and implementing it properly in a particular receiver design.

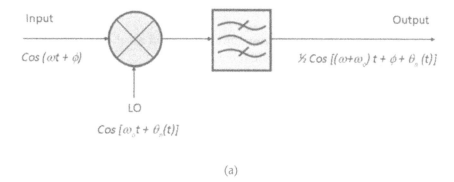

(a)

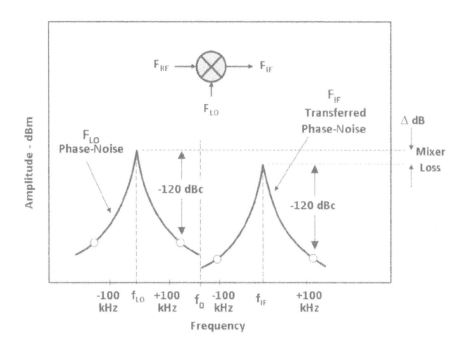

(b)

Figure 15.6 (a) Mathematics of the LO phase noise transfer to the IF in a mixer. (b) Typical transfer of LO phase noise into the IF in a mixer. The phase noise of a synthesizer can transfer directly, dB-per-dB (minus the insertion loss of the mixer), into the IF passband of a receiver. If the synthesizer is noisy, it can impact the MDS of the receiver and contribute to complex intermodulation distortion problems.

15.7 Conversion Loss and Noise Figure of Diode Mixers

Conversion loss in a diode mixer is usually referred to as the single-sideband conversion loss, and is simply the power difference (expressed in dB) between the input RF level at the R port (expressed in dBm) and the output IF level (expressed in dBm) at the I port, of the desired sideband as shown in (15.6) where L_c is the conversion loss.

$$L_c(\text{dB}) = P_{RF(\text{dBm})} - P_{IF(\text{dBm})} \tag{15.6}$$

For example, if a −10-dBm number is used for the RF input, and the desired IF sideband measures −16 dBm, then:

$$L_c = -10 - (-16) = 16 - 10 = 6 \text{ dB}$$

Because only one sideband of the frequency translation is used as the IF, with the other sideband being filtered out by the IF filter, there is an automatic 3-dB conversion loss in a mixer. The additional losses come from product energy being inadvertently produced and from mismatches between the balanced elements of the mixer, as well as heat losses in the diodes and junctions.

Although single-sideband conversion loss is usually specified by the mixer manufacturer, (typically 6 to 9 dB for a DB mixer) its importance is crucial to the design of a radio receiver because of its noise figure. As we previously discussed, the noise figure of a lossy part such as a mixer is its loss in decibels.

Referred to as the single-sideband noise figure (SSB NF), this value is a direct function of the conversion loss plus the estimated contribution in electrical noise caused by the diode junctions, a value that is typically 0.5 to 1 dB as shown in

$$SSB\ NF_{(\text{dB})} \leq L_{c(\text{dB})} + 1_{\text{dB}} \tag{15.7}$$

Knowledge of the SSB NF of a mixer dictates not only the amount of preamplification needed to make up for it, but more importantly, it can set the noise figure of an entire radio receiver as we have previously discussed in Chapter 13.

15.8 Two-Tone Intermodulation Performance in Mixers

The *two-tone intermodulation* ratio intercept method for the overall performance of an entire radio receiver described earlier in Chapter 13 can be applied to all or any parts that make up an RF system, including amplifiers, filters, and mixers. The total IMD performance of a system depends on the individual performance of the many parts, and specially the mixer that is often the most important contributing element to this performance. We can use the previous setup shown in Figure 13.6 to perform a two-tone test on a mixer.

When two $F_{(RF)}$ signals are simultaneously applied to the input of a mixer they will mix with the $F_{(LO)}$, producing the sums and the differences as well as the undesired intermodulation products, as discussed earlier in Chapter 12. The largest magnitude of undesired intermodulation product level is the third-order product. This is why it was chosen as a means of performance specification for mixers as well as entire RF systems.

The two-tone intermodulation ratio is the ratio of the third-order intermodulation (IM) product to an IF output level at a specified power level for the two RF inputs.

If the output products versus RF input power are plotted on a log-log graph, the third-order IM products present a 3:1 slope while the desired IF outputs present a 1:1 slope. The two slopes meet at the third-order intercept point as previously shown in Figure 13.5. The intercept point is then specified in terms of its RF input power level and is the means of expressing the mixer's performance.

A rule of thumb is that the higher the intercept point is, the better the mixer performance is. Generally, improved performance can be obtained with higher LO drive levels, which will switch the diodes further into the linear regions at both ends of their I/V curves while reducing the amount of time spent in the nonlinear region centered around $V = 0$. While the forward current through each diode pair is almost unlimited, the reverse voltage is limited by the forward voltage drop of the nonconducting diode pair. This can present a reliability problem.

The solution for this problem is adding more series diodes, series resistance to each diode, or choosing diodes having a higher breakdown voltage. This in turn increases the requirement for LO drive levels, which increases DC power consumption.

15.9   Compression Point (–1 dB) in Mixers

When a single RF frequency is applied to the RF port of a mixer, along with a large LO signal, the conduction transfer characteristic of the diodes changes as the RF input level is increased beyond a point, and the output will eventually start saturating until no increase is obtained no matter how much the RF input level is increased.

The point at which the output level drops 1 dB from following the input is the –1-dB compression point. This point was evident when looking at the previous plot in Figure 13.5. This point plays an important role in determining the upper limit of the linear dynamic range of a mixer or radio receiver. Its importance was discussed in detail in Chapter 13 (Figures 13.5 and 13.6).

Sometimes, the compression point is not listed in mixer specifications, but a third-order intercept point is. As a rule of thumb, the third-order intercept point (IP3) is approximately 15 dB above the 1-dB compression point in diode mixers. In FET mixers, the rule of thumb is 10 dB. Subtracting either of these two numbers from the given IP3 number gives us an estimate of the compression point.

15.10   Desensitization Level and Isolation

Another receiver parameter that depends on the mixer performance is the 1-dB desensitization created by a mixer. This level is the RF input power of an interfering signal that causes the small-signal conversion loss to increase by 1 dB. This level is typically 2 to 5 dB above the compression point level and should not be confused with it.

Isolation between ports is a good measure of the internal circuit balance of a mixer. It is an important element in receiver design, as we do not want LO energy to spill back out through the antenna port or impact the weak IF signal energy.

Isolation in mixers is a measure of insertion loss between any two of the R, L, and I ports. However, because the LO usually has the highest level, specifying and measuring isolation in mixers is most important between L-to-R and L-to-I ports, with the second being the most critical.

Isolation is measured in dB and is specified for a given bandwidth, over a specific temperature range, and at a certain LO level. The L-to-R isolation is measured as the difference between the actual LO level and the amount of LO power seen at the RF port with the IF port terminated in 50 ohms. The L-to-I isolation is measured as the amount LO power attenuated at the IF port, with the RF port terminated in 50 ohms. A spectrum analyzer is used to make these measurements.

SB mixers usually have good isolation between L and I, as well as L and R, ports because of the good balance achieved in transformer baluns and near perfect diodes matching.

Although most DB mixers have good isolation (typically 50 dB) and can be operated over wide frequency ranges, some demanding receiving systems might require additional filters and amplifiers for improving this characteristic, especially at the microwave frequencies.

15.11   Commutative Mixers, FET, and H-Mode Mixers

While the DB diode mixer remains the most popular type, there are other types of mixers in use today. Among them is the DB commutative mixer implemented with FETs, which can offer superior IP3 performance over class 3; Schottky-barrier diode type; DB mixers; and triple-balanced mixers. Commutative mixers are those mixers whose operation depends on switches implemented with FETs and are usually driven with quadrature digital signals as LOs.

A well-designed commutative FET mixer can offer an IP3 in excess of +38 dBm with less LO drive than a class 3 diode-type mixer, which can only offer a maximum of +30 dBm IP3 using a higher LO drive.

Commutative FET mixers are usually suitable for HF, VHF, and UHF receivers, but newer models have been manufactured at frequencies of up to 2 GHz. The schematic diagram of a basic DB commutative FET mixer is shown in Figure 15.7(b). Looking at this figure, the switches (FETs) are activated via their gates in push-pull via a well balanced transformer, and signals are collected in a series, source-drain arrangement.

H-mode mixers are also members of the commutative mixers family. Instead of the series switching arrangement of the previous commutative mixers, they are set up to switch the source terminals to ground instead. The high third-order intercept is achieved by arranging the switches to switch the signal to ground when they close, rather than to the output load as in the previous models.

This provides a more positive switching effect, which in turn prevents instability at the gate voltage, thus creating less intermodulation distortion than a regular DB FET mixer. Consequently, better IP3s in excess of +50 dBm can be realized with H-mode DB mixers at the lower IF frequencies. A +60 dBm H-mode mixer has been reported recently when used in conjunction with special IF filters.

The IP3 of the upconvert Star-10 H-mode mixer previously presented in Chapter 14 is +45 dBm using a +27 dBm LO drive.

H-mode mixers are used mainly in HPOI HF receivers. They are usually driven with quadrature square wave digital signals obtained via high speed edge triggered D flip-flops. The quadrature square waves have to obey by rigorous matching rise times between the quadrature signals, which could present a problem if we are

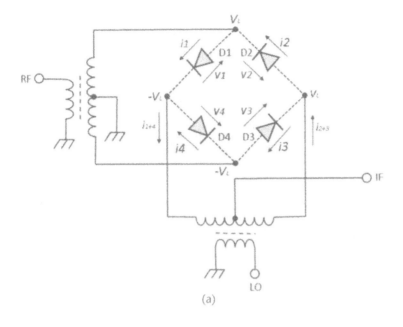

(a)

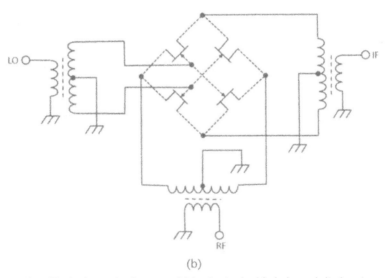

(b)

Figure 15.7 Simplified schematic diagram of (a) a basic double balanced diode mixer and (b) an FET mixer.

trying to use such mixers in wideband general coverage receivers. Consequently, such mixers are usually found in single-band or chanalized radios in order to guarantee the narrow tolerance rise time matching. In the general coverage Star-10 transceiver, the quadrature drive has been administered through transformers as the only way to obtain wideband operation. This method worked well over the four octaves frequency coverage.

The H-mode mixer was invented by Colin Horrabin in Great Britain and was improved by several others. Figure 15.8 shows the basic operation of an H-mode mixer.

15.12 Integrated Circuit Mixers—Gilbert Cell Mixers

When power consumption requirements are important, high LO drive passive mixers have been replaced by active Gilbert cell mixers. This is particularly true in battery-operated VHF/UHF applications such as cellular telephones. In addition to being low-power devices, Gilbert cell mixers offer conversion gain instead of conversion loss. A typical Gilbert cell mixer model is shown in Figure 15.9(a).

The mixer consists of an emitter-coupled transistor pair Q1 and Q2 and four collector-crossed coupled transistors, Q3 through Q6. As can bee seen, Q2 is RF grounded while RF is applied to Q1. The LO is applied at Q3.

With the RF applied to Q1, a 180° differential path is created in the two branches. The LO is applied at Q3 through Q6 where it mixes with the RF from Q1 and Q2 operating in reverse, and the IF is output is at Q7, which is an emitter follower that sets the output impedance via a resistor. The conversion gain in a Gilbert cell mixer is provided by the emitter-coupled amplifier configuration as shown in Figure 15.9(a).

Additional voltage and current sources are used throughout to ensure proper balance, which is very important in this kind of design. Consistency in manufacturing

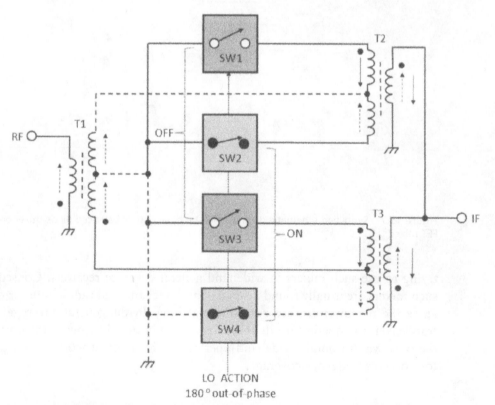

Figure 15.8 Signal modeling in an H-mode mixer.

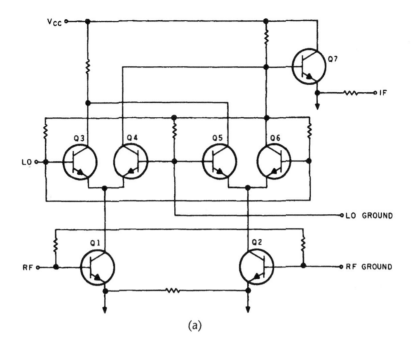

(a)

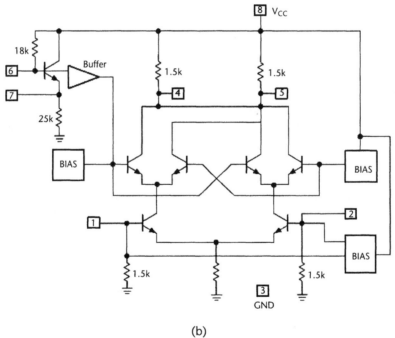

(b)

Figure 15.9 (a) Gilbert cell mixer concept. (b) Actual implementation of a SA602A Gilbert cell mixer showing biasing sources.

is key to Gilbert cell mixer performance and is only possible through precise circuit integration. A more accurate schematic diagram of an actual Gilbert cell mixer product is shown in Figure 15.9(b).

A good representative of the Gilbert cell mixer family is the SA602A. It is a low-power VHF monolithic DB Gilbert cell mixer with an onboard oscillator, voltage regulator, and a temperature compensated bias network. It is intended for low-power communication systems. This IC provides approximately 18-dB conversion gain and has a built-in local oscillator, which can be set up to operate at up to 200 MHz. Current consumption is typically 2.4 mA. Its noise figure is approximately 5 dB, while its IP3 is −17 dBm.

15.13   Image-Reject Mixers

The image problem discussed earlier in receivers with relatively low IFs can be improved with the help of image rejection mixers.

This is particularly true in simple data communications or cellular telephone receivers, which, because of their tight packaging requirements, cannot afford double-conversion approaches using complex preselectors. DB mixers are combined with other wideband devices to accomplish this type of mixer. Figure 15.10 shows a pair of gain- and phase-matched mixers arranged to provide image rejection by cancellation in the hybrids. A low-IF radio receiver employing such a mixer will exhibit a typical image rejection of 20 to 30 dB without preselection.

15.14   Image Recovery Mixers

We previously discussed how mixers respond to signals at the image frequency; however, mixers can also generate energy at this frequency. This phenomenon is accomplished through two mechanisms. The second harmonic of the LO can mix with the incoming RF signal creating energy at the image frequency as shown in

$$F_{Image(internally\ generated)} = 2_{FLO} \pm F_{RF} \tag{15.8}$$

Image energy can also be created by the IF signal being reflected back into the mixer (due to mismatching at the IF port) and remixing with the LO energy as shown in

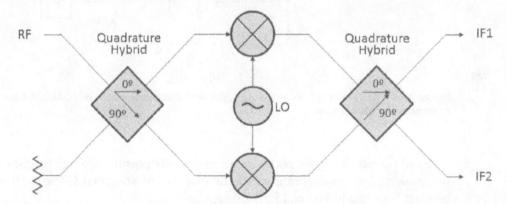

Figure 15.10   Internal configuration of an image-reject mixer.

$$F_{Image(internally\ generated)} = 2_{LO} \pm F_{IF} \qquad\qquad (15.9)$$

To verify these mechanisms, the reader is encouraged to use the image models previously presented at the beginning of this book.

In these cases, the image-created energy can be recovered and used so as to create additional power at the IF frequency, reducing the conversion loss of the mixer.

15.15 Mixer Technology Conclusions

Several types of mixers were described and a few technologies were analyzed. Today, the DB diode mixer is the most widely used. Despite its relatively high conversion loss (6 to 7 dB) it provides a typical dynamic range approaching 100 dB with high output intercepts (> +30 dBm). The balanced mixer disadvantages are that it is susceptible to odd-order harmonic mixing and it requires a proper termination.

At up to 200 MHz, mixers use ferrite cores for wire-wound transformers. This type of technology covers several octaves.

Beyond 200 MHz, electric coupling is used utilizing various combinations of microstrip designs on different substrates with soldered-in miniature quad diode ring packages. At the higher frequencies (beyond 3 GHz), coplanar waveguide and slot line technologies are being used.

New recent developments in DB mixer design indicate third-order intercept point achievements in excess of +50 dBm (at HF) with LO drives in excess of +28 dBm when using FET DB mixers. The JFET mixer, while relatively more expensive, provides superior noise figures (typically 4 dB), combined with gain and high intercept (> +36 dBm). The bipolar transistor in an active DB form will continue to provide a low-cost approach, while experiments with MOS FETS look promising from a performance standpoint. The H-mode mixer provides high IP3s of as much as +60 dBm at HF frequencies.

Reference

[1] Sabin, W. E., and E. O. Schoenike, (eds.), *Single-Sideband Systems and Circuits*, New York: McGraw-Hill, 1987.

Selected Bibliography

Bakner, B. D., "Specifying High-Rel Mixers and Amplifiers," *Watkins Johnson Tech-Notes*, Vol. 7, No. 6.

Calogic Corporation, "SD-8901 and SD-5000 Wideband Ring Demodulator Data."

Cartoceti, S., "A Doubly Balanced 'H-Mode' Mixer for HF," *QEX*, July/August 2004.

Henderson, B. C., "Mixers, Part 1 and 2," *Watkins Johnson Tech-Notes*, Vol. 8, No. 2 and 3.

Henderson, B. C., "Mixers in Microwave Systems, Part 1 and 2," *Watkins Johnson Tech-Notes*, Vol. 17, No. 1 and 2.

Horrabin, C., "Technical Topics, High Performance Mixer," *RadCom*, October 1993.

Makhinson, J., "A High Dynamic Range MF/HF Receiver Front End," *QST*, February 1993.

Marki, F. A., "The Coplanar Mixer," *Watkins Johnson Tech-Notes*, Vol. 1, No. 4.

Mini-Circuits, U.S. Patents 5,416,043 and 5,600,169.

Mini-Circuits Engineering Department, "Novel Passive FET Mixers Provide Superior Dynamic Range," http://minicircuits.com.

Oxner, E., "Designing a Super-High Dynamic Range Double-Balanced Mixer," Siliconix.

Philips Semiconductors, "SA602A, Double-Balanced Mixer and Oscillator Product Specification Datasheet."

Signatron, "Nonlinear System Modeling and Analysis with Applications to Communications Receivers," prepared for Rome Air Development Center, AD-766 278, NTIS, U.S. Department of Commerce.

Signetics Corporation, "SA/NE-602 Double Balanced Mixer and Oscillator Data."

Volterra, V., *Theory of Functionals*, 1880.

Smith, A., "Notes on the Basic Operation of Commutative Mixers," http://g4oep.atspace.com/mixers/notes_on_the_basic_operation_of_.htm.

Tayloe, D. R., "Product Detector and Method Therefore," U.S. Patent 6,230,000.

Watanabe, G., H. Lau, and J. Schoepf, "Integrated Mixer Design," *Proceedings of the 2nd IEEE Asia Pacific Conference on ASICs*, 2000.

Waugh, R. M., T. -Z. Chen, and M. Kumar, "Broadband Monolithic Balanced Mixer Apparatus," U.S. Patent 4,896,374.

Waugh, R. M., and M. Kumar, "Monolithic Double Balanced Mixer with High Third Order Intercept Point Employing an Active Distributed Balun," U.S. Patent 5,060,298.

Weiner, D. D., and J. E. Spina, *Sinusoidal Analysis and Modeling of Weakly Nonlinear Circuits, with Application to Nonlinear Interference Effects*, New York: Van Nostrand Reinhold, 1980.

Weiner, S., D. Neuf, and S. Sphrer, "2 to 8 GHz Double Balanced MESFET Mixer with +30 dBm Input 3rd Order Intercept," *IEEE Symposium on Microwave Theory and Techniques*, 1988.

Wholey, J., I. Kipnis, and C. Snapp, "Silicon Bipolar Double Balanced Active Mixer MMIC's for RF and Microwave Applications up to 6 GHz," *Microwave and Millimeter-Wave Monolithic Circuits Symposium*, 1989.

Williams, F. D., "Gilbert Active Mixers," *Communications Quarterly*, Spring 1993.

Wilson, M., (ed)., "Mixers, Modulators and Demodulators," Chapter 11 in *The ARRL Handbook*, ARRL, 2007.

Wong, A. K., S. H. Lee, and M. G. Wong, "Current Combiner Enhanced Active Mixer Performance," *Microwaves & RF*, March 1994.

Wyse, R. D., "Active Commutated Double Balanced Mixer," U.S. Patent 6,230,001.

Young, J. P., "Dual Gate FET Mixing Apparatus with Feedback Means," U.S. Patent 4,814,649.

Yuen, G. W. M., "Transformer-Coupled Mixer Circuit," U.S. Patent 5,821,802.

CHAPTER 16

Frequency Synthesizers

16.1 Introduction

This chapter provides an overview of all the available synthesis methods, technologies, and their relationship. More importantly, it shows how they can be used together harmoniously. While an entire book can be written on synthesis, we will focus on the relationship of the different synthesis methods available for complex HPOI receivers. We will then discuss circuit design topics backed by practical examples. By the end of this chapter, you should have a good idea of how to approach diverse and complex synthesis applications as they apply to particular receiver designs. We will begin with some of the key terminology required when dealing with local oscillators.

16.2 Definitions

16.2.1 Leeson Oscillator Noise Model

In an oscillator, noise behaves differently than in other circuits. It tends to congregate mainly near the oscillation frequency, which in turn can impact a superheterodyne or direct sampling receiver MDS (as well as other dynamic range elements) through the mixing or sampling process, as discussed in Chapter 15.

A perfect oscillator is described mathematically by a perfect sinusoidal wave:

$$\cos\left[\omega_0 t\right]$$

A practical oscillator will exhibit both, an amplitude noise modulation component, $n(t)$, and a phase noise modulation component, $\theta n(t)$:

$$\left[1 + n(t)\right]\cos\left[\omega_o t + \theta_o t + \theta n(t)...\right]$$

where $n(t)$ and $\theta n(t)$ are random processes and resulting LO phase noise will limit the ultimate signal-to-noise ratio.

The noise fractional bandwidth was first defined by Edson in 1960 [1]. The terminology of "noise spectrum density" was first introduced by Leeson in 1966 [2]. He described heuristically the noise distribution of an oscillator, which resembled practical observations. In the 1990s, Craninckx, Steyaert [3], and Razavi [4] described formulas for LC-tank and CMOS oscillators. Mathematical dependencies in an oscillator according to Leeson are shown in Figure 16.1.

Based on these assumptions, a simplified graph of phase noise contributions was created and is shown in Figure 16.2(a) [5]. It gives a visualization of phase noise distribution in the frequency domain along with some ideas on how to modify its characteristics in the distribution of the output spectrum. This graph reflects a modified version of Leeson's equation that depicts an oscillator's SSB phase noise behavior.

In Figure 16.2(a), there are three key regions of interest in the phase noise distribution of an oscillator, as shown. At up to f_0, the phase noise decreases by 30 dB per decade. Beyond this point and up to the resonator's half bandwidth $(f_0/2Q)$, the phase noise decreases at 20 dB per decade. Beyond this point, the phase noise content is mainly determined by the available RF power level and the active device's thermal noise properties. This region is called the *noise floor* of the oscillator.

This noise floor can be improved by using low flicker noise devices and very high-Q frequency resonators. In addition, the entire phase noise curve can be improved by increasing the oscillator signal-to-thermal noise ratio (increasing power ahead of the resonator), thus reducing the active device noise factor, just as in the receiver noise analysis previously discussed. In synthesizers, the classic model from Figure 16.2(a) is modified further due to many other factors such as power supply ripple or phase detector noise (in the case of phase-locked loops), as we will see later in this chapter. Fundamental Laplace transform phase noise relationships are shown in Figure 16.2(b).

16.3   Long-Term and Short-Term Frequency Stability

Frequency stability is important in quantifying local oscillator (LO) characteristics. It is defined as the degree to which a local oscillator (LO) produces the same frequency throughout a specified period of time.

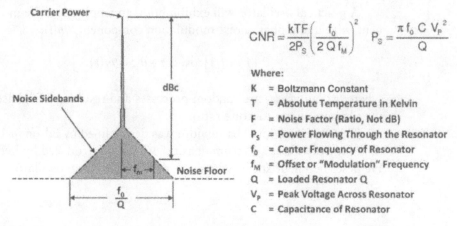

$$CNR = \frac{kTF}{2P_S}\left(\frac{f_0}{2Qf_M}\right)^2 \qquad P_S = \frac{\pi f_0 C V_P^2}{Q}$$

Where:

K = Boltzmann Constant
T = Absolute Temperature in Kelvin
F = Noise Factor (Ratio, Not dB)
P_S = Power Flowing Through the Resonator
f_0 = Center Frequency of Resonator
f_M = Offset or "Modulation" Frequency
Q = Loaded Resonator Q
V_P = Peak Voltage Across Resonator
C = Capacitance of Resonator

Figure 16.1 Phase noise in an oscillator according to Leeson.

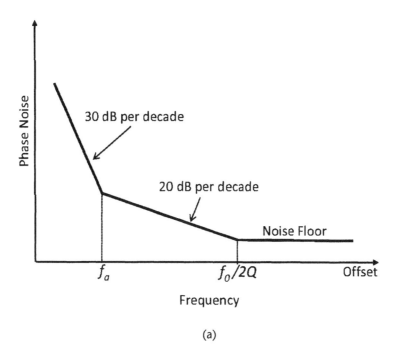

Figure 16.2   (a) Classical phase noise density profile in a local oscillator. (b) Fundamental Laplace transforms phase noise relationships.

Every LO source has a certain amount of frequency instability that can be broken down into two components, long-term stability and short-term stability. Long-term stability refers to the frequency variations that occur over relatively long periods of time, expressed in parts per million per hour, day, month, or year. In a synthesizer, long-term stability is ensured by a master reference oscillator, which is usually quartz-based. This will be discussed in Section 16.21.

By contrast, the short-term stability contains all discrete elements, causing frequency changes about the nominal frequency with a duration of less than a few seconds.

The short-term stability pertains to the complex set of phenomena known as phase noise and spurious performance that are impacted by the model presented earlier and by many other causes depending on the synthesizer type and technologies used.

Before going into the details of various synthesis methods as well as the master reference unit, we will discuss the mathematical conventions behind the phase noise and its counterpart, the jitter, as well as the spurious performance, two key frequency stability issues.

As discussed in Chapter 15, phase noise transfers directly decibel per decibel into an IF of a superheterodyne or a direct sampling receiver within the passband of interest. Spurious performance is equally important, as spurs can mix with products and create additional intermodulation distortion performance problems. This in turn can impact the MDS and other dynamic range elements of an HPOI receiver.

Phase noise is the result of short-term frequency instability. Ideally, a sine wave can be expressed mathematically as in:

$$V_{(t)} = V_0 \sin 2\pi f_0 t \tag{16.1}$$

where V_0 is nominal amplitude, $2\pi f_0 t$ is the linearly growing phase component, and f_0 is the nominal frequency.

A better model of a sinewave signal is expressed by

$$V_{(t)} = |V_0 + \varepsilon(t)| \sin |2\pi f_0 t + \Delta_\Phi(t)| \tag{16.2}$$

where $\varepsilon(t)$ is the amplitude fluctuation and $\Delta_\Phi(t)$ is the random fluctuating noise.

The random fluctuating noise ($\Delta_\Phi(t)$) can be further broken down into two categories. The first category is comprised of discrete signals that appear at distinct places in the spectral plot. They are usually connected with a power supply or digital glitches along with mixer-created products, called *spurious products*. The second category is caused by random fluctuations such as flicker noise, thermal agitation, or shot noise in the actual circuits. This category is called phase noise. Its integral is called jitter and plays an important role in understanding analog-to-digital, digital-to-analog, and DSP requirements.

Phase noise is observed in a per-hertz basis on a single sideband of a double-sideband signal as a Laplace ($\mathcal{L}(f)$) transform function. It behaves as in

$$\mathcal{L}(f) = \frac{1}{2} S \Delta_\Phi(f) \tag{16.3}$$

where $\mathcal{L}(f)$ is P_{SSB}/P_s and P_s equals the total signal power. $\mathcal{L}(f)$ is plotted logarithmically directly in decibels as offset from the carrier and is expressed in dBc/Hz. This is shown in Figure 16.3.

16.4 Residual Phase Noise and Absolute Phase Noise

Two kinds of phase noise are usually measured and specified in synthesizers. For brute force synthesizers (e.g., multipliers, dividers, mixers), the residual phase noise is the device's inherent phase noise regardless of the reference's phase noise. In residual phase noise measurements, the effects of external noise sources such as power

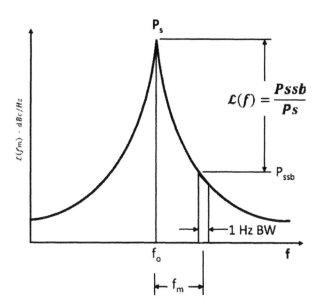

Figure 16.3 Defining the single-sideband phase noise. $\mathcal{L}(f)$ is plotted logarithmically directly in decibels as offset from the carrier and is expressed in dBc/Hz.

supply ripple or reference feed through the problems are canceled out. In the absolute phase noise measurements, these effects are included in the measurements [6].

16.5 Allan Variance

The Allan variance (named after Dr. David W. Allan) is a proven way of characterizing the frequency stability of oscillators (short-term and long-term) as averaged in the time domain [7, 8]. It is related to $\mathcal{L}(f)$ and $\mathcal{L}(fm)$. This characterization is based on the sample variance of fractional frequency fluctuations. Averaging differences of consecutive sample pairs with no dead time in between yields the Allan variance. The Allan variance is expressed by $\sigma_y^2(\tau)$ and is given by

$$\sigma_y^2(\tau) = \frac{1}{2(M-1)} \sum_{k=1}^{M-1} \left(\overline{y}_{k+1} - \overline{y}_k\right)^2 \tag{16.4}$$

where $\overline{y}_k$ is the average fractional frequency difference of the kth sample measured over sample time τ. The samples are taken with no dead time between them. Figure 16.4 shows an example of using the Allan variance equation.

Because the Allan variance depends on the time period τ used between samples, it is a function of the sample period, as well as the distribution being measured, and is usually displayed as a graph rather than a single number as shown above. A low Allan variance is a characteristic of an oscillator with good stability over the measured time period. There are also a number of variants around the Allan variance measurements, primarily the modified Allan variance, the total variance, the moving, the Hadamard variance, the modified Hadamard variance, the Picinbono variance, the Sigma-Z variance, and others. Allan deviation is sensitive to a

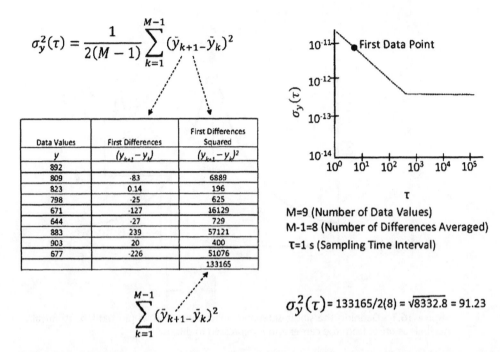

Figure 16.4 Example of using the Allan variance $\sigma_y^2(\tau)$ equation. In this example the data values are parts in 10^{12}.

constant drift, while the Hadamard deviation is unaffected by a constant drift and instead is sensitive to changes in drift. All these variances can be categorized into the same form of stability variances, mainly as mean-square averages of the output of a finite-difference filter acting, not on the phase or frequency samples, but on their cumulative sums.

16.6 Phase Noise and Jitter Concepts

RMS clock jitter is calculated by integrating the $\mathcal{L}(f)$ SSB phase noise curve from the low-frequency limit to the high-frequency limit. The calculation adds 3 dB for dual sideband integration. The seconds of jitter are calculated by dividing the radian jitter by 2π radian per cycle and the A/D clock frequency. Additional information is shown in Figure 16.5(a, b), while jitter calculations using three methods are shown in Figure 16.5(c) [9, 10].

$$\theta_{noise} = 2pf_{clk}T_j \qquad T_j = \theta_{noise}/2pf_{clk}$$

16.7 Defining Coherency in Synthesizers

We will discuss next the complex topic of synthesized local oscillators (LO) that are used to feed mixers in superheterodyne HPOI receivers or serve as reference oscillators in direct sampling digital receivers.

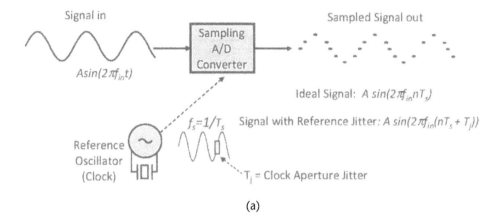

(a)

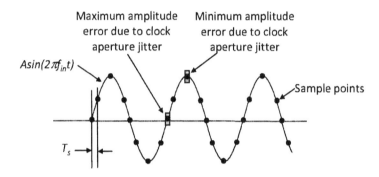

Ideal sampled signal = $A \sin(2\pi f_{in} nT_s)$ [T_s = Sample Interval]
Signal RMS = 0.707A

Ideal signal with jitter = $A \sin(2\pi f_{in}(nT_s + T_j))$[$T_j$ = aperture jitter in units of RMS]

Max error = $A \sin(2\pi f_{in} T_j)$ [$2\pi f_{in} nT_s$ = 0 or multiples of π, zero crossings]
    = $A \, 2\pi f_{in} T_j$ [Small angle approximation]

Error RMS = 0.707A2$\pi f_{in} T_j$ [Error varies as a sinusoidal function]

SNR = signal RMS^2/error RMS^2 SNR in dB = $20 \log_{10}[1/(2\pi f_{in} T_j)]$
   = $1/(2\pi f_{in} T_j)^2$

(b)

Figure 16.5 (a, b) Reference (clock) aperture jitter as it impacts a direct sampling A/D converter. Aperture jitter is caused by the phase noise performance of the reference oscillator. (c) Three methods of converting phase noise to jitter. (*From:* [9].) The methods contain information about periodic jitter components, but do not take into consideration "single" events.

In the past, variable frequency oscillators (VFO) or permeability tuned oscillators (PTO) combined with fixed frequency-quartz oscillators have been used alone or in combination as local oscillators in HF receiving systems. When a fixed number of channels were required, fixed frequency LOs have been used. Channel tuning was achieved by switching in various quartz crystals using the same or separate oscillator circuits.

Today, coherent frequency synthesizers are being utilized almost exclusively to generate local oscillators (LO) frequencies into the multiple gigahertz range for all mixers in single, double or multiple conversion receivers. The term coherent

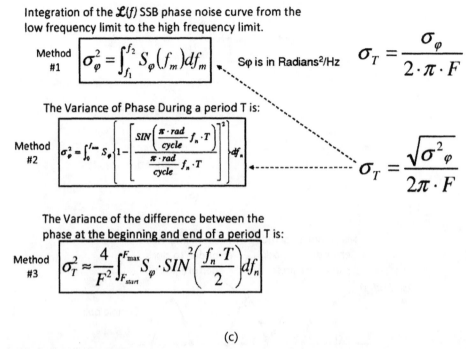

Integration of the $\mathcal{L}(f)$ SSB phase noise curve from the low frequency limit to the high frequency limit.

Method #1
$$\sigma_\varphi^2 = \int_{f_1}^{f_2} S_\varphi(f_m) df_m$$

$S\varphi$ is in Radians²/Hz

$$\sigma_T = \frac{\sigma_\varphi}{2 \cdot \pi \cdot F}$$

The Variance of Phase During a period T is:

Method #2
$$\sigma_\varphi^2 = \int_0^{f_{max}} S_\varphi \left\{ 1 - \left[\frac{SIN\left(\frac{\pi \cdot rad}{cycle} f_n \cdot T \right)}{\frac{\pi \cdot rad}{cycle} f_n \cdot T} \right]^2 \right\} df_n$$

$$\sigma_T = \frac{\sqrt{\sigma^2 \varphi}}{2\pi \cdot F}$$

The Variance of the difference between the phase at the beginning and end of a period T is:

Method #3
$$\sigma_T^2 \approx \frac{4}{F^2} \int_{F_{start}}^{F_{max}} S_\varphi \cdot SIN^2 \left(\frac{f_n \cdot T}{2} \right) df_n$$

(c)

Figure 16.5   (continued)

indicates that all frequency sources (fixed and/or variable) are derived from a single, high-quality, stable reference source, which is usually quartz-based, but could also use other technologies as we will discuss later in this chapter. While coherency usually means "phase coherent," this does not necessarily apply in synthesizers. Coherent synthesizers do not have to be phase coherent except in some very special applications where single cycle synchronization is required to establish absolute physical position of an object in space, for example.

Designing coherent synthesizers for receivers and transceivers requires complex system design skills such as paying attention to all mixer ratios over wide bands (as we previously discussed), and understanding how realizable technologies can be for a certain frequency scheme. In multiple conversion receivers, the phase noise performance of the first LO has to be matched by the performance of all other LOs. If this cannot be done, the performance of the worst LO can spoil the receiver's total performance regardless of how good the first LO's performance is.

16.8   Open Loop Systems: Mixing VFOs with Crystal Oscillators

In some receivers, an open loop mixing scheme such as the one shown in Figure 16.6 has been used successfully to provide injection to the first mixer and the product detector of a single conversion 9-MHz IF-HF receiver.

As can be seen, the LOs in Figure 16.6 are not coherent or synthesized since the variable frequency oscillator (VFO) is free-running and can be impacted by temperature changes. In addition, the various quartz crystal oscillators can change frequency independently of each other in time due to aging and can upset the calibration when switching from band to band.

This receiver does not use an up convert-down convert approach and is intended as a channelized-bands only receiver since the IF frequency is in the band of the entire RF spectrum, creating numerous products if such receiver was to be used for general coverage. The switched crystal oscillator output mixes together in M3 using both, the plus or the minus outputs as shown in Figure 16.6.

On the other hand, if we perform a careful ratio analysis on all mixers (as we discussed in Chapter 12) for the frequency bands of interest, we can see that the intermodulation distortion performance of all mixers (M1, M2, and M3) is relatively good if we stay inside the given RF bands and adhere to the limited frequency coverage offered by this scheme. This can be proven by finding the ratios and using the IMD Web chart from Figure 12.2 or Tables 12.1 and 12.2.

An additional benefit to this design is that it is a single conversion which means less intermodulation distortion (IMD) if we adhere to the channelized scheme. Also, the 9-MHz IF allows for the right percentage bandwidth to realize an ultimate bandwidth of 2.4 kHz (bandwidth required for SSB voice communications) at the first and only IF, an advantage in signal processing.

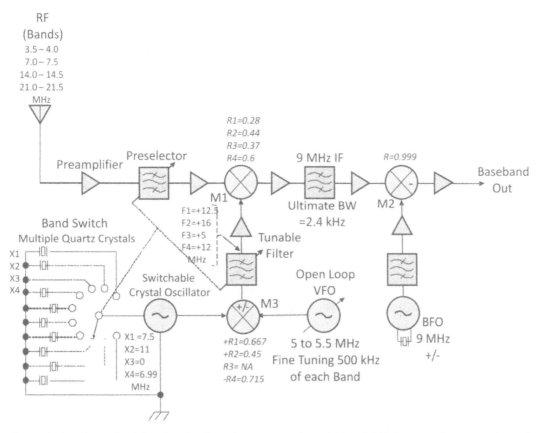

Figure 16.6 Channelized, bands-only HF receiver system using a 9-MHz IF. This is not a coherent synthesized system as the variable frequency oscillator (VFO) runs open loop and each quartz crystal is independent of the other's frequency stability and accuracy. Intermodulation distortion is well behaved in the particular bands due to careful choice of the frequency plan, but the system cannot be used as a general coverage receiver due to the many in-band products.

Table 16.1 shows what could happen from an IMD point of view if this receiver was to be used as a general coverage receiver. Because the IF frequency falls in the middle of the RF band, many products can be generated. This design is not suitable for general coverage.

The local oscillator (LO) in the limited channelized receiver works as follows. The VFO operating at 5.0 to 5.5 MHz is mixed with several crystal oscillators (one for each 500-kHz band), as shown in our example. The mixing products are carefully filtered out at the output of the mixer M3. The output filter to this mixer is also tuned to the required frequency as shown in Figure 16.6.

This scheme provides good results over a direct-switched free-running VFO system by getting its phase noise performance and stability primarily from the quartz crystal oscillators. A mixing scheme such as the one shown combined with fixed frequency crystal oscillators is actually a combination of two forms of brute force synthesis as we will see later in this chapter. However, the use of the free-running VFO in this system spoils the idea of coherent synthesis.

As we discussed previously, a mixing scheme should be chosen with extreme care since the additional mixing process generates a multitude of products that could interfere with the receiver's performance. One of the easiest mistakes to make in an RF system design is to choose LO frequencies that fall right in the IF passband

Table 16.1 Products Generated in a General Coverage 9-MHz IF Receiver Using the Discussed Frequency Plan

Operating Frequency (MHz)	Local Oscillator Frequency (MHz)	BFO Frequency (MHz)	Lowest-Order Spurious Product Equal to i-f	Typical Carrier Level (μV) for 10 dB S+N/N		
				Adjacent Frequencies		Spurious Frequency, Typical
				Maximum	Typical	
3.0025	12.0025	9.0025	3LO–3BFO	0.3	0.2	2.0
3.6030	12.6030	9.0025	5LO–6BFO	0.3	0.2	2.0
5.4035	14.4035	9.0025	5LO–7BFO	0.3	0.2	6.0
6.0033	15.0033	9.0025	3LO–4BFO	0.3	0.2	30.0
9.0025	18.0025	9.0025	LO–BFO	0.4	0.2	Receiver blocked
11.2550	20.2550	9.0025	4LO–8BFO	0.4	0.2	8.0
12.0050	21.0050	9.0025	3LO–6BFO	0.4	0.2	300 for 3 dB S+N/N
13.5000	4.5000	8.9975	2LO	0.5	0.3	Receiver blocked
14.99916	5.99916	8.9975	3LO–BFO	0.5	0.3	8.0
18.0000	9.0000	8.9975	LO	0.5	0.3	Receiver blocked
20.9975	11.9975	8.9975	3LO–3BFO	0.6	0.4	0.5
21.5970	12.5970	8.9975	5LO–6BFO	0.6	0.4	2.0
23.3965	14.3965	8.9975	5LO–7BFO	0.6	0.4	8.0
23.99667	14.99667	8.9975	3LO–4BFO	0.6	0.4	2.0
25.1960	16.1960	8.9975	5LO–8BFO	0.6	0.4	4.0
26.9975	17.9975	8.9975	LO–BFO	0.6	0.4	30.0
29.2450	20.2450	8.9975	4LO–8BFO	0.6	0.4	6.0

without realizing it. This actually happens more often than we would like. When it does, no amounts of shielding can prevent the leakage of the undesired frequency into the system.

A full system analysis should be performed prior to the circuit design by using the methods and tools previously discussed. The problem gets worse in synthesized systems as mathematical decisions may result in improperly chosen solutions.

The open loop approach to LO generation can also be used in upconvert receivers. A local oscillator system for an upconvert HF receiver with a first IF of 46 MHz is shown in Figure 16.7. The 46-MHz IF frequency was chosen on purpose so that the 45-MHz product of mixer M3 cannot fall in the IF passband. This is still not a coherent synthesizer due to the free-running VFO, which can drift. However, the system comes closer to a coherent synthesizer approach than the previous method because it involves a phase-locked loop.

16.9 Synthesizer Forms and Classifications: Brute Force, Direct and Indirect, and Nonbrute Force, Direct and Indirect

The real solution for solving the stability problem in open loop systems is the coherent frequency synthesizer, but the design becomes more complex. As mentioned before, a coherent frequency synthesizer means that all frequency sources are slaved

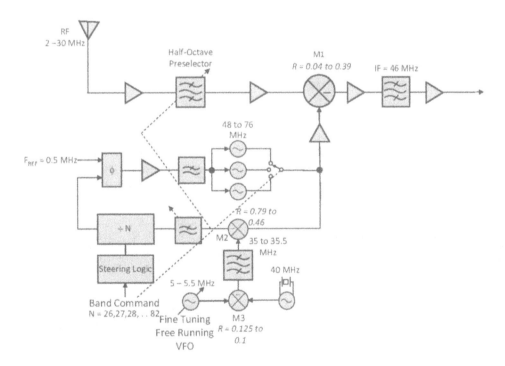

Figure 16.7 Evolution of open loop LO systems. This system uses a free-running VFO in conjunction with a phase-locked loop in an upconvert HF receiver with an IF of 46 MHz. The phase-locked loop, having a reference of 500-kHz steps in half megahertz increments (for each integer N change), and the VFO frequency, which is mixed with the fixed 40-MHz oscillator, perform fine tuning over each of the half megahertz coverage.

to a single stable reference source using one or more frequency synthesis methods known.

While some believe that the only form of synthesizer is the phase-locked loop, the truth is that there are many forms of synthesizers, all with their advantages and disadvantages. Depending on the application, several of these methods can be carefully chosen and used in concert to achieve high-performance, high-resolution coherent systems. Choosing the right combination of technologies requires experience and a good understanding of how to take advantage of each one of the methods along with a full awareness of all topics we have discussed so far, including the use of mixers, mixer ratios, noise figures, gain, compression points, and intermodulation distortion (IMD) as applied to local oscillators.

Some engineers who have not been intimately involved with the design of complex synthesizers tend to use simplistic PLLs when faced with frequency synthesis design assignments regardless of the application. Although the PLL in its simplest form can be easily understood, it is not an exclusive solution to all synthesis problems.

In addition, complex, high-resolution, fast-switching requirements over wide bands can involve several forms of synthesis, which, if not properly chosen, can degrade results.

Choosing a synthesizer system and a circuit design go hand-in-hand with the receiver system architecture and circuit design. Oversimplification of the requirements or ill-defined system designs can result in the wrong technologies being used for the wrong reasons. With several forms of PLLs and many other synthesizer forms available, the receiver/synthesizer designer is faced with making quick decisions that may not necessarily be the right ones. Choosing the right technology or combination of technologies is needed in a unique receiver application. This is as important as the circuit design, particularly when low phase-noise, high-resolution coherent schemes are needed.

What is frequency synthesis? It is defined as the process or processes through which mathematically related frequencies can be created or derived from one or more stable reference frequencies. This definition includes, but is not limited to, the more simplistic (and inexact) definition of frequency synthesis sometimes found in published material that defines it as "the process providing a finite number of frequencies, all equally spaced." Coherent frequency synthesis uses a single stable reference frequency for all synthesis components.

Although it is impossible to do a total classification of frequency synthesizers, we will discuss proven methods that produce results. The classification will emphasize certain properties of synthesizers that will make us understand when and how to use them collectively.

Shown in Figure 16.8 is the classification of most known synthesizer forms. As can be seen, frequency synthesizers can be categorized in two major classes: brute-force direct and indirect, and nonbrute-force direct and indirect. Further classification is possible as shown.

16.9.1 Brute Force

A brute-force direct synthesizer is usually used in applications where several (mostly up to 10) frequencies are required. It is defined as a process of deriving related

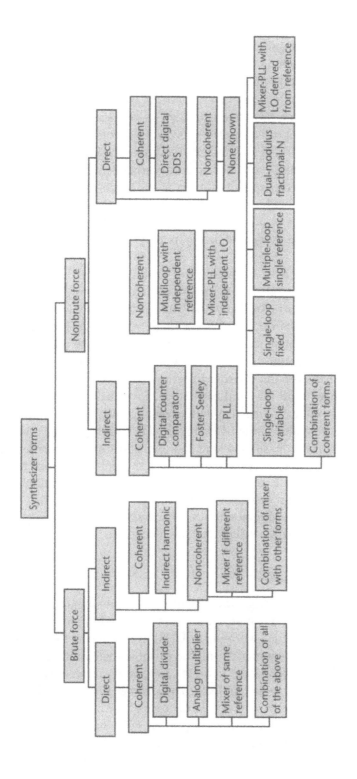

Figure 16.8   Classification of frequency synthesizer forms.

frequencies by using intentionally created products in mixers or harmonic generation in frequency doublers, triplers, and the like, and/or by simple digital logic division.

It is important to note at this point that brute force synthesis by multiplication degrades phase-noise performance by a $20\log N$ value (expressed in decibels) as added to the reference's phase noise performance (a higher number meaning a worse performance). In this simple equation, N is the multiplication number. Conversely, synthesis by brute force synthesizers using division improves phase-noise performance by using the same equation, as impacted only by the ultimate noise floor of the digital part used in the process. Consequently, the digital division process is always a preferred synthesis method over the multiplication method.

In Figure 16.8, the brute-force direct synthesizer can be noncoherent or coherent (fully synthesized) depending on how the various processes use the reference frequencies. The noncoherent synthesizer simply means that the derived frequencies are the result of two or more reference frequencies, each impacting the final stability and resolution independently as we previously discussed at the beginning of this chapter.

A simple example of a noncoherent synthesizer is the mixer type where independent crystal controlled LOs can be combined to produce new frequencies as we have seen in Figures 16.6 and 16.7. By contrast, a coherent or fully synthesized process derives all its outputs from a single reference. Thus, the outputs will be coherent with the reference. (Note: A small out-of-phase relationship between the reference and the output is acceptable within this definition.)

Full synthesis becomes harder to achieve when the designer is trying to obtain total coherence for all the local oscillators in a signal processing system using several forms of synthesis. Such a system is usually much harder to design than the alternatives because all resulting LOs required in a multiple conversion receiver have to have comparable phase noise and spurious performance in order to maintain the entire receiver's dynamic range performance through all its mixers. Knowing how to combine the diverse methods and technologies is of paramount importance in such a complex design case, as we will see later in this chapter.

Most brute-force direct synthesizers, with the exception of the mixer type (when used with separate references), are coherent.

16.9.2 Nonbrute Force

The other two general categories of synthesizers from Figure 16.8 are the nonbrute-force direct and the nonbrute-force indirect. Within the context of the first category, we find the direct digital synthesizer (DDS) as a coherent form (fully synthesized). The Foster Seeley discriminator and the various single-loop PLLs have been identified within the nonbrute-force indirect category as coherent types. They are called indirect because they indirectly lock to a reference rather than being used as direct products of a reference process, as in the case of the brute force-direct synthesizers.

In addition, various multiple-loop PLLs with independent references and the mixer type PLL when used with an independent (not reference-derived) LO have been classified as noncoherent types within the same class. All proven known synthesizer types have been included in Figure 16.8.

16.10   The Mixer as a Synthesizer

A good example of a brute-force, noncoherent synthesizer is the mixer type shown in Figure 16.9. In this example, two or more crystal oscillators are mixed in a certain mathematical arrangement to produce new frequencies intended as LO injection for channelized receivers, transmitters, or transceivers. This technique is relatively inexpensive and has been used extensively in simple-fixed channel receivers. When designing such a synthesizer, attention should be paid to the mixer ratios as explained in Chapter 12.

The noncoherent mixer synthesizer can become coherent when the reference is mixed with its own products and when used with other coherent forms of brute force synthesizers, such as the digital/analog divider type or the harmonic multiplier, shown in Figure 16.10(a, b).

Figures 16.9 and 16.10 reveal that, although relatively simple at first glance, brute-force synthesizers, particularly the mixer type synthesizer, require special care in system design. When designing such synthesizers, it is important to take into consideration the in-band, compounded mixer products created in the intermediate frequency (IF) chains of the synthesizer that can be detrimental to the purity of the final outputs. The design tools from Chapter 12 for predicting intermodulation distortion in mixers should be used. Only frequency ratios that produce seventh-order and higher distortion products should be considered if at all possible, and the process should be carried through all the mixing chains much like in multiconversion receiver design. The mixer type synthesizer tends to use many filters. Because of this and the shielding involved, the seemingly simple mixer synthesizer can sometimes become a cumbersome piece of equipment. Among its advantages are concurrent parallel outputs that can translate into fast switching without the lockup characteristic of other forms of synthesis, and straightforward implementation. Historically, the mixer synthesizer dates back to the 1940s and is probably the first form of synthesizer ever developed. Its use has diminished with the recent

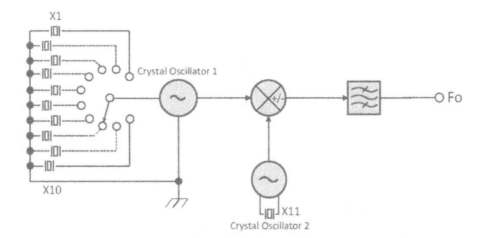

Figure 16.9   A simple brute-force direct, noncoherent mixer type synthesizer using independent quartz crystals.

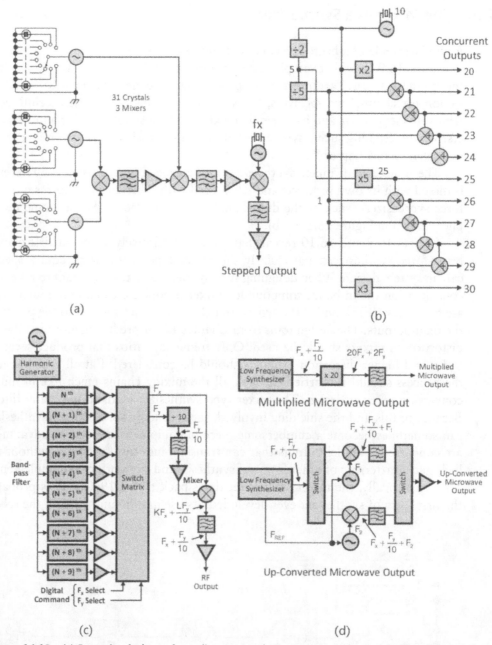

Figure 16.10 (a) Example of a brute-force direct, noncoherent mixer synthesizer. Frequency resolution and stability depend on different quartz crystals. (b) Coherent brute-force direct synthesizer that uses mixers, dividers, and multipliers. Its resolution and frequency stability are derived from the same reference. (c, d) Additional brute-force synthesizers using harmonic generator and mixer synthesis methods.

introduction of other forms of synthesizers. However, the mixer synthesizer is finding renewed use when combined with other forms of synthesis in complex hybrid systems.

16.11   Digital and Analog Regenerative Dividers

Frequency division is another form of brute force direct-coherent synthesizer. It can be realized by using digital or analog means. A static digital divider is usually implemented using D flip-flops with the not Q output fed back to the D input as shown in Figure 16.11. It has a large bandwidth, from a few hertz to the frequency capability of the logic family involved.

The digital divider is a good choice as a component in brute force frequency synthesis as it offers phase noise improvements over other methods. It is also used in the feed back circuits of phase-locked loops. As mentioned before, the phase noise improvements come from the $20\log N$ equation where N is the division number. This improvement, however, is limited by the divider's noise floor as we previously discussed.

At the lower frequencies (up to 2 GHz) digital dividers implemented in ECL technology have been used extensively in the past as fixed divide-by prescalers. They divided down high VCO frequencies by fixed numbers or dual-modulus numbers in order to be handled by lower frequency TTL or CMOS dividers. They are usually used in integer or fractional phase-locked loops (PLL) as we will see later in this chapter. More recently, such prescalers have been incorporated in NMOS or CMOS technologies directly on the synthesizer PLL chips.

Very high-frequency microwave binary dividers (up to 18 GHz) are readily available commercially as independent components. Noise floors are usually –150 dBc/Hz. Relatively simple to implement, the digital divider can use any of the logic families available to the designer (TTL, CMOS, ECL, NMOS), depending on the frequencies of interest. However, only the lowest frequency logic family required to satisfy the particular application should be used, in order to keep the harmonic-rich

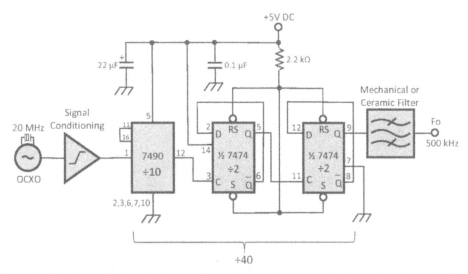

Figure 16.11 Example of a simple brute force direct coherent synthesizer using the static digital divider. This example uses a divide-by-40 digital divider made of a divide-by-10, 7,490 TTL IC and two divide-by-2 D flip flops from a single 7,474 TTL logic IC. In a divide-by-2 D flip flop divider, the not Q signals are fed back to the D input in the two cascaded units. Waveform is 50% duty cycle. Conditioning of the input and output waveforms is necessary for proper operation. This idea can be used at much higher frequencies by changing the logic family and the filter/conditioning setup.

square waves from wandering into the RF signal processing circuits. A combination of slow rise times, Schmitt triggering, and analog filtering are often used to keep such effects down. One-shots are avoided due to their nonsynchronous nature and reliance on temperature-dependent RCs. Large-scale integration (LSIC) of synchronous counters can be easily implemented and is highly recommended.

The relative simplicity of digital division makes this method of synthesis very attractive for brute-force synthesizers at up to 18 GHz. Any binary division number combination from 2 to 16 is available commercially, usually implemented in InGaP GaAs HBT technology. Divisions by 5 and 10 are also available. An HMC-C005, divide-by-2 digital RF divider that operates at up to 18 GHz is shown in Figure 16.12.

This divider is a low noise divide-by-2 static digital divider utilizing InGaP GaAs HBT technology packaged in a miniature, hermetic module with replaceable SMA connectors. It operates from a 0.5–18-GHz input frequency from a single +5V DC supply. The low SSB phase noise floor of –150 dBc/Hz at a 100-kHz offset helps the designer maintain excellent system noise performance.

Shown in Figure 16.13 is the HMC494LP3 divide-by-8 static digital RF divider and its surface mount implementation. This divider also uses InGaP GaAs HBT technology and is packaged in a leadless 3 × 3 mm QFN surface mount plastic package. This device operates from almost DC to an 18-GHz input frequency using

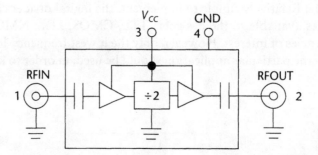

Figure 16.12 The HMC-C005 is a high-performance divide-by-2 static digital RF divider operating from 0.5 to 18 GHz and using +5V DC. It is implemented in InGaP GaAs HBT technology and exhibits an SSB phase noise floor of –150 dBc/Hz at a 100-kHz offset. (Courtesy of Hittite Microwave Corporation © 2000–2008 Hittite Microwave Corporation, All rights reserved.)

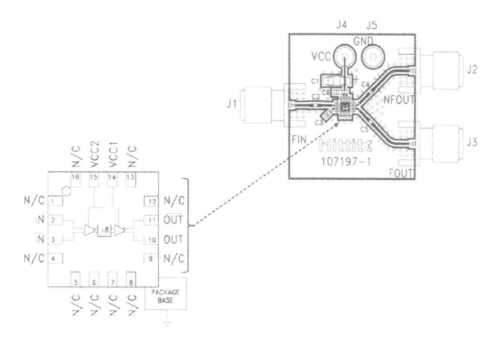

Figure 16.13   The HMC494LP3 is a high-performance divide-by-8 static digital RF divider operating from almost DC to 18 GHz and using +5V DC supply. It is implemented in InGaP GaAs HBT technology and exhibits an SSB phase noise floor of –150 dBc/Hz at a 100-kHz offset. (Courtesy of Hittite Microwave Corporation © 2000–2008 Hittite Microwave Corporation, All rights reserved.)

a single +5V DC supply. The phase noise floor is also –150 dBc/Hz at a 100-kHz offset.

For superior phase noise performance at the VHF/UHF frequencies or at frequencies beyond 18 GHz, the regenerative analog frequency divider is being used. In contrast with the static digital divider, the regenerative divider can operate at higher frequencies, but has a narrower bandwidth and is consequently frequency-specific.

The regenerative divider was first introduced by R. L. Miller in 1939 [11]. Modern regenerative dividers operate at frequencies of up 65 GHz. They offer a very low phase noise performance because of their ability to utilize clean sine wave signals. This eliminates the requirement for conditioning circuits, which is necessary with the digital dividers. A block diagram of a regenerative divider model is shown in Figure 16.14(a). An actual implementation of a regenerative divider is shown in Figure 16.14(b).

In Figure 16.14(a), the way the regenerative divider works is as follows. Assume that a frequency of 10 GHz is input to the mixer M1. An initial residual $F_{in}/2$ noise present in the circuit triggers the regenerative process. This 5-GHz component is then lowpass filtered and amplified to be output. A small portion of it is coupled back to the mixer's L-port after it has been tweaked for phase coherence via a microstrip delay line. Then it is mixed again with the input frequency generating a reinforced 5-GHz signal. Once started, the process repeats. It can be seen that the process has a start-up time delay that can be measured in microseconds depending on the frequency used and the layout of the system. Because of its layout-sensitive nature, a regenerative mixer has a narrow bandwidth of operation (as opposed to the static digital divider). Because of this, regenerative dividers are usually

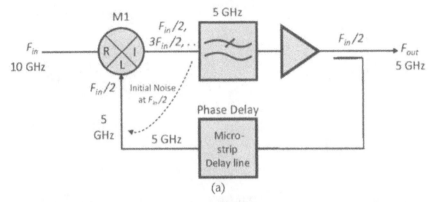

(b)

Figure 16.14 (a) Block diagram of a regenerative divider. The mixer generates outputs at $F_{in}/2$, $3F_{in}/2$, and more. The filter allows the $F_{in}/2$ component to pass. The regenerative divider operation is based on the assumption that a leaked subharmonic component of F_{in}, ($F_{in}/2$), is present as noise in the loop at start-up, which triggers the regenerative process. (b) An LNRD4 regenerative divider manufactured by Wenzel. The LNRD4 is a 500-MHz regenerative divider per the Wenzel Web site: http://www.wenzel.com/pdffiles1/Dividers/LNRD4.pdf. (Courtesy of Wenzel Associates.)

custom-made to a particular frequency part. Designing a regenerative divider for production at the higher frequencies can be demanding, especially if mil-spec or industrial temperature ranges are required.

Regenerative frequency divide-by-2 devices have been manufactured commercially at up to 60 GHz. At the higher frequencies, LSICs are usually utilized to ensure consistent performance. At the lower frequencies, the construction of these devices is hybrid. A 500-MHz LNRD4 regenerative divider, which is a 500-MHz divider manufactured by Wenzel, is shown in Figure 16.14(b). The regenerative divider has three outputs, two at 1/2 frequency and one at 3/2 frequency input. More on this product can be found on the Wenzel Web site at http://www.wenzel. com/pdffiles1/Dividers/LNRD4.pdf.

These modules use low noise mixers and amplifiers yielding exceptional phase noise performance. The typical phase noise performance has been up to −170 dBc/Hz at the higher frequencies (e.g., 10 GHz) and −177 dBc/Hz at the lower frequencies.

The phase noise of a divide-by-2 regenerative divider is given by the phase modulation (PM) noise equation [12]:

$$\mathscr{L}_{Regen\,Div(f)} = \frac{\sum \mathscr{L}_{Devices(f)}}{4} \tag{16.5}$$

where $\mathscr{L}_{Regen\,Div(f)}$ is the SSB phase noise power spectral density of the divider, and $\mathscr{L}_{devices(f)}$ is the SSB phase noise power spectral density of all the devices within the divider loop.

These limits are dictated by careful setting of the phase shift within the divider loop parameters at the expense of the output power and loop stability.

The resulting phase modulation (PM) noise is due mainly to the mixer and amplifier noise figures as they perform within the loop parameters. Consequently, (16.5) can be rewritten as:

$$\mathscr{L}_{LNRD}(f) = \left[\mathscr{L}(1/f)_{mixer}(f) + \mathscr{L}(1/f)_{amp}(f) \right]$$
$$+ \mathscr{L}_{thermal\,mix}(f) + \mathscr{L}_{thermal\,amp}(f) \tag{16.6}$$

where the first and second terms are the flicker noise of the mixer and amplifier, the third term is the conversion loss (noise figure) of the mixer, and the last term is the thermal noise floor (or noise figure) of the amplifier.

If the flicker noise of the mixer is less than the amplifier's noise figure, then the close-in phase noise of the divider can be represented by the phase noise of the active amplifier in the loop as divided by 4.

In addition, the noise floor of the entire mechanism is dictated primarily by the noise figure of the amplifier, the mixer conversion loss, and the available RF power to the R-port. Consequently, the SSB phase noise of the regenerative divider is given by:

$$\mathscr{L}_{RD(thermal)}(f) = \mathscr{L}_{Amp(thermal)}(f)$$
$$= \tfrac{1}{4} KTF_{(amp)} / 2P_{(in\,amp)} \tag{16.7}$$

As an example, if $T = 300K$, $K = 1.38 \times 10^{-23}$ J/K, the amplifier power is $P_{(in\ amp)} = -2$ dBm and $F_{amp} = 6$ dB, and then the phase noise floor of this divider example is calculated as -175 dBc/Hz.

Regenerative dividers have been used in combination with static digital dividers in coherent frequency synthesizers to achieve division by 2 and 4 at input frequencies exceeding 60 GHz. At the higher frequencies they have been recently implemented in 0.18-μm SiGe BiCMOS and other technologies in order to keep the loop mechanics stable. In addition, to ensure no self-oscillation, the loop gain must be reduced below 1 when the RF input signal is absent.

We will discuss next an application of regenerative and static digital dividers in a millimeter-wave coherent synthesizer used in a 60-GHz wireless receiver [13].

Recent demand for portable high-definition (HD) video has called for networks at the 57–64-GHz band. This band has been recently introduced by the IEEE 802.15.3 standard as the new high-speed band.

At these frequencies the percentage bandwidth is such as to allow for very wide bandwidths to be transmitted and received with precision despite the limited range characteristic of these frequencies, which obey interesting atmospheric physics laws. The 60-GHz band has been known for a long time to provide intersatellite secure communications or point-to-point covert communications due to the high oxygen absorption in the atmosphere at this frequency, which protects it from ground eavesdropping. In addition, extra secure and limited range communication can be obtained because of the ability to point very narrow antenna beams at each other at these frequencies. Data rates in excess of 5 Gbps have been realized. This competes favorably with fiber optic communications and is fast enough to allow downloading wirelessly a 2-hour high-definition (HD) movie in roughly a minute. Many other advantages besides percentage bandwidth and ultrafast data rates exist at these frequencies. Because the wavelength is very short, the antenna can be incorporated in the transceiver chip. In actuality, an array of antennas can be incorporated in each transceiver allowing for phasing and selective directivity which in turn can allow for thousands of ultrawide bandwidth transmissions to be deployed concurrently in geographically confined areas without the fear of interference. The oxygen absorption property only enhances this multitude of channels, which can allow for many smaller ultrawide bandwidth networks to coexist smartly in the same area competing favorably with fiber optic networks and without the need for physical connections. A basic block diagram of a 60-GHz receiver is shown in Figure 16.15.

In Figure 16.8, we can see that this receiver uses a fully coherent synthesizer utilizing brute-force direct coherent components (both regenerative and digital dividers) as combined with a nonbrute-force indirect coherent synthesis form, a phase-locked loop. The common reference runs at 6 GHz and can be obtained from a multiplied quartz or SAW master reference oscillator. We will discuss master reference oscillators in Section 16.21.

The voltage controlled oscillator (VCO) runs at 48 GHz to mix with the 60-GHz RF signal to produce a 12-GHz first IF. The VCO could use a dielectric resonator oscillator (DRO) or a waveguide cavity oscillator on MMIC technologies [14]. The oscillator and additional circuitry can be readily integrated to form a phase-locked system implemented on InP HBT or SiGe BiCMOS processes.

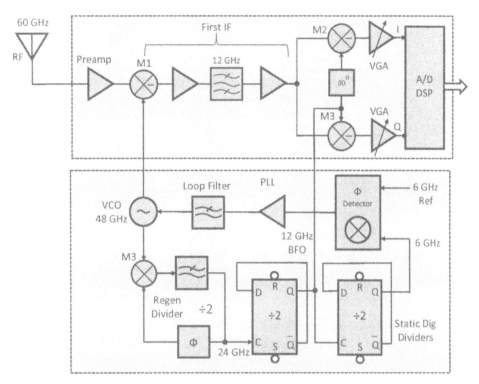

Figure 16.15   Block diagram of a 60-GHz receiver using a brute-force coherent synthesizer utilizing a regenerative divider operating at 48 GHz, and two static digital dividers for a total division of 8. A phase-locked loop is also used in conjunction with the dividers [13].

As can be seen in Figure 16.15, the very high frequency of the VCO dictates the use of a regenerative divide-by-2 divider. Following this divider is a divide-by-4 static digital divider using two divide-by-2 *D* flip-flops to produce a 6-GHz signal for the phase-locked loop phase-frequency detector. This signal is then compared with a 6-GHz reference to close the loop via the VCO control voltage, which keeps the loop in lock at 48 GHz. The 12-GHz signal obtained after the first digital divider is used as a beat frequency oscillator (BFO) and is mixed with the IF to produce baseband signals to be further processed through the A/D and DSP section. Special care is exercised in isolating the 12-GHz LO frequency from the first IF.

An even more sophisticated brute-force coherent synthesizer using regenerative and digital static dividers is shown in Figure 16.16. This synthesizer offers a clever plug-and-play approach to a multiband millimeter and submillimeter generator. It takes advantage of two regenerative dividers used in conjunction with a static digital divider to provide a variable reference to a 1-GHz direct digital synthesizer (DDS), which in turn closes the loop against a 50-MHz reference. The fractional division feature of the DDS is fully exploited here by using specially selected reference frequencies for each band, which in turn allows for a step resolution of 23 Hz in any of the bands of interest.

Plug-in DROs are used as VCOs for the required band of interest. A DRO is a mechanically and/or electrically tunable fundamental oscillator that can be designed to operate in bands of approximately 40 MHz from 1 GHz to 100 GHz. It includes a custom-designed voltage-controlled-oscillator (VCO) monolithic

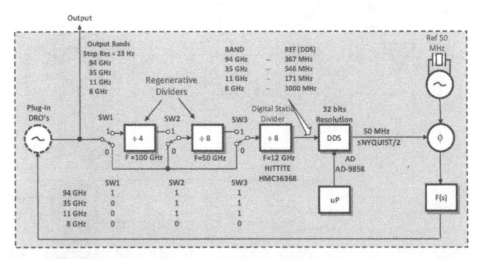

Figure 16.16 Multiple-band plug-and-play microwave synthesizer uses regenerative dividers together with static digital dividers and a direct digital synthesizer (DDS) to achieve a 23-Hz step resolution in any of the four microwave bands. Plug-in DROs are used as VCOs.

microwave integrated circuit (MMIC), a dielectric resonator disk ("puck"), and two varactor coupling circuits, all laid out on a thick alumina substrate. The dielectric resonator disk (ceramic in nature) varies in diameter depending on the frequency. The oscillator is usually mounted in a cavity metal housing. Frequency can be adjusted (locked) over +/– 20 MHz typically.

The synthesizer from Figure 16.16 offers the advantage of using a single assembly to address all four bands with high resolution, which is a departure from the customary independent dedicated synthesizer used for each band.

With the proper DRO plugged in, the synthesizer uses three toggle switches and a truth table to select the proper divisor combination in order to generate the right reference for the DDS. In any of the selections, the DDS reference does not exceed the 1 GHz frequency limit of the Analog Devices AD-9858 DDS part as shown.

16.12 Harmonic Multipliers

The harmonic approach to frequency synthesis can be used whenever wide channel spacing between adjacent frequencies is desired and an integer multiple of the input frequency is required, such as in a complex synthesizer using several combined synthesis methods. This approach uses the harmonic content of a reference frequency which equals the step required. This is also referred to as a comb generator because of the resemblance with a comb when the output signal is viewed on a spectrum analyzer. Each individual output can be selected with the help of a variable bandpass filter as shown in Figure 16.17(a). Another approach is presented in Figure 16.17(b). In this example, there is no variable bandpass filter as used at A, as selectivity is provided by the fixed frequency crystal bandpass filter used as an IF. Selection of the proper harmonic is obtained through the process of double mixing provided by the tuned oscillator.

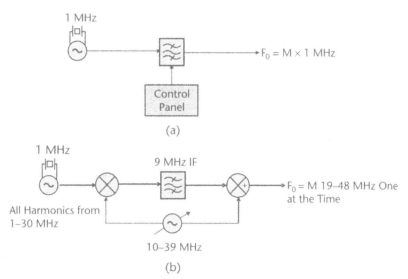

Figure 16.17 (a) Harmonic synthesizer using a variable bandpass filter to select multiples of the 1-MHz reference. (b) Harmonic synthesizer using a tuned oscillator in a double mixing scheme with a narrowband IF.

This system can offer superior spurious performance over the previous one because of the IF type processing offered by the crystal filter; however, an image problem develops because of the superheterodyne approach.

Frequency multipliers can use simple silicon diodes at the lower frequencies or step recovery diodes (SRD) at the higher frequencies. Schottky diode multipliers are used for narrowband or broadband, low-phase noise applications. Microwave bipolar transistors and PIN diodes are also used for high-efficiency, low-order multipliers from 500 MHz to 10 GHz. GaAs FETs, Impatt diodes, and varactors are used for multipliers above 10 GHz. SRD multipliers are used for high multiplication ratios sometimes beyond 100 GHz. An SRD multiplier model is shown in Figure 16.18.

The circuit of Figure 16.18 uses a transmission line resonator to couple the harmonic energy out of the diode. The network L1, C1 is a highpass filter. This filter is needed, because the SRD diode may enter oscillation if it becomes resonant below the input frequency. This filter has a cutoff frequency of about 20% below the input frequency. This will isolate the diode from any low input frequency. The resistor R is the bias resistor for the SRD diode. The value of R is determined

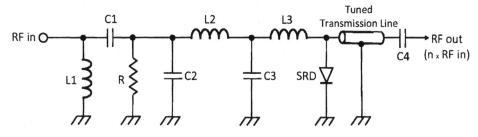

Figure 16.18 Step recovery diode (SRD) multiplier model.

experimentally. C2 and L2 are an impedance matching network, while C3 and L3 form an impulse generator.

The diode conducts on the positive-going half of the input wave, but it does not stop conducting on the negative-going slope until all charge carriers flow out of the diode's inert region. Upon the total depletion of carriers, the diode opens up abruptly, collapsing the magnetic field in inductor L3, which in turn generates an avalanche of impulse signals with fast rise times and short pulse width. This pulse train is rich in harmonic content. The circuit is followed by a transmission line resonator, which is tuned to the multiplied frequency of interest.

Traditionally, SRDs have been used in multipliers. Newer harmonic generators have been using nonlinear transmission lines (NLTLs). A nonlinear transmission line is comprised of a transmission line periodically loaded with varactors, where the capacitance nonlinearity arises from the variable depletion layer width, which depends both on the DC bias voltage and on the AC voltage of the propagating wave. Figure 16.19(a) shows a scanning-electron microscope (SEM) photo of the sampling circuit of an NLTL. Figure 16.19(b) shows an equivalent circuit model of an NLTL. Figure 16.19(c) shows the phase noise performance plot of a comb generator using NLTL with a 200-MHz input and an 8-GHz output. This data shows that the residual phase noise performance of NLTL multipliers is usually below the noise floor of the measurement system.

Straightforward analog multiplication is seldom used alone in frequency synthesis unless utilized at the very high microwave frequencies where no other means of synthesis exist.

Multipliers are usually used in combination with other synthesis forms. Properly combining dividers and multipliers with the mixer synthesizer can result in new, powerful, brute-force coherent synthesizers. Further combining such synthesizers with direct-digital and/or fixed phase-locked loops can provide a solid foundation for designing versatile, agile structures, as we will discuss later.

While many variations of the harmonic synthesizer exist, this form of synthesizer has been utilized in a less-known HF application, the Barlow-Wadley loop, which was briefly discussed in Chapter 9. This design has been commonly used in communications receivers until the recent introduction of the phase-locked loop and can still be found in new applications. The concept is so appealing that engineers have used it repeatedly. The block diagram in Figure 16.20 shows the implementation of a Barlow Wadley synthesizer in a HF communications receiver. One of the most interesting features of the Barlow Wadley scheme is the drift-canceling mechanism provided by the double mixing of a free-running oscillator (the megahertz oscillator) as shown in Figure 16.20.

In Figure 16.20 a 1-MHz reference oscillator generates LO markers at exactly 1-MHz intervals anywhere between 3 and 32 MHz. The 29 harmonics are fed to one side of a loop mixer as shown. A separate free-running variable oscillator called the megahertz oscillator generates frequencies between 55.5 and 84.5 MHz and runs an open loop. The output of this oscillator is fed simultaneously to the loop mixer and the first mixer of the receiver where it combines continuously with the incoming RF signal, generating a 1-MHz-wide IF centered at 55 MHz. The signal information, which has a bandwidth of 1 MHz (54.5 to 55.5 MHz), is then fed to one side of the second mixer. The other side of this mixer is fed from the loop mixer through a narrow bandpass filter centered at 52.5 MHz. In this approach the

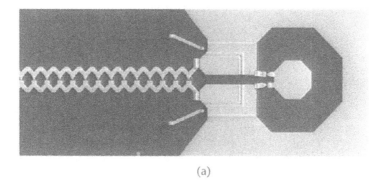

(a)

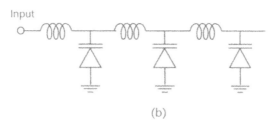

(b)

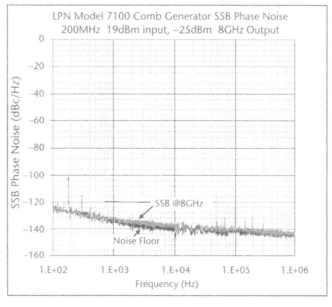

(c)

Figure 16.19   (a) An integrated NLTL and sampler structure. (b) Circuit diagram of nonlinear transmission line network. (c) Phase noise measurements for Picosecond Labs Model 7100 comb generator. (Courtesy of Picosecond Labs.)

52.5-MHz signal will only be true at exactly 1-MHz intervals as a result of the selective mixing process which takes place in the loop mixer, as previously discussed in Chapter 9.

The drift of the megahertz oscillator is completely eliminated by the subtraction process in the second mixer, providing a stable synthesized conversion for the second IF of the receiver.

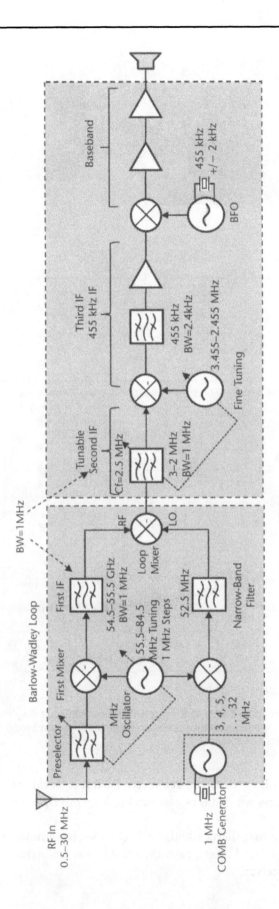

Figure 16.20 A good example of indirect harmonic synthesis is the Barlow Wadley loop. A stable receiver injection LO is synthesized in 1-MHz steps from a 1-MHz comb reference. Despite the open-loop nature of the megahertz oscillator, any drift is canceled automatically due to the double mixing process.

The advantage of this synthesizer is obvious: simple circuitry that can provide many relatively wide and equally spaced frequencies derived from a stable 1-MHz reference. The disadvantages are due to the harmonics of the reference oscillator being heard at the beginning and end of each band, a phenomenon that can only be minimized through the extensive shielding of the stages and by locating the harmonic generator as far away from the RF processing circuits as possible.

An interesting satellite communications receiver for frequencies above 10 GHz, using a drift and phase noise cancellation process, has been introduced by the patent 4,918,748 [15] (see Figure 16.21).

Although using a phase-locked loop, the invention utilizes a 1-GHz comb generator at 14 through 17 GHz to mix and cancel out drift and phase noise produced by relatively noisy gallium arsenide components. The noise-canceling loop is used with the tuner section of a communications receiver, which includes a gallium arsenide voltage controlled oscillator (VCO) for upconverting the received signal. Phase noise and posttuning drift associated with the phase-locked oscillator are canceled by the signal supplied via a second oscillator.

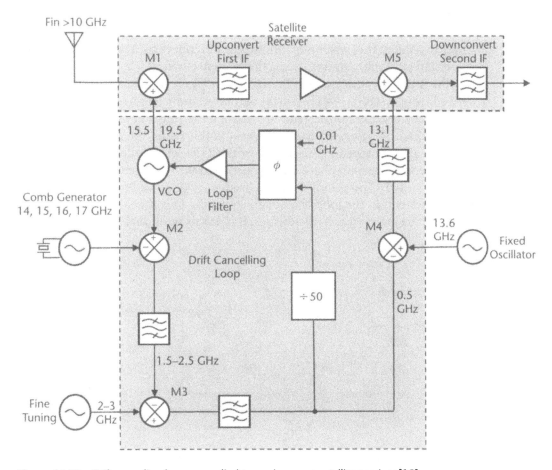

Figure 16.21 Drift canceling loop as applied to a microwave satellite receiver [15].

16.13   Single-Loop Integer Phase-Locked Loop (PLL)

We will next turn to the phase-locked loop (PLL). To better understand the various forms of PLLs, we will have a discussion intended to see how the reference frequency affects the step resolution in the integer PLL.

Referring to Figure 16.8, the PLL is defined as a coherent type synthesizer under the nonbrute-force indirect category. It uses phase/frequency comparison with a reference to correct the frequency of a VCO in closed loop, a simple concept that has many ramifications. The PLL is actually a frequency multiplier in which the divide-by N number is the multiplier number. This generally tends to spoil the phase noise performance of the VCO by the proverbial $20logN$ formula, where N is the division number. Although several things can be done to improve the situation, we will use an exaggerated example to make a point.

Figure 16.22 shows a block diagram of a simple PLL synthesizer. This scheme could be used to generate the entire range of frequencies necessary for the local oscillator injection in our previous 9-MHz IF communication receiver example from Figure 16.6. However, there are some problems.

Unlike the previously discussed brute-force synthesizers, this requirement calls for a high-resolution frequency source in order to replace the continuous VFO feel. A step resolution of 10 Hz would be required of the PLL over the frequency range. In order to achieve this resolution, an integer PLL obeys by a simple rule that says that the step resolution always equals the reference frequency, in our case, 10 Hz.

In the simple hypothetical synthesizer from Figure 16.22, the 10-Hz reference is derived from a stable 1-MHz crystal oscillator and is fed to one side of a phase and frequency comparator/detector. To change the frequency, the integer divider number is changed by one or more integers. At this point, the entire loop searches within the capture range of the synthesizer until the second input line to the phase detector becomes equal in phase and frequency to the reference frequency, which is 10 Hz. The near DC output of the phase/frequency detector bounces up and down until it settles within the loop filter characteristics. This is called settling time and depends on the loop filter characteristics. It is at this point that the near DC output

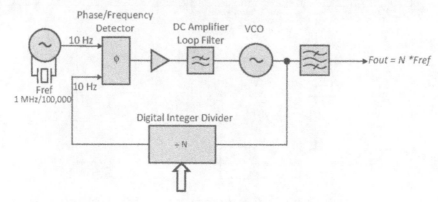

Figure 16.22 A hypothetical high-resolution single-loop integer phase-locked loop (PLL) synthesizer. Step resolution obeys the general rule of frequency resolution in a phase-locked loop (PLL) in which the step resolution equals the reference frequency, in this case, 10 Hz. This example was used only to describe a concept. In reality it has problems connected with phase noise being impacted by the reference feedthrough.

of the phase detector (which is the control line to the voltage controlled oscillator) ensures that the desired frequency is developed at its output.

The loop tracks continuously so that the stability of the output is always based on the stability of the reference frequency. The lowpass filter (the loop filter) is intended to smooth out the control voltage to the VCO for better control signal purity. Too little damping translates into worse phase noise. Too much damping produces a cleaner voltage, but it increases the settling time, not necessarily a desirable feature.

If a change in frequency is desired, all that needs to be done is to change the division number N and the synthesizer will search for whatever VCO frequency divided by N will give a 10-Hz signal at the second input to the phase detector to match the reference frequency and phase. This relationship can be expressed by

$$F_{out} = N * F_{ref} \qquad (16.8)$$

where F_{out} is the desired output frequency, N is the integer division number, and F_{ref} is the reference frequency.

It can be seen from (16.8) that the integer phase-locked loop approach can provide stable injection frequencies to the first mixer of a communication receiver in integer steps, simulating the VFO's continuous coverage. To approach this continuous resolution, the synthesizer bases itself on a stable reference frequency, but cannot tune continuously. The reference frequency determines the switching step resolution, in our case, 10 Hz. Every time the division number N is changed digitally, a new 10-Hz step is obtained. This synthesizer is coherent.

In reality, this synthesizer is not practical. This exaggerated example was used only to demonstrate step resolution dependency on reference and some of the PLL problems, as we will see next. First, with the PLL problems, the 10-Hz reference would require a very dampened lowpass filter that would slow down the system lockup time. This, in turn, would make the switching time impractical. It may take seconds to achieve locking in such a system, which would be a long time to retune from an operator standpoint.

Second, because the control voltage to the VCO is not pure DC, we are faced with a new phenomena called phase detector reference feedthrough. The 10-Hz reference would bleed into the phase detector's output modulating the VCO on each sideband of Fo and create a noisy local oscillator. Because these sidebands are only 10 Hz away from the carrier, they are actually in band and cannot be filtered out. They would blend together with the phase noise and contribute to it. This, in turn, would intermix with the incoming signals of the receiver and create intermodulation distortion. Third, such a system would be affected by hum since the reference frequency is kept low.

What we would like to do is to increase the reference frequency to a much higher frequency, say, 10 MHz. This would move the reference feed through sidebands out by 10 MHz on each side of the VCO carrier, and could be easily filtered out.

However, increasing the reference frequency affects the basic law of integer PLL resolution from (16.8). Consequently, the step resolution would change to 10 MHz and the reference feedthrough would be far enough out to be filtered, but we would loose the fine resolution property of our PLL synthesizer (see Figure 16.23).

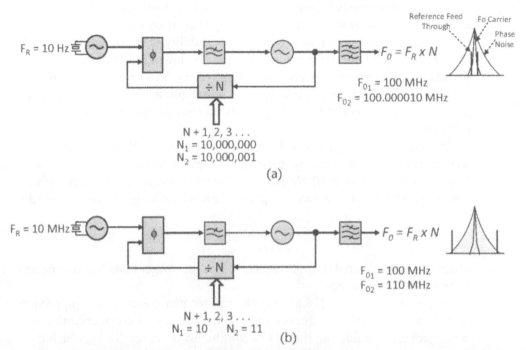

(a)

(b)

Figure 16.23 The reference feedthrough in a PLL. (a) When reference is low in order to achieve fine resolution (e.g., 10 Hz), it inadvertently shows up in each sideband of the VCO's output frequency (usually –40 dBc), offset by the 10-Hz reference. It cannot be filtered out and contributes to the phase noise degradation. (b) By changing the reference to a higher frequency (e.g., 10 MHz), the reference feedthrough sidebands are moved out from the carrier, allowing for filters to be used effectively. However, by changing the reference frequency, the step resolution of the integer PLL also changes from 10 Hz to 10 MHz according to (16.8) ($F_{out} = N*F_{ref}$).

To remedy this problem, various methods have been devised over the years to obtain high resolution while using high reference frequencies, as will be discussed next.

16.14 Multiple-Loop, Phase-Locked Loop (PLL)

Because of the issues presented above, single-loop integer synthesizers have been used mainly as fixed frequency multipliers. A 12.8-GHz integer phase-locked loop (PLL) circuit design using Hittite parts is shown in Figure 16.24. The divider (multiplier) number is 128 and the reference frequency is 100 MHz. The digital phase detector used is state-of-the-art and allows for references of up to 1.3 GHz. Even higher output frequencies are possible with this design if the VCO can be further multiplied. Other frequencies are also possible by changing N.

We have seen that the reference frequency dictates the frequency step resolution in a integer PLL. We have also seen that relatively low frequencies would have to be used in order to achieve high resolution in such a PLL. However, the approach in Figure 16.22 is impractical.

So how do we achieve fine resolution using higher reference frequencies? One of the methods to satisfy this requirement is the multiple-loop PLL design.

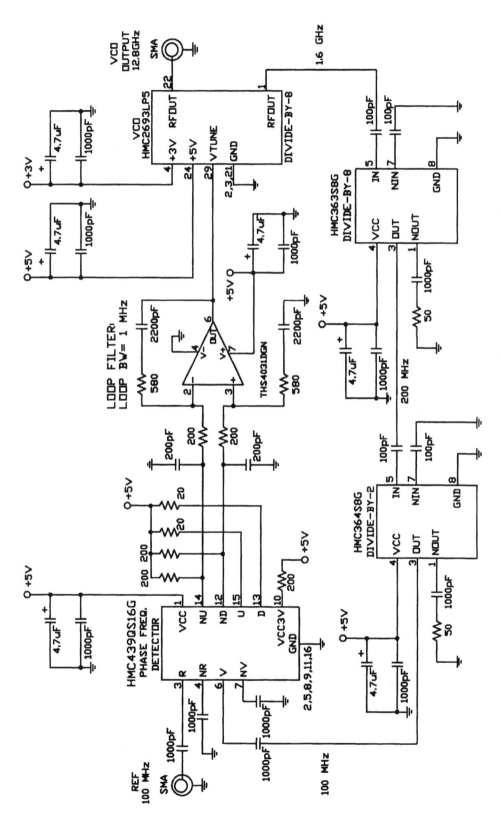

Figure 16.24   A 12.8-GHz integer phase-locked loop using a 100-MHz reference, a 1.3-GHz phase/frequency detector, and a chain of cascaded static digital dividers. $N = 128$ for a multiplied frequency of 12.8 GHz. (Courtesy of Hittite Microwave Corporation © 2000–2008 Hittite Microwave Corporation, All rights reserved.)

A good example of such synthesizer is the approach shown in Figure 16.25. In this approach, high resolution is achieved through the use of two relatively high-frequency references that have been coherently derived from the same reference. These references have been chosen such that when the two integer PLL outputs are subtracted from one another in a mixer, their final resolution is determined by the delta frequency between the two references, in this case, 100 Hz.

Also shown in Figure 16.25 is a small part of the tedious algebraic manipulation of the two numbers (N1 and N2) necessary to tune a small portion of the frequency range.

Multiloop synthesizers are a fact of life in today's technology. As many as four or five loops are sometimes used in multiconversion receivers. Complicated steering mechanisms provide fast switching over wide frequency ranges, while lockup times have been reduced to milliseconds and sometimes to microseconds. A complex multiple loop synthesizer is shown in Figure 16.26. In this design, the output of the first phase-locked loop is combined to a second phase-locked loop in combination with a mixer type approach. Dual-modulus prescalers are used throughout.

The introduction of the PLL has at least temporarily obscured many of the other forms of synthesizers. Because of its apparent simplicity, the PLL has been viewed as a solution to all frequency generation problems without a real understanding of the new problems it created and its obvious limitations. While the PLL can offer great versatility, it is far from perfect and cannot be used by itself except in simplistic applications. Many forms of synthesizer technologies need to

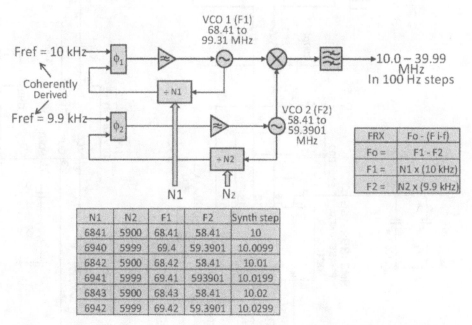

N1	N2	F1	F2	Synth step
6841	5900	68.41	58.41	10
6940	5999	69.4	59.3901	10.0099
6842	5900	68.42	58.41	10.01
6941	5999	69.41	593901	10.0199
6843	5900	68.43	58.41	10.02
6942	5999	69.42	59.3901	10.0299

Figure 16.25 A dual-loop synthesizer provides high resolution and relatively fast switching using higher reference frequencies than the synthesizer step resolution requirement. The references that are derived from the same crystal oscillator are 10 kHz and 9.9 kHz, respectively. The 100-Hz resolution is obtained by subtracting the two VCO outputs and consequently their references in a mixer-type synthesizer. This moves the reference feedthrough problem further away from the close-in phase noise of the output at the cost of possible new intermodulation distortion problems created by the mixer.

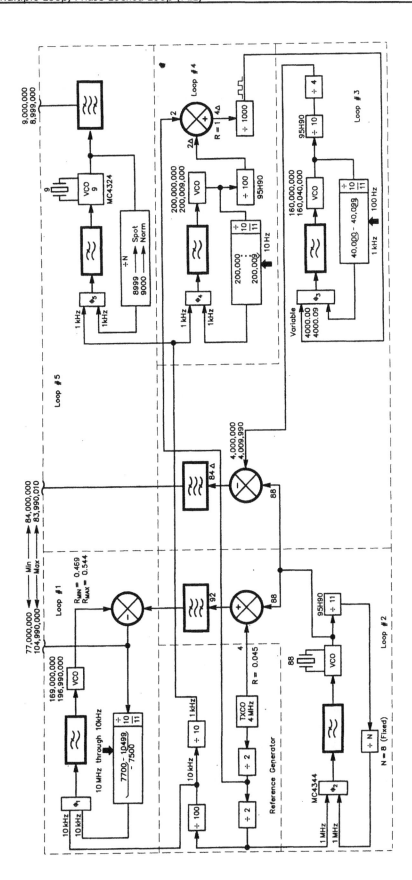

Figure 16.26 Block diagram of an early multiple-loop coherent synthesizer combining brute-force synthesis with the dual-modulus approach.

be combined to produce superior results, especially when high-resolution, fast-switching parameters are required.

Designing high-performance, multiple-loop high-resolution switching synthesizers that can generate thousands of frequency steps from a single reference at microwave frequencies is a very different task than designing a brute-force synthesizer or a master reference oscillator using dividers and multipliers and mixers. In the case of high resolution, the phase noise performance usually suffers when compared with the fixed brute-force methods because the required resolution implies more noise.

Today practical high-resolution full synthesizers combine all the coherent brute-force and nonbrute-force types to achieve fast switching and low phase noise. The experienced designer will mix technologies such as to take advantage of the various properties of the different techniques explained here in order to derive all frequencies from a single reference. We will turn next to other methods of obtaining high-resolution synthesis.

16.15 Digital Counter/Comparator and Digiphase Synthesizer

One of the less-known synthesizer forms is the digital counter/comparator (not to be confused with the Digiphase approach), which is defined as a coherent indirect type. It is characterized by using the reference of a high-resolution digital counter and cascaded static digital comparators to correct the frequency of a VCO in closed loop. The locking mechanism is achieved by activating an up/down counter connected to a D/A converter through the greater than, less than, or equal outputs of a cascaded static digital comparator chain, which constantly compares the VCO's output frequency to a preset number. Consequently, fast-locking transitions can be achieved with a "sliding effect" caused by the speed of the steering counter. This effect can be useful in certain applications. Because of the relatively large number of cascaded counters required for high resolution, this synthesizer type seldom exceeds four decades of resolution. Consequently, the output frequency is open loop between the steps (the ambiguous digit problem), and simple frequency discriminators can be used in mixing schemes to achieve final stability.

Among its advantages are the fast switching with relatively low noise content. Its true digital nature makes it a robust candidate for very producible and reliable signal processors without the problems usually associated with PLL manufacturing. Large-scale integration (LSI) is feasible and recommended. This synthesizer is shown in Figure 16.27(a).

In Figure 16.27, an accumulator increases its number (N_A) by the number presented at its control lines (N) every cycle of the reference frequency (F_{REF}). The output number is the desired number of phase cycles for the synthesizer as seen at some point in the reference cycle. It is this digital number that is latched and compared with a digital number produced by a frequency counter that is continuously monitoring the VCO's output (N_{dc}). The comparison is performed by a static digital comparator that switches a current I_p into the loop filter whenever the accumulator number (N_A) is greater than the counter number (N_{dc}).

The digital counter frequency-locked loop approach does well from a reliability point of view since there is no phase detector reference feedthrough or capture

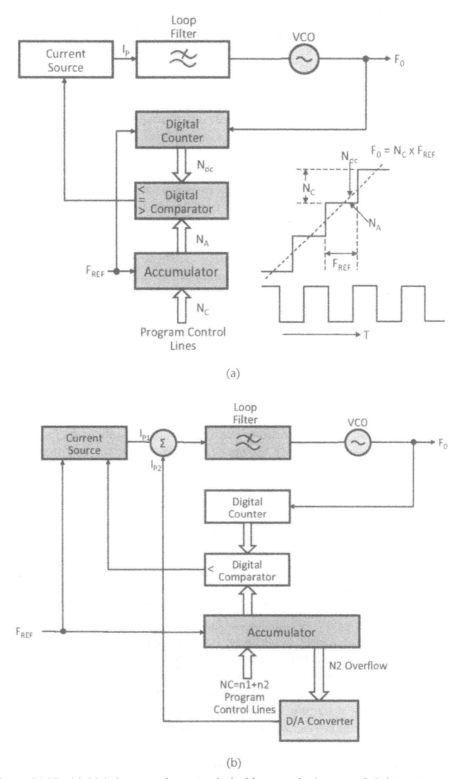

Figure 16.27   (a) Digital counter frequency-locked loop synthesizer uses digital comparator and counter. This is a completely digital approach which does away with phase-locked loop problems such as out of capture range or locking due to temperature changes. (b) The Digiphase approach to synthesis.

range problems. Its resolution is based on the number of cascaded programmable static digital counters (e.g., 74192 for TTL frequencies) and comparators (7485) involved, which have a practical limit of about four digits in order to keep integration manageable.

The basic Digiphase technique, which allows high resolution independent of the reference frequency, is shown in Figure 16.27(b). In this approach, the accumulator register is intentionally built longer than the digital counter or the digital comparator registers. The control number N_c is comprised of a main number n_1 and a fractional number n_2, which represents the resolution. Although the total number N_c is not compared with the content of the digital comparator because of the lack of resolution in it, the overflow of n_2 in the accumulator results in a current I_p, which is summed together with I_{p1} allowing for small current increments to be presented to the loop filter and provide fine resolution for the VCO.

Using this technique, a resolution of 1 Hz can be achieved with a reference frequency of 100 kHz. The Digiphase synthesizer was introduced by G. C. Gillette under the U.S. Patent 3582810 in June 1971 [16, 17] and was implemented in a variety of instruments and receivers. Further improvements to the Digiphase method of synthesis have been introduced by Racal and others. This consists of a further reduction of the low-frequency components at the output of the phase detector by using digital techniques. For example, the output of the accumulator in Figure 16.26(b) is further integrated in a similar size accumulator. When an overflow occurs from the second accumulator, a pulse is removed from N_1. By adding this pulse to the next count, a better phase noise performance is obtained for one reference period, balancing out the continuously decreasing phase due to $(Fo/N_1) > F_{ref}$.

The fractional-N synthesizer achieves basically similar high-resolution results as the Digiphase approach, except that it is more like the traditional PLL.

16.16   Fractional-N and Dual-Modulus Divider Phase-Locked Loop (PLL)

We have seen that in order to obtain fine steps in an integer phase-locked loop, we would have to use low phase detector frequencies, which means in-band reference feedthrough problems spoiling the phase noise performance. This, in turn, necessitates very narrow loop bandwidths that slow down PLL settling times.

What is needed is a method of achieving high resolution while using a high reference frequency. While this can be achieved by using multiple-loop PLL systems combined with a brute-force mixer synthesis, this method can suffer from additional mixer intermodulation distortion problems and can be mathematically complex.

Thus, the fractional-N phase-locked loop was invented. The fractional-N synthesizer belongs to a group of synthesizer forms that allow the minimum frequency step to be a fraction of the reference frequency. In a fractional-N synthesizer, achieving a noninteger VCO frequency divide ratio of $N < N_f < N + 1$ is performed by switching between the N and the $N + 1$ states to average out a division ratio N_f.

In addition, when the synthesizer is designed for VHF/UHF frequencies, simple programmable dividers may not always be available. The solution to this problem is to combine the properties of a high-speed divider with the functional properties of the slower programmable divider along with a programmable dual integer divider function. This higher-frequency device is referred to as the dual-modulus prescaler and the technique used in impleementing it was named the pulse swallow-ing technique. A PLL synthesizer using this approach will usually be referred to as a dual-modulus prescaler-PLL synthesizer.

The rationale behind the fractional-N approach is as follows. If it was pos-sible for N to take on fractional values, the output of a PLL could also be changed in fractional increments. Because fractional division is not possible with standard integer digital dividers, a different mathematical approach has to be adopted. By dividing the output frequency by $N + 1$ every M cycle and dividing by N the rest of the time, an effective division ratio function would result as shown by

$$F_o = (N + 1/M)f_{ref} \qquad (16.9)$$

This expression shows that F_o can vary in fractional increments of the reference frequency by varying M. This is implemented through a fractional phase value, which is added to the phase detector's output as shown in Figure 16.28.

Dual-modulus counters allow the division ratio to be changed under logic con-trol swallowing of one or two pulses from the total train of pulses provided by the VCO in a closed loop. Direct programmability over the entire range of a VCO operating at frequencies beyond the capabilities of conventional counters can be achieved through the dual-modulus $(P/P + 1)$ prescaling process. The process is usually implemented in ECL technology. To understand this technique, refer to Figure 16.29 and the following explanation.

Assume that an output frequency (F_o) of 150.020 MHz is expected with a ref-erence of 5 kHz. From the simple phase-locked loop equation $F_o = F\,r \times N$.

Then

$$N = Fo/Fr = 150.020 \times 10\,\text{Hz}/5 \times 10\,\text{Hz} = 30,004$$

Assume the dual-modulus values P and $P + 1$ are the numbers 20 and $20 + 1 = 21$. If N is divided by the P modulus, the resulting quotient is the N program value and the A program value in the remainder.

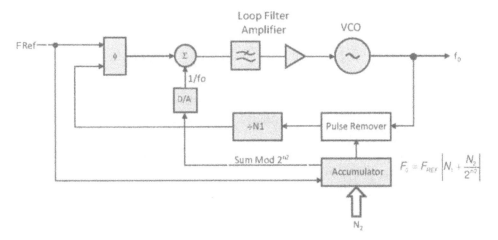

Figure 16.28   Fractional-N synthesizer approach.

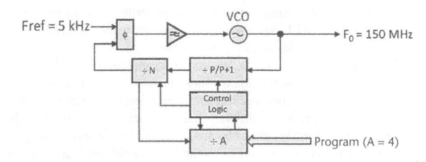

Figure 16.29   Dual-modulus PLL approach.

$$N/P = 30,004/20 = 1,500 \text{ and a remainder of } 4$$

The key to the operation of this synthesizer is the $P + 1$ function of the dual-modulus prescaler. If the above value for N and A are programmed into the synthesizer, the loop will perform in the following manner: The $P/P+1$ divider is initially set up to divide by 21 and will output a pulse to the dlvide-by-n counter as well as to divide-by-A counter for every 21 pulses it receives. Since divider A is programmed to divide by 4, this cycle will continue for another three times (each cycle equals 21 VCO pulses it receives at the $P/P + 1$ counter), at which time divider A outputs a command (count of 4) to the control logic and instructs divider $P/P + 1$ to change the divisor to $P = 20$. This change is executed and an inhibit signal is fed back to the divider A, preventing it from any further count until the total cycle is repeated.

In the meantime, divider N which has already counted to 4, continues to receive pulses from divider $P/P + 1$ at a ratio of 20:1. After the remaining 1,496 pulses have been received (1,500 − 4), divider N finally outputs a pulse to the phase comparator as well as to the control logic which resets all counters and the process repeats. It can be seen from this simple explanation how the swallow terminology was born. To check the math involved, use the following equations.

$$N = (P+1)A + (n - A)P$$
$$\text{If } N = 30,004, \text{ then}$$
$$30,004 = (2 \times 4) + (20 \times 1,496)$$

Various division ratios such as 5/6, 8/9, 10/11, 15/16, 20/21, 32/33, 40/41, 64/65, 80/81, 100/101, and 128/129 are available in ECL technology operating at frequencies exceeding 2 GHz.

Predicting and reducing spurs in fractional-N synthesizers has been a constant battle. These spurs are usually generated by the divider switching mechanism. They have been especially troublesome near the fundamental frequency component and have been combated via several methods, more noticeably by using high-order delta-sigma ($\Delta\Sigma$) modulators, which have the property of providing a noise notch while enhancing the desired carrier signal. We will discuss delta-sigma modulators in more detail in Section 24.9.

The Digiphase and the fractional N synthesizers can achieve high resolution with a relatively high-frequency reference. They have been used in receiver synthesizer design for a long time. The dual-modulus fractional approach is still being used today, but has been recenly surpassed by the DDS-driven PLL.

16.17 The Mixer Phase-Locked Loop (PLL)

When the dual-modulus prescaler approach is not economically practical, or the frequency of the VCO is so high that there is no digital component available to perform this function, as in the case of some microwave synthesizers, the mixer phase-locked technique is usually used, as shown in Figure 16.30.

This approach allows the operation of the VCO at frequencies, several orders of magnitude higher than the speed capability of the programmable divider, providing for inexpensive parts to be used in this part of a system. In a HPOI receiver, this advantage is twofold due to the inherently quiet nature of the lower-frequency CMOS counters that can be used in this type of synthesizer.

Referring to Figure 16.30(a), the VCO's output operating at VHF or UHF frequencies is mixed in M1 with the fixed frequency F1, which is greater than the maximum frequency of the VCO. This injection frequency is usually derived by means of frequency multiplication from the crystal controlled reference oscillator. Another way of obtaining this frequency would be to use a fixed PLL (multiplier), which is locked to the same reference.

The output of the mixer that operates in a subtracting mode is fed to the programmable divider, and the loop is finally closed at the phase detector.

The mixer PLL approach has the advantage of allowing the use of conventional counters without VHF/UHF prescaling. An important advantage of this synthesizer is the possible mathematical reversion of N due to the mixer subtraction mode, which, in turn, improves the loop dynamics. On the negative side, the reference multiplied input required by the mixer needs to be derived from the reference frequency, which spoils the phase noise performance of the reference by the $20\log N$ law (14 dB in our case). This implies starting with a better phase noise performance reference oscillator than would normally be required or using another method of generating the ×5 frequency. Although an attractive solution in certain applications, only a limited loop bandwidth is possible with this approach. The mixer PLL can be used in conjunction with all other forms of synthesizers.

16.18 Direct Digital Synthesizer (DDS)–Driven PLL

A relatively new form of direct coherent synthesis, and probably the most promising type of synthesizer implementation today, is the direct digital synthesizer (DDS), also called the numerically controlled oscillator (NCO). Its concept is simple and effective. In effect, the DDS is a fine resolution programmable divider that provides phase noise improvements that behave by the $20\log N$ rule explained earlier.

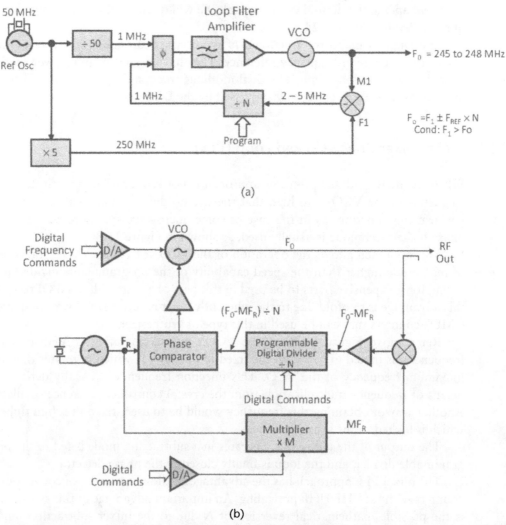

Figure 16.30 (a) The mixer PLL synthesizer is used to eliminate VHH/UHF prescalers, but limited frequency coverage results. (b) A presteerable version of the mixer PLL synthesizer.

The DDS uses a digital binary adder to increment a preprogrammed constant number representing a phase increment at several arbitrary points in time over a 360° cycle. (Note: Some designs sample over several cycles to keep spurious content down at the cost of higher harmonics.) Stored digital amplitude values representative of a cycle are sampled every phase increment (the Nyquist criteria calls for a minimum of two samples). These are selected from a look-up table that takes the form of a read-only memory (ROM) and are then presented to a digital-to-analog (D/A) converter, which translates them into a sine wave output. Lowpass or bandpass filtering is used to attenuate harmonics and some of the spurious products caused by the digital switching.

The concept can be viewed as a wheel in a gear train that turns continuously since the cycle repeats every 360° due to digital overflow. When a frequency change is required, a different gear ratio is engaged with the transition keeping the new frequency in phase with the old one. The block diagram of a DDS is shown in Figure 16.31(a).

An important limitation of a DDS is that the output frequency is limited by the Nyquist criteria, which say that the sample rate of the output frequency must be at least twice the output frequency. Higher sampling ratios are recommended, especially when wideband frequency ranges are required. Up to eight times the highest output frequency is recommended for better spurious free response. The expected output of a DDS operating at 30 MHz and using a reference clock of 100 MHz is shown in Figure 16.31(b). As can be seen, the output follows a $\sin(x)/x$ response with zeroes at the clock frequency and multiples afterwards. This is known as aliasing in DDS systems.

One of the biggest advantages of this synthesizer type is its ability to achieve fine step resolution because of the sampling technique. In addition, fast, coherent switching of frequencies is possible, which, in turn, lends to easy frequency hopping or frequency/phase shift keying (FSK or PSK). The phase accumulator width is designed such as to accommodate the smallest possible phase (frequency) step. The number of bits can be as large as 32 or even 64.

The phase noise performance of a DDS synthesizer can be superior to other synthesis methods. Since the DDS is a programmable fractional divider, its phase noise performance will be improved by the $20\log N$ rule, where N is the division number. The phase noise will be present as a result of the data converter's sample clock. The typical improvement to phase noise in a programmable frequency divider such as a DDS is expressed by:

$$S_{\Phi DDS}(f) \approx S_{\Phi, Fs}(f)Z^2 \tag{16.10}$$

where F_s is the sample clock and $F_{out} = Z*F_s$ (typically, $Z = X/Y$) and where X and Y are prime integers.

Consequently, a $20\log 10\ (Z)$ (dB) improvement is obtained, but limited by the noise floor of the digital divider. There are many other contributors to the phase noise and spurious performance of a DDS. This is because of the switching in the circuitry plus the flicker noise and white noise contributors. A typical phase noise performance in a DDS can be expressed by [8]:

$$S_{\Phi DDS}(f) \approx S_{\Phi, Fs}(f)Z^2 + \frac{10^{-10\pm 2}}{f} + 10^{-15\pm 1} \tag{16.11}$$

Problems usually associated with phase-locked loop settling characteristics and, consequently, phase-noise components are virtually not existent in this form of synthesizer, making it a design choice attempting to replace the PLL.

However, DDS synthesizers (even by recent standards) have been found to have spurious problems. Contrary to popular belief, the limiting factor in this type of synthesizer is not the speed of available digital-to-analog products, but the spurious performance of this technology when used in wideband systems. This is mainly caused by the "glitch energy" present in the switching.

It is not unusual to find DDS implementations that incorporate RC and LC lowpass filters in the switching lines to reduce the spurious content. Other

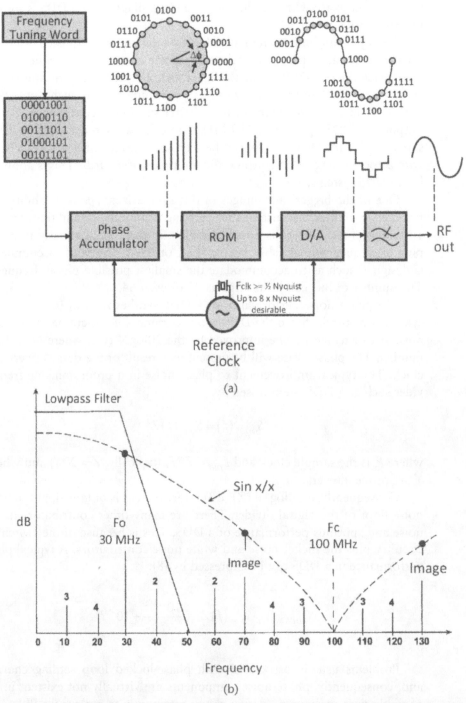

Figure 16.31 (a) The concept of DDS is based on sampling amplitude values at various points over a full-wave cycle of 0 to 360° (and sometimes over several cycles). The number of sampled points can vary but cannot be less than the Nyquist rate of 2. Once completed, the process rolls over to 0° and repeats. Implementation is a sequential-parallel, digital-to-analog (D/A) conversion. (b) Aliasing in a DDS. The output follows a sin(x)/x response with images and aliasing as shown. A reference of up to eight times the maximum output frequency and a lowpass filter are used to remedy these problems.

canceling techniques, including noise phase inversion, are being used to improve on the performance.

Recent developments in gallium arsenide (GaAs) D/A technology allow for low-spurious DDSs at very high frequencies. Today's state-of-the-art DDS has reached practical reference frequencies of 3 GHz (with up to 12 GHz in the near future) and high resolutions of fractions of 1 Hz. The EUVIS DS872 is such an example. It is a high-speed direct digital synthesizer (DDS) with a frequency tuning resolution of 32 bits, a ROM phase resolution of 13 bits, and a D/A amplitude resolution of 11 bits. The analog outputs of D/A can be selected between a normal-hold mode (for the first Nyquist band) and a return-to-zero mode (for the first, second, and third Nyquist band) operation. Sine waves can be generated up to the first Nyquist band near 1.5 GHz (at a 3-GHz clock rate) with a normal-hold mode of D/A or up to the third Nyquist band around 4.5 GHz with a return-to-zero mode of D/A. A block diagram of the DS872 DDS is shown in Figure 16.32.

Judicious use of this technology in combination with the phase-locked and mixer approaches can make for cost-effective, multioctave, high-resolution synthesizers into the microwave frequencies. As much as 90 dB of spurious attenuation has been realized in such high-resolution hybrid approaches.

The solution for improving on the spurious performance of the DDS has been the DDS-driven PLL. It was early realized that the fractional nature of the DDS resolution could be used to replace the fractional-N approach [18]. After all, a high-frequency DDS with a resolution of a fraction of 1 Hz could serve as a programmable reference for a fixed integer PLL, thus moving out the reference feedthrough problem while providing a small step resolution to the PLL. This idea came about at the same time with the introduction of the DDS and was introduced as a practical

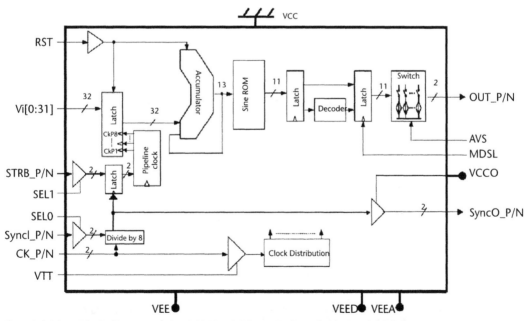

Figure 16.32 Block diagram of the DS872 DDS from Euvis. It features a 32-bit frequency tuning word, a 13-bit ROM phase address resolution, an on chip 11-bit DAC, and a clock rate of up to 3.2 GHz. (Courtesy of Euvis Corporation, http://www.euvis.com/index.html.)

solution for the first time by the author at the 1988 RF Expo Symposium in February 1988 [18]. It was presented as a means of improving performance and obtaining high resolution in a PLL without the use of fractional or multiple loops.

Figure 16.33 shows the evolution of the DDS to the DDS-driven PLL. Prior to this, only multiloop fractional-N PLL synthesizers were the rule to achieve high resolution. The fractional nature of the DDS used as a variable reference in a PLL allowed for a much more simplistic and better-performing high-resolution synthesizer. Figure 16.34 shows a block diagram of the DDS-driven PLL synthesizer as it was introduced at the 1988 RF Expo Symposium. We will return to the DDS-driven PLL in much more detail in Section 16.26 with a circuit design of a microwave DDS-driven PLL synthesizer for an HPOI receiver. Design and performance will be presented.

Today's DDS can provide phase-noise performance that can exceed that of the PLL. Latch synchronization can be important. As previously stated, the spurious content of a DDS is directly proportional to the D/A's speed and can be degraded if the system's latches are not synchronized with the reference as noticed in some of the published designs.

The prediction of spurious response for a DDS is primarily based on the "image" frequency, which is determined by the reference frequency and its harmonics. Due to the Nyquist rate, this spurious product is located at least 1 octave away from the highest-frequency output expected and should be greatly attenuated by the synthesizer's lowpass filter. The mixing effects of this product can be attenuated by the half-octave filters usually present at the input of a receiver. As much as 90 dB of spurious rejection has been observed in narrowband synchronized low-frequency DDSs, while as little as 40 dB of spurious rejection has been reported in wideband VHF-DDS systems.

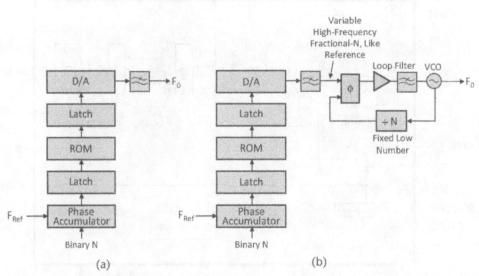

Figure 16.33 The evolution of the DDS into the DDS-driven PLL. (a) The DDS can generate high frequencies with high fractional resolution. (b) A PLL uses a DDS in its reference (DDS-driven PLL) to achieve high resolution and fast switching with a relatively high-frequency reference. The phase-noise performance and the spurious performance of such a synthesizer can be superior to the DDS only synthesizer.

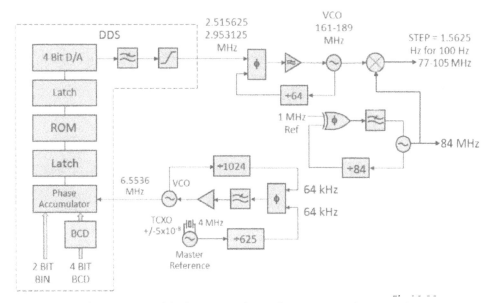

Figure 16.34 Block diagram of the first practical DDS-driven PLL introduction at the 1988 RF Expo Symposium [18] by the author. The DDS provides a high-resolution reference to the PLL, eliminating the need for complex multiple loop or fractional-*N* synthesizers.

Modern DDSs use binary arithmetic inputs, which make for fractional reference frequencies that cannot be easily derived from even frequency reference oscillators (e.g., 4, 5, or 10 MHz). Additional translation, which can take the form of a microprocessor, is used if binary-coded decimal (BCD) inputs are required.

16.19 Foster Seeley and Digital Frequency Discriminators

As previously mentioned, Foster Seeley and ratio detectors have been used in FM demodulators and automatic frequency controls (AFC). A Foster Seeley analog discriminator is shown in Figure 16.35(a). The operation of analog discriminators is usually based on the vector sum of two rectified voltages created across an inductor. One of the voltages is 180° out of phase from the input signal, while the other voltage is only 90° out of phase. The vector combination of the two, after being filtered, can be used in a feedback circuit of other forms of synthesizers, usually through a mixing process. Quick response and limited range are the primary characteristics of this form of synthesis.

An example of a digital frequency discriminator that operates linearly over a wide bandwidth of 2 MHz centered at 4 MHz is shown in Figure 16.35(b). This approach involves a delay line in the form of a static shift register. The analog input signal is conditioned and applied to the 64-bit shift register, as well as to the exclusive-OR detector, and is compared against a fixed 8-MHz stable reference. The output is inversely proportional to the phase difference of the two signals. Because the time delay is fixed, the change in phase imposed by the delay line is directly proportional to the frequency change. Consequently, when phases are compared in the exclusive-OR comparator, the output is inversely proportional to the frequency

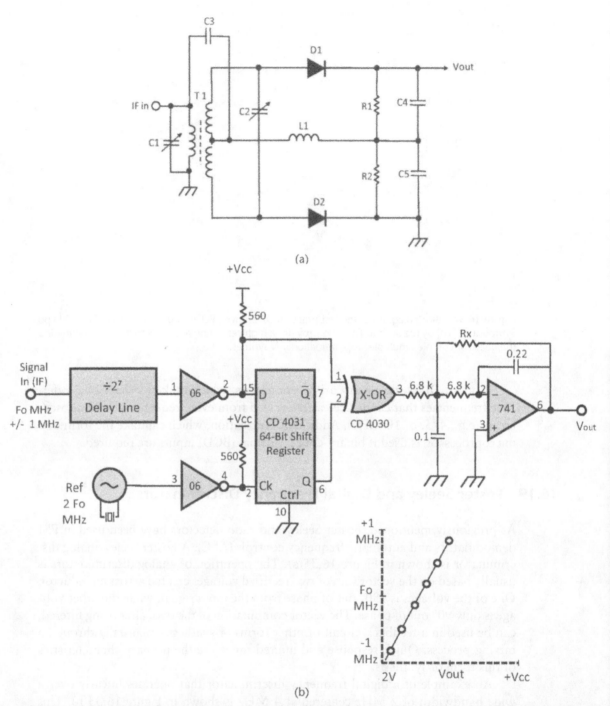

(a)

(b)

Figure 16.35 (a) Foster Seeley discriminator. It works on the principle of the vector summation of two voltages after the recti-fication. A voltage is created across the inductor L1, which will be 180° out of phase from the input voltage. A different voltage is created across the tuned circuit and will be 90° out of phase from the input at the center frequency. It will have a greater angle at higher frequencies and a smaller angle at the lower frequencies. The resulting baseband output will be the difference between the absolute sum of these two voltages. (b) The digital frequency discriminator uses a CMOS logic circuit that produces a 2 to 6 VDC linear output over a 3–5-MHz range. Using a faster logic family can increase the operational frequency. Such a system can be used to compensate for the ambiguous digit phenomena experienced in the digital comparator synthesizer from Figure 16.26(a).

change. For a Vcc of 10 VDC, a linear voltage curve is obtained as shown. The delay is 64/8 MHz or 8 μs.

Judicious use of such circuits can help achieve coherent fine tuning in many forms of synthesizers, as it was mentioned in the digital/counter comparator example from Figure 16.27(a).

16.20   Phase-Locked Loop (PLL) Key Components

A PLL is a control loop. It senses the phase and frequency of the VCO and compares it to a fixed reference frequency. If there is a frequency/phase error, it "tunes" the VCO either up or down to minimize it. It uses a charge pump to achieve this. We will now turn our attention to key elements in the design of phase-locked loops. The basic PLL synthesizer contains five functional blocks.

1. Master reference oscillator/unit (MRU);
2. Phase/frequency detector;
3. DC amplifier/loop filter;
4. Voltage controlled oscillator (VCO);
5. Programmable divider.

We will analyze each of the five blocks and give some practical design examples.

16.20.1   Master Reference Oscillator/Unit (MRU) Technology Classifications: Quartz (TCXO, OCXO, MCXO), SAW, Photonic, Rubidium, and Caesium (Cesium) Hydrogen Maser

The reference oscillator can be thought of as the heart of a synthesizer because it establishes the stability and sometimes the phase noise performance characteristics of the synthesizer output. The reference oscillator is usually called the master reference unit (MRU). It can take the form of a simple quartz crystal oscillator using one of the classical approaches to oscillator design, such as the Hartley, Colpitts, or Clapp. A Colpitts approach is shown in Figure 16.36.

The reference oscillator can take the form of a highly stable, temperature-compensated frequency source that, in turn, could be locked to atomic clocks or slaved to time/frequency standards such as WWVB or GPS.

Depending on the technology used, a practical high-quality master reference unit (MRU) can provide frequency accuracies from 1×10^{-8} to 1×10^{-14} of the nominal reference frequency per year.

Commercially available quartz-based reference oscillators usually operate in the frequency range of 1 to 150 MHz. The 10-MHz and 100-MHz models are probably the most widely used today.

Frequency schemes using phase-locked crystal oscillators (PLXO) are also common. For instance, a 10-MHz oven-controlled quartz crystal oscillator (OCXO) can be used in an integer phase-locked loop to provide frequency stability to a high-quality, 100-MHz phase-locked quartz crystal oscillator (PLXO), which, in turn, can drive a 1-GHz SAW oscillator and a dielectric resonator oscillator (DRO) operating at 10 GHz. The 10-MHz OCXO provides superior stability, while the

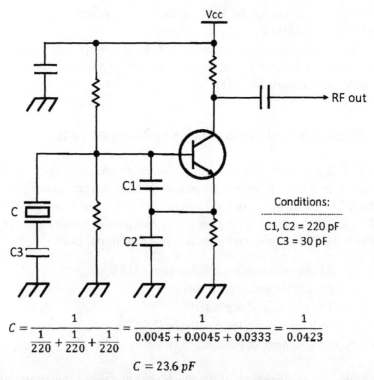

$$C = \cfrac{1}{\cfrac{1}{220} + \cfrac{1}{220} + \cfrac{1}{220}} = \frac{1}{0.0045 + 0.0045 + 0.0333} = \frac{1}{0.0423}$$

$$C = 23.6 \, pF$$

Figure 16.36 A quartz crystal Colpitts oscillator circuit used as a reference oscillator. Its output frequency is offset from the crystal's series resonance point to provide oscillation. Third and fifth overtone crystals can be used with this design. The load capacitance (the capacitance presented to the crystal by the circuit) for the quartz crystal can be determined from the three series capacitors using the formula as shown. This data is needed when ordering crystals from manufacturers. Additional stray capacitance caused by the layout and other parts in the circuit can usually account for an additional 7 to 10 pF to the load capacitance, bringing C to ~31 pF.

100-MHz phase-locked quartz crystal oscillator (PLXO) provides improved phase noise performance at the higher frequency for the following stages. Temperature-compensated crystal oscillators (TCXO) and microprocessor-controlled crystal oscillators (MCXO) are also used. Photonic master reference units using laser interferometry generating high-quality 10-GHz fundamental frequencies directly have also been used in HPOI receivers. In addition, rubidium, cesium, and GPS-disciplined oscillators have been used in very demanding applications such as radio astronomy receivers. A general classification of the most popular master reference sources and some of their key properties are shown in Table 16.2.

The technology choice and the requirements for a master reference unit (MRU) originate with a total receiver system design that takes into consideration all pertaining elements including how much power consumption is allowed for this function, how much volume is available, the type of information or modulation to be received, jitter requirements in the case of phase modulation or radar, its phase noise requirements in view of the minimum discernable signal (MDS), and the total coherency requirements for the synthesizer system.

The phase noise requirements are driven mainly by where the received information fits on the SSB phase noise slopes of the synthesizer (LO) outputs and how they are degraded by various multipliers in the entire system. Furthermore, if used in

Table 16.2   Various Forms of Master Reference Units (MRU) and Their Characteristics

Technology	Frequency (MHz)	Stability (Allan Variance)	Acquisition (Warm-Up) Time	Phase Noise	Aging Per Year	Power Consumption (Watts)
TCXO	1 to 25	1×10^{-8}	1 minute	Fair	5×10^{-7}	0.05
OCXO	10 to 150	1×10^{-12}	2 minutes	Very Good	6×10^{-9}	6 at start, 2 after
MCXO	10 to 150	1×10^{-10}	1.5 minutes	Average	2×10^{-10}	0.5
SAW	150 to 1,200	1×10^{-8}	2 minutes	Good	5×10^{8}	0.5
Rubidium	6,834.682,612	1×10^{-12}	1.5 minutes	Good	NA	20 at start, 8 after
Caesium (Cesium)	9,192.631,770	1×10^{-14}	20 minutes	Good	NA	25 at start, 10 after
Photonic	10,000	1×10^{-11}	30 to 120 minutes depending on performance level	Excellent	NA	20 to 80 depending on performance level

multiple conversions, all LO outputs from a synthesizer will have to have a similar performance to guarantee the receiver's entire performance.

Although many technologies are available today, the quartz oscillator is usually the technology of choice when designing MRUs for synthesizers. TCXO, OCXO, or MCXO MRUs are commonly used.

Quartz as applied to oscillators was first investigated at Bell Labs in the 1920s. Large-scale production of quartz-based radio equipment spanned from 1939 to 1945. The first precision quartz-based clock operating at 2.5 MHz with a stability of 1×10^{-10} per day was demonstrated at Western Electric in 1958. Low power and low drift commercial oscillators were first produced in the 1960s. Large-scale production of watch quartz crystals started in 1970. Today, OEM quartz oscillators for synthesizer applications are produced by many companies.

Quartz technology was previously discussed in detail in Chapter 4. The equivalent circuit for a quartz crystal was shown in Figure 4.31. The same key parameters used in the design of the quartz filter from Chapter 4 can be used to design a quartz oscillator. At the higher frequencies, crystals designed for third and fifth overtone operations are being used.

The TCXO adds temperature compensation circuits to the oscillator circuits to guarantee stability, while the OCXO uses a thermally insulated heater element that brings the quartz crystal temperature to a constant high temperature (usually over 90°C), where outside temperature impacts on the quartz element do not matter anymore. OCXOs have relativelly long warm-up time periods, so they are usually left running continuously to maintain stability. When the requirement is for a very quick warm-up to the required stability, the MCXO or a preamble method is usually used. The MCXO uses a microprocessor and preset data to correct against an anticipated drift curve in order to speed up the warm-up process, including aging.

Master reference units (and quartz crystals in general) come with specific aging guarantees. Aging refers to a cumulative process that lends to a gradual and permanent deterioration of the initial quartz tuning conditions. Because of imperfections and impurities in the process of cutting, polishing, and handling the quartz crystals, a certain amount of drift over time in the resonant frequency of the quartz elements exists. This phenomenon takes place mostly during the first year after a crystal has been cut. Although the phenomenon is not fully understood, an aging frequency shift of 5×10^{-7} per year is typical of overtone crystals. This means that if the MRU shifts in frequency in time, an entire coherent synthesizer can shift in frequency along with it. To prevent against this, MRUs are designed with mechanical and/or electrical setting features that can be periodically adjusted to calibrate the MRU against more stable sources. Typical aging specifications call for a life cycle of 10 to 15 years. In addition, if low aging rates are required, quartz crystals can be preaged through a high-temperature baking process at approximately 105°C for 7 to 15 days before they are implemented in circuits. A typical MRU implementation showing its antiaging mechanical calibration set screw is shown in Figure 16.37.

16.21   Designing a High-Performance MRU for an HPOI Receiver

We will next look at a practical design implementation of a high-performance MRU. It is a phase-locked crystal oscillator (PLXO) for the Star-10 transceiver. In the block diagram in Figure 16.38, the DDS-driven PLL frequency synthesizer used in the Star-10 transceiver uses a reference frequency of 84 MHz. This is required for DDS-1 and 2 and is eight times the highest frequency generated by DDS-1 for good Nyquist sampling and superior spurious performance. The 84-MHz reference doubles as a fixed local oscillator (LO) for the second conversion from 75 MHz to 9 MHz. This reference is, in turn, locked to a 10-MHz OCXO, which is disciplined to WWV signals for good stability (1×8^{-10}).

As previously explained, Star-10 is a *fully coherent* system. This means that all local oscillator frequency sources in the double conversion superheterodyne implementation are locked to a single high stability–high spectral purity source whose

Figure 16.37   A 100-MHz TCXO master reference unit (MRU) has mechanical setting to compensate for aging.

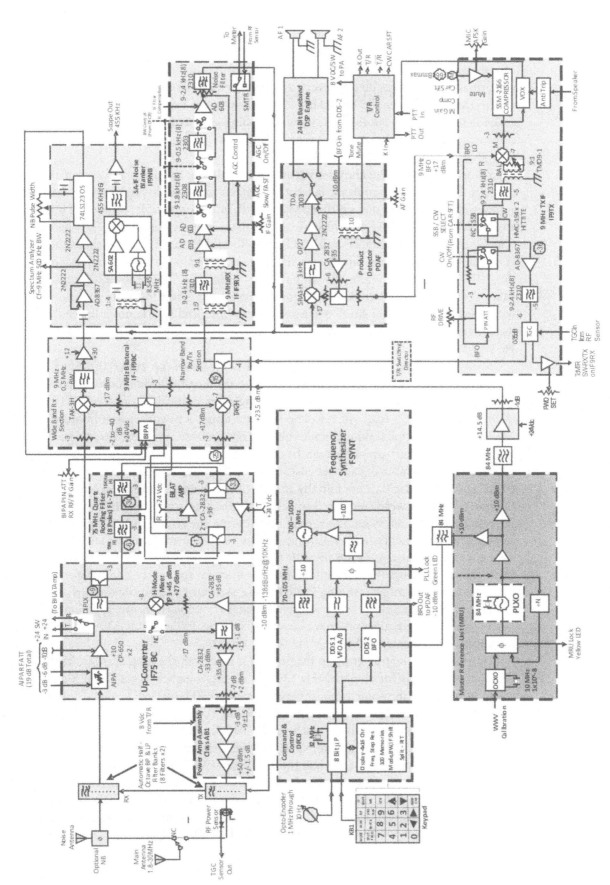

Figure 16.38 Block diagram of the transceiver showing the master reference unit (MRU) assembly.

performance is reflected in the total long-term stability and phase noise behavior of the system and is directly translated into the receiver's performance. A high-performance master oscillator is required to provide reference frequencies for the synthesizer LOs. Thus, the 84-MHz MRU LO serves as both, a high-quality reference for the two DDSs in the FSYNTH and a high-quality fixed LO for the second conversion. A fully coherent synthesizer system with equally good phase noise performance at all mixer LO input ports results.

Before going into the circuit description, calculating the phase noise contributions of all LOs in a complex coherent RF system such as this requires a careful synthesizer analysis to ensure that they are all fully compatible with each other and with the receiver MDS, phase noise, and spurious performance. This means that synthesizer performance at all LO ports should be equally good throughout as phase noise translates directly decibel-per-decibel (minus the mixer loss) into the MDS of the radio within the IF bandpass of interest as previously discussed. Many receiver and transceiver designers do not take this fact into consideration as witnessed by the receiver MDS being sometimes obscured (phase noise limited) by the converted phase noise. The Star-10 MRU schematic diagram is shown in Figure 16.39 [19, 20].

In Figure 16.39, the MRU utilizes an 84-MHz precision-cut (0.001%) fifth overtone quartz crystal (X1) with one side of the crystal grounded in a phase-locked series resonant Colpitts oscillator arrangement comprised of Q1 (2N5179), L2, L3, C3, C4, R2, R3, and R4. The Colpitts approach was chosen because of its well-known circuit stability, while the 0.001% quartz crystal cut was chosen to guarantee initial start-up almost on frequency before locking occurs. C3 and C4 are high Q, silver mica capacitors customary of the Colpitts implementation. The 84-MHz Colpitts oscillator is initially tuned within its narrow resonance range via L2 and L3, which were calculated to resonate the circuit on the fifth overtone of the crystal. Additional tweaking is required to bring the circuit into resonance due to board stray elements, with L3 being the key-tuning element. For 84 MHz, this inductor is air-wound using seven turns of a #20 wire on a 1/4-inch molded plastic form and using a high-Q aluminium core as an initial frequency control element. L2 is wound in a similar fashion. The 84-MHz oscillator is digitally divided down by 84 for a 1-MHz square wave reference signal to be phase compared against a 1-MHz precision frequency signal obtained from the 10-MHz OCXO, which, in turn, is compared against the 10-MHz WWV signal.

Upon turning the power on to the Star-10 transceiver, the loop searches within the first 30 seconds for the 10-MHz OCXO signal, which is forced into a quick warm-up mode at the same time (the oven heats up). The 84-MHz signal follows and searches back and forth quickly at a decreasing frequency of approximately 10 Hz and decreasing to 0 until the oven in the 10-MHz OCXO reaches its internal temperature (over 100°F) and an exact 10-MHz reference signal is obtained and heard by the receiver beating against the 10-MHz WWV signal. A front panel yellow LED reports to the operator the lockup process. At this point (MRU locked), the receiver can be redirected to the frequency of interest via the keypad or via the opto-encoder using the proper digit underscore marker on the main dial. All LOs in the Star-10 are now coherent with the MRU and WWV within 1×8^{-8} and operation can begin. The receiver is guaranteed to be exactly on frequency regardless of where it is tuned within the 2–30-MHz range. There is no drift.

At the circuit level, the high-performance, low phase noise Colpitts quartz oscillator starts up almost on frequency due to its 0.001% precision cut. The clean sine wave generated is further amplified by Q2, a 2N5109 transistor. The signal is then filtered using a similar 84-MHz quartz crystal at X2 and is presented to the divide-by-2 digital divider U1, a UPB1509. This chip has analog-to-digital conditioning circuits that allow for a clean 42-MHz signal to be produced. The 42-MHz analog signal is further conditioned/filtered in the IC for extremely low jitter, only to be filtered again via a 42-MHz tubular quartz crystal filter at X3. The signal is finally presented to U2, a MAX999 low jitter comparator. The threshold of this chip is adjusted via a 10-turn potentiometer, R10.

The MAX999 comparator is billed to guarantee a 4.5-ns propagation delay time at 100 MHz. This implies low jitter performance with rise times in the range of 2 ns or less, but experiments comparing the 84 MHz directly through this device showed a relatively noisy signal. Thus, the divide-by-2 conditioning resulted. Using a 42-MHz signal into the comparator showed superior and stable (low jitter) results. The clean 42-MHz square wave signal at U2 pin 1 is further presented to a digital divider string comprised of U3 and U4, two 74161 chips for a divide-by-42 (7 and 6) function. A clean 1-MHz square wave results from the 84-MHz Colpitts oscillator, which is further presented to U9A, an exclusive-OR phase detector. Further signal conditioning is achieved using U6, a high-speed 54S00 used as sequential gates. The other side of the phase detector is presented with a highly accurate 1-MHz comparison signal derived from the 10-MHz OCXO through another MAX999 comparator and a 50% duty cycle: divide-by-10 IC at U8, a 74LS290 part.

It should be noted that the exclusive-OR phase detector was chosen on purpose because of its narrow $\pi/4$ capture range, which is exactly what is needed to lock a 0.001% deviation quartz crystal over the operating temperature range of interest. Using a wider phase detector would result in more searching and additional jitter translating into inferior phase noise performance. Exclusive-OR phase detectors need 50% duty cycle digital signals, and a locked condition is achieved when the two reference signals are out of phase by 90°. These conditions are fully achieved in this design. The loop correction voltage is obtained at U9A pin 3. The signal is applied through the loop filter comprised of R11, R12, and C19 to the varactor CR1, a BB109. The loop correction signal (2.5-V DC when locked) is fed back to the 84-MHz Colpitts oscillator at the point between C3 and C5, which also serves as the output point of the locked 84-MHz oscillator to be amplified by class A amplifiers at Q3, Q4, Q6, and filtered by X4, X5, over two channels to provide +10-dBm to +13-dBm fixed reference signals to FSYNTH and serve as a second LO drive in IF9BC. Two additional fifth-order tubular narrow bandpass filters help reduce harmonic and spurious content. It was initially feared that the 1-MHz reference square waves would generate multiple markers at every megahertz in the receiver throughout the HF range. Because of the comprehensive filtering and shielding used, this has not been the case. The actual implementation of the MRU is shown in Figure 16.40(a). The measured phase noise performance of the 84-MHz MRU is shown in Figure 16.40(b).

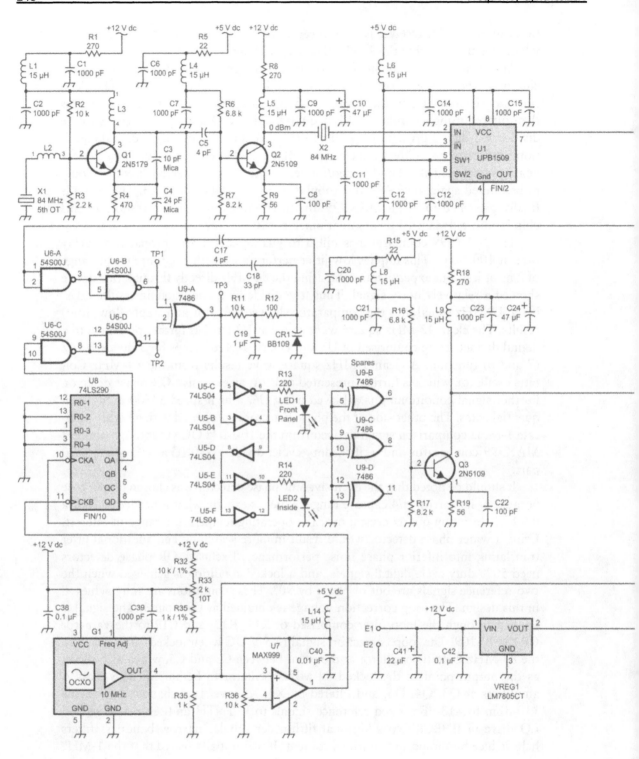

Figure 16.39 Schematic diagram of an 84-MHz MRU using a fifth overtone quartz crystal locked to a 10-MHz OCXO which is disciplined to WWV. Stability after 30 seconds is 1×10^{-8}. (*From:* [21]. Graphics courtesy of ARRL.)

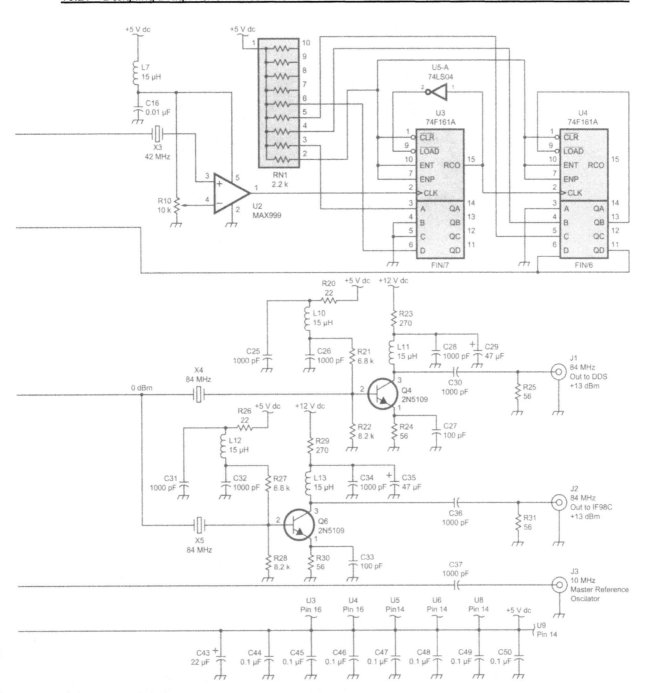

Figure 16.39   (continued)

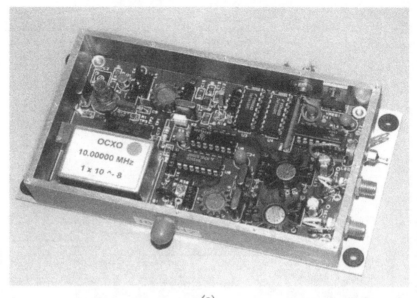

(a)

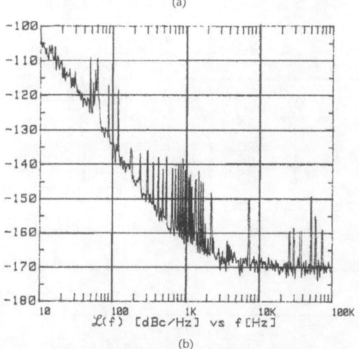

$\mathcal{L}(f)$ [dBc/Hz] vs f [Hz]

(b)

Figure 16.40 (a) Actual board and assembly implementation of the 84-MHz MRU. An 84-MHz PLXO is locked to a 10-MHz OCXO, which, in turn, is disciplined to WWV. (b) Phase noise performance of the 84-MHz MRU (Note that the various spikes are customary artifacts of the measurement process and not spurious signals.)

16.21.1   Photonic Master Reference Unit (MRU)

We have seen that master reference oscillators at up to 150 MHz use high Q, quartz resonators to set the frequency of oscillation. The process is sustained through an active feedback loop and the quality of the LO signal is mainly based on the Q of the quartz. Beyond this frequency, surface acoustic wave (SAW) devices using quartz or other substrates take over at up to approximately 1.5 GHz.

More often than not, quartz oscillators use multipliers to achieve higher MRU frequencies at the cost of degrading the overall phase noise performance of synthesizers.

A new technology has recently evolved from the world of photonics using a fundamentally distinct approach for generating signals. This technology circumvents the inherent limitations of conventional quartz technology and is insensitive to frequency impacts as it is based on laser interferometry. Consequently, MRUs up to 10 GHz have been developed exhibiting phase noise (jitter) performance independent of the frequency. This performance has been found to surpass quartz technology phase noise performance by at least three orders of magnitude. With this technique, a 10-GHz signal with a performance of –163 dBc at 10 kHz from the carrier was demonstrated. It is claimed to be unmatched by any other oscillator [22, 23].

Laser interferometry is not a new science. Ring lasers have been used in ring laser gyros for many years. The new MRU is based on a unidirectional optical ring laser, an optical amplifier to sustain a constant light field, isolators to restrict light from circulation around the ring, lenses, an optical delay line, a photodetector, and an RF bandpass filter tuned to the output frequency. The new method has been awarded the U.S. Patent 5,723,856 [24]. A block diagram of the photonic MRU is shown in Figure 16.41(a). The phase noise performance is shown in Figure 16.41(b).

16.21.2 Hydrogen Maser, Caesium (Cesium), and Rubidium Master Reference Units (MRUs)

Atomic frequency standards are used as MRUs in critical receiver applications. The hydrogen maser is an atomic clock used in radio astronomy receivers, deep space communications, precision tests of gravitation, relativity, and quantum mechanics experiments. The cooled atomic hydrogen maser is the most stable time and frequency source known. Hydrogen masers are active oscillators that operate at 1.420 GHz. The frequency stability of these devices is typically better than 1×10^{-15}.

A caesium (cesium) atomic clock is a device that uses as a reference the exact frequency of the microwave spectral line emitted by atoms of the liquid metal element caesium (cesium). Caesium bears the symbol Cs and the atomic number 55. It is a soft, silvery-gold alkali metal with a melting point of 28°C (83°F).

A caesium oscillator operates at 9.192,631,770 GHz with an accuracy of 1×10^{-14}. It works by exposing caesium atoms to a microwave field until they vibrate at one of their resonant frequencies and then counting the corresponding cycles as a measure of time. The frequency involved is that of the energy absorbed from the incident photons when they excite the outermost electron of the atom to transition from a lower to a higher atomic orbit.

The International System of Measurements decided in 1967 to base the unit of time, the second, on the properties of cesium. Consequently, one second has been defined as counting 9,192,631,770 cycles of the radiation, which corresponds to the transition between two hyperfine energy levels of the ground state of the 133 Cs atom.

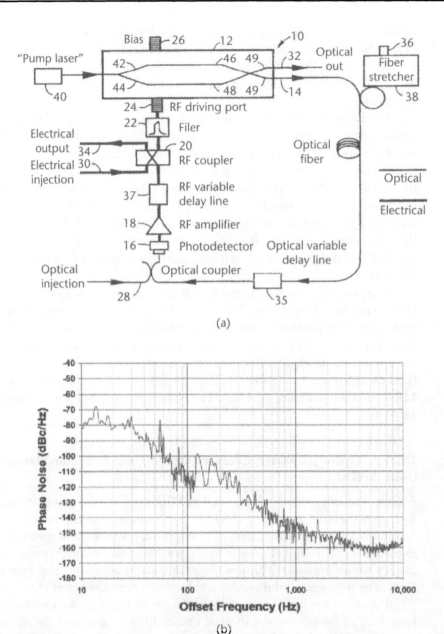

Figure 16.41 (a) Block diagram of the opto-electronic oscillator as it appears in the U.S. Patent 5,723,856 [22]. (b) Phase noise performance of a 10-GHz opto-electronic oscillator. (Courtesy of OEwaves Inc [24].)

Cesium MRUs have been used in conjunction with rubidium MRUs in GPS and other satellite applications.

Commercial rubidium oscillators are the next atomic standard in the arsenal. They operate typically by disciplining a 10-MHz crystal oscillator to the hyperfine transition at 6.834682612 GHz in rubidium. The amount of light from a rubidium discharge lamp that reaches a photodetector through a resonance cell will drop by about 0.1% when the rubidium vapor in the resonance cell is exposed to microwave power near the transition frequency. The 10-MHz crystal oscillator is usually stabilized to the rubidium transition by detecting the light dip while sweeping an RF frequency synthesizer (referenced to the crystal) through the transition frequency. The frequency stability of rubidium oscillators is typically better than 1×10^{-12}. Rubidium MRUs are used in conjunction with cesium MRUs in GPS and other satellite and communications applications.

A commercial rubidium-disciplined quartz crystal oscillator, the PRS10 from Stanford Research Systems, is shown in Figure 16.42.

16.22   Phase Detectors

An important element necessary in the design of a phase-locked loop is the phase/frequency detector. It is considered a key part of the system because it senses errors and outputs proper commands to the VCO in order to lock onto the reference frequency. The phase detector uses the reference frequency from the MRU to compare

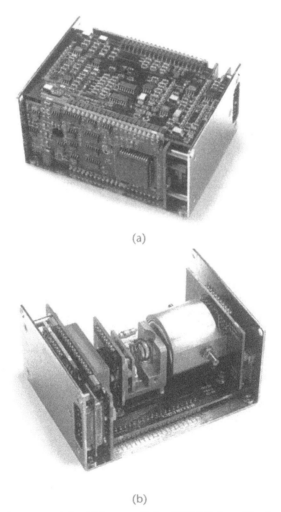

(a)

(b)

Figure 16.42   (a. b) A commercial rubidium oscillator (PRS-10) from Stanford Research Systems. (Courtesy of Stanford Research.)

against the signal from the VCO, as modified by the programmable dividers, and outputs a pumped up or down DC-like signal to the VCO proportional with the correction necessary to bring it to the required frequency. A number of factors influence how quickly a phase detector corrects the phase/frequency error. To better understand how the phase detector works, consider the electromechanical analogy illustrated in Figure 16.43.

As the chopper switches between the two sides of the transformer at the feedback frequency rate, the reference signal appears at the armature alternately at 0° or 180° phase points. The reference signal (F_r) and the feedback signal (F_f) are not equal in frequency; the armature will actually see the sum and the difference of the two signals ($F_R + F_f$) and ($F_R - F_f$), making the circuit act much like a mixer. The lowpass filter will reject the sum ($F_R + F_f$), allowing only the difference ($F_R - F_f$) to pass through with increasing amplitude as the reference frequency is approached by the feedback frequency. When the two are exactly the same, a near-DC voltage is generated as the error signal. The level of the signal is directly proportional to the phase difference (Φ) between the two signals. This value is usually expressed in electrical degrees as shown in

$$\Delta\Phi = \frac{t_2 - t_1}{T} \times 360° \qquad (16.12)$$

where $\Delta\Phi$ is the phase difference in electrical degrees, t_1 is the event time in seconds from the start of one cycle on the reference frequency, t_2 is the corresponding event time expressed in seconds in the feedback signal, and T is the total time required for one cycle (360°) of the reference frequency.

In the example wave form in Figure 16.44, the reference frequency (F_R) has a cycle time (T) equal to 1.8 seconds. The event point (t_1) was chosen to be on the rising edge of (F_R) and is equal to zero seconds. The corresponding edge of the feedback frequency (t_2) is delayed by 0.3 second, as shown.

Then:

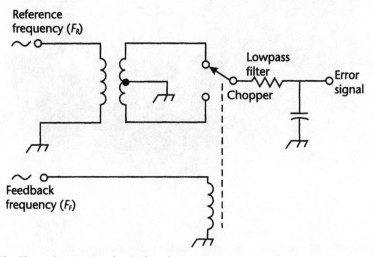

Figure 16.43 Phase detector mechanical analogy.

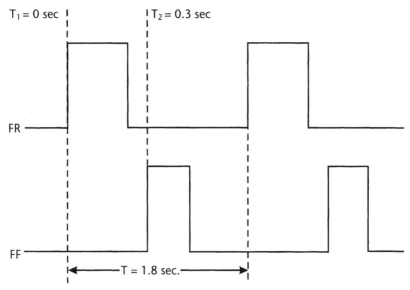

Figure 16.44   Calculating the phase difference (Φ) between two signals, F_r and F_f.

$$\Delta\Phi = \frac{0.3}{1.8} \times (360°) = 60°$$

It can then be said that the reference frequency in this example leads the feedback frequency by 60° or that the feedback frequency lags the reference frequency by the same amount. In a practical phase-locked loop, the error voltage generated by the phase detector is dependent on this value, as shown by

$$V_o = K\Phi \times \Delta\Phi \tag{16.13}$$

where V_o is the average output voltage of the phase detector in volts, $K\Phi$ is the phase detector conversion gain in volts/radian, and $\Delta\Phi$ is the phase difference expressed in radians (1 radian=180°/π = 57.295).

Practical phase detectors may be implemented using various technologies depending on capture requirements. Various phase detector implementations are shown in Figure 16.45. They vary from the simple mixer type to the exclusive-OR approach and the edge triggered D flip-flop type.

The analog mixer phase detector is used where phase noise performance is very important. It achieves the best noise transfer characteristics, but could become unstable if it rolls over its capture range. The exclusive-OR digital phase detector is used for input waveforms with a 50% symmetrical duty cycle. Lock is achieved at a 90° shift from the reference as shown in Figure 16.45, and the capture is only $\pi/4$. Simple, exclusive-OR phase detectors can be used with good results where capture range is relatively narrow, such as in a quartz-based master reference oscillator as discussed in the Star-10 MRU design. This method is sensitive to harmonic multiples of the feedback frequency and also any duty cycle changes on either of the inputs. Consequently, erroneous locking can result.

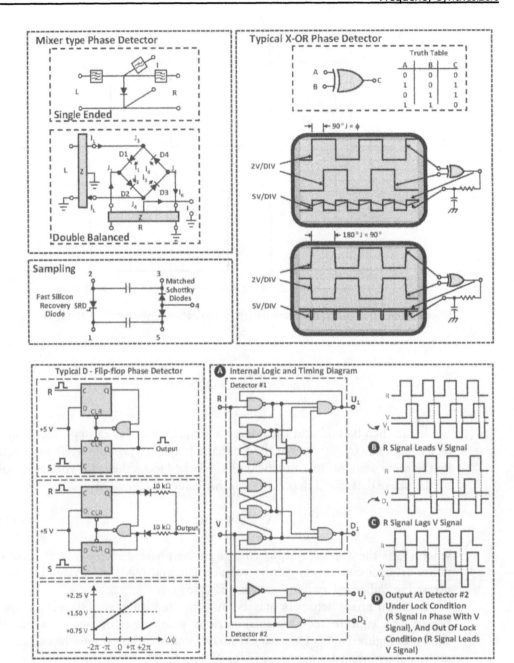

Figure 16.45 Typical phase detectors used in phase-locked loops. The mixer phase detector is used for best phase noise performance. It uses sine wave signals and therefore gets better phase noise performance. Any DC-coupled mixer type will work as a phase detector (even a simple diode). In practice, DB (double-balanced) types are the best because of good isolation and DC offsets. Digital, (Type 1) X-OR, implemented in TTL, ECL technologies is used for narrow capture requirements ($\pi/4$). The lock condition is achieved when two 50% duty cycle input signals are out of phase by 90°. (Type 2) D flip-flop or combinational logic, implemented in TTL or ECL, is used for up to 4π capture range, uses square wave edges for locking, and is duty cycle insensitive.

To avoid these inconveniences, special edge triggered D flip-flop phase detectors in an IC form have been produced by various manufacturers. Edge triggered flip-flop phase detectors are used for wider capture ranges of up to 4π.

Figure 16.45 shows the diagram of such a digital phase detector, which consists of two digital phase detectors, a charge pump, and an amplifier.

In this example, phase detector 1 is intended for systems using negative edge triggered circuitry, where zero frequency and phase difference are required at lock. The total transfer characteristic curve of phase detector 1 (with the charge pump connected) is a sawtooth with a wide linear range of 4 radians as shown. The typical conversion gain ($K\Phi$) for this detector is 0.12 volt/radian.

A second phase detector is usually connected in parallel with phase detector 1 and can be used in a quadrature lock if desired. Its outputs can also be used to drive a lock indicator for phase detector 1. Timing diagrams are also shown in Figure 16.45.

A point to be made in phase detectors choices is that the designer should use the best phase detector for the application at hand. For instance, it would be a mistake to chose one of the more elaborate phase detectors that would provide locking over 4π instead of using the simple X-OR $\pi/4$ phase detector in a phase-locked quartz crystal oscillator (PLXO) because the quartz element guarantees to start almost on frequency and the initial oscillator signal will always be within the capture range of the much simpler, exclusive-OR phase detector. The phase noise performance will also be better with the tighter capture range.

16.23 Amplifier/Loop Filter Trade-Offs

In Figure 16.45, a loop lockup occurs when both outputs U1 and D1 are high. Otherwise, a pulsed waveform will appear on either one of the outputs, depending on the phase-frequency relationship of the R and V inputs.

The U1 and D1 outputs are usually fed to a self-contained charge pump, which is usually an integral part of the PLL integrated circuit.

The pulsed waveform present on either PU (pump up) or PD (pump down) will make the charge pump generate the right voltage in the proper direction for phase differences between R and V inputs of $\pm 2\pi$ radians, as shown in Figure 16.45.

Figure 16.45 shows that a "no-pump" condition is achieved with a phase difference ($\Delta\Phi$) of zero π radians corresponding to an output voltage at UF and a DF of 1.5 volts, which is the lockup condition. It is this output that is used to further steer the VCO in the right direction. Ideally, this signal should be a fast "sliding," pure DC voltage, in order for an inherently good VCO to remain clean from a phase-noise point of view.

In reality, this signal is made of a multitude of AC signals and a DC component. The high-frequency components are generated in the digital part of the phase detector. The charge pump can also add noise components to the output. In order for the phase-locked loop to perform properly, a loop filter is introduced between the output of the phase detector and the VCO's input. This lowpass arrangement has the role of suppressing AC components from the output of the phase detector while still maintaining the response requirement for the entire phase-locked loop. This contradictory requirement usually calls for a design having a loop response at about 1% of the reference frequency.

For example, if a 10-kHz reference was considered, the 3-dB cutoff point of the loop filter would be 100 Hz, meaning that the noise performance of the VCO can be improved only between 0 to approximately 100 Hz, leaving all other AC components to pass through and appear as sideband noise in the output of the VCO.

As previously mentioned, one of the worst problems encountered in such a system is the feedthrough of the reference frequency and its harmonics, which appear as angle modulation sidebands on both sides of the VCO's output. These sidebands, if not filtered, are about 40 dB down from the fundamental VCO output and are located on either side of the VCO's output by the amount of the reference frequency and its harmonics. They can contribute to the phase noise performance and greatly hamper the performance of a receiver.

The twin-T notch network, shown in Figure 16.46, will provide an additional 20-dB attenuation for the reference frequency or its harmonics, if introduced in the output of the phase detector. This filter is sometimes used to reject both the reference frequency and its second harmonic. The example in Figure 16.46 shows two twin-T networks tuned to the 10-kHz reference frequency and its second harmonic at 20 kHz.

The loop filter is also responsible for the phase-locked loop's capture range, bandwidth, lockup time, and transient response. This filter is usually of the low-pass active design, as shown in Figure 16.47. Its cutoff frequency is expressed by (16.14).

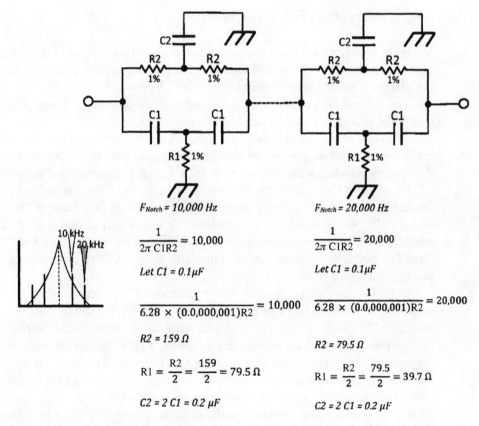

$F_{Notch} = 10,000 \text{ Hz}$

$$\frac{1}{2\pi \, C1R2} = 10,000$$

$Let \; C1 = 0.1\mu F$

$$\frac{1}{6.28 \times (0.0,000,001)R2} = 10,000$$

$R2 = 159 \, \Omega$

$$R1 = \frac{R2}{2} = \frac{159}{2} = 79.5 \, \Omega$$

$C2 = 2 \; C1 = 0.2 \; \mu F$

$F_{Notch} = 20,000 \text{ Hz}$

$$\frac{1}{2\pi \, C1R2} = 20,000$$

$Let \; C1 = 0.1\mu F$

$$\frac{1}{6.28 \times (0.0,000,001)R2} = 20,000$$

$R2 = 79.5 \, \Omega$

$$R1 = \frac{R2}{2} = \frac{79.5}{2} = 39.7 \, \Omega$$

$C2 = 2 \; C1 = 0.2 \; \mu F$

Figure 16.46 The T-notch filter network can be introduced after the loop filter of a PLL to reduce reference feedthrough. In this example, two T networks are used with 10-kHz and 20-kHz references to reduce this effect.

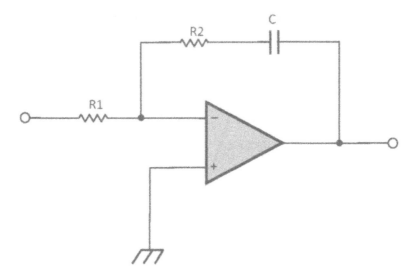

Figure 16.47 A typical inverting lowpass filter using an operational amplifier can be used for the loop filter in a PLL.

$$\omega LPF_{(RAD/s)} = \frac{1}{R1C} \qquad (16.14)$$

where ωLPF is the cutoff frequency for a lowpass filter.

The loop natural frequency and the damping factor can be found from (16.16).

$$\omega n_{(RAD/s)} = \left(K_\Phi K_0 \omega_{LPF}\right)^{1/2} \qquad (16.15)$$

$$\zeta = \left(\frac{R2C}{2}\right)\omega n \qquad (16.16)$$

where ωn is the loop natural frequency, $K\Phi$ is the phase detector conversion gain in volts/radians, K_0 is the VCO conversion gain in radians/seconds/volts, and ζ is the damping factor.

The total transient response of the phase-locked loop is controlled by this filter. Upon a change in command (÷N), the loop will search until the VCO travels from one frequency, F1, to the other frequency, F2. This shifting is not encountered suddenly as in a DDS, for instance, but when frequency F2 is reached, the output of the VCO will oscillate around its value until finally settling down, as shown in Figure 16.48.

The time that has passed between the given command and the output of the VCO to settle within some certain limits (usually within 10% of F2) determines the switching speed of the phase-locked loop and is usually referred to as the lockup time.

The amount of damping (ζ) determines how fast this process is completed. In designing a fast-switching synthesizer for an HPOI receiver, Figure 16.49 should be

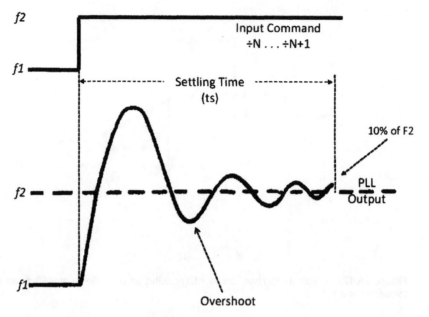

Figure 16.48 Transient response of a phase-locked loop.

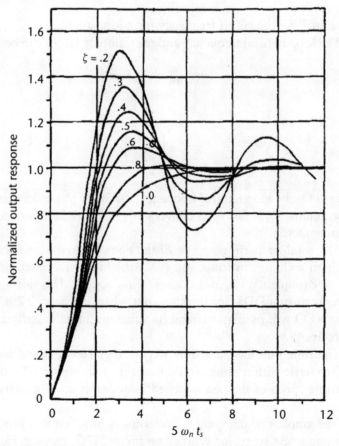

Figure 16.49 Normalized transient response for a second-order system.

used for determining the value ω_n for a given damping factor ζ in order to apply it in the lowpass equations presented earlier. A rule of thumb in using this graph would be to design for a certain amount of overshoot within a given settling time.

Another factor controlled by the loop filter is the lock-in range (ω_c). With the filter shown in Figure 16.47, (16.17) can be used to find this value:

$$\omega_c = \omega_1 \left(\frac{R2}{R1} \right) \qquad (16.17)$$

where ω_c is the lock-in range in radians/second and ω_1 is the hold-in range in radians/second (this is defined as to how far the input frequency to the phase detector can deviate from a given VCO frequency ω_o; this number is numerically one-half the lock range).

An example of a modern digital phase detector is the HMC439QS16G from Hittite Microwave. It operates with reference inputs from 10 MHz to 1.3 GHz and has an ultralow SSB phase noise floor of −153 dBc/Hz at a 10-kHz offset. Its application, in conjunction with a very quiet second-order loop filter, is presented in a 12.8-GHz PLL example shown in Figure 16.50.

16.24   Voltage Controlled Oscillator (VCO)

The next circuit required in a PLL is the *voltage controlled oscillator* (VCO). This circuit takes the voltage commands from the loop filter and outputs the required sine wave frequency necessary for the receiver's conversion. This signal is also used to feed the programmable counters, which, in turn, close the loop at the second input to the phase detector, as previously discussed.

Ideally, the VCO supplies an output frequency directly proportional to its input DC voltage control over the required temperature range. A VCO model is shown in Figure 16.51. Its transfer function characteristics can be expressed by

$$\omega_o = K_o V_F \qquad (16.18)$$

where ω_o is the VCO output in radians/second, V_f is the VCO control voltage provided by the loop filter, and K_o is the VCO conversion gain in radians/seconds/volts.

The output frequency transfer function of a VCO expressed in hertz is given by

$$F_o = \frac{\omega_o}{2\pi} \qquad (16.19)$$

where F_o is the VCO frequency in hertz, ω_o is the VCO frequency in radians/second, and $\pi = 3.14$.

From a practical standpoint, the VCO resembles VFO technologies and follows typical oscillator topologies, with the tuning being achieved through the use of voltage-variable capacitors (varactors) in the oscillator tank circuit. Low-noise VCOs can be designed today by using field-effect transistors at frequencies of up to

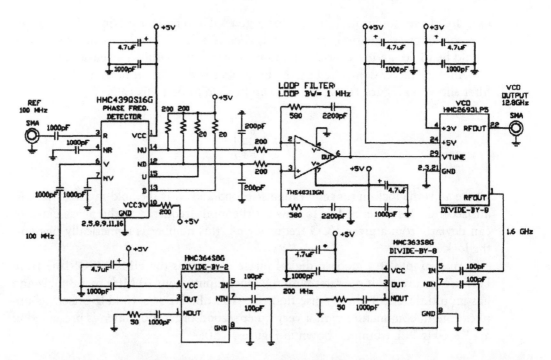

Figure 16.50 Application of a modern 1.3-GHz digital phase detector in the 12.8-GHz integer PLL using a 100-MHz reference. The second-order loop filter is implemented with a low noise operational amplifier. (Courtesy of Hittite Microwave Corporation © 2000–2008 Hittite Microwave Corporation, All rights reserved.)

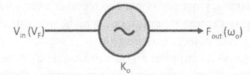

Figure 16.51 Basic model of a VCO.

500 MHz. A practical VHF VCO design for frequencies of 110 MHz through 165 MHz is shown in Figure 16.52.

Figure 16.53 shows the linearity test results of the design over the temperature range of −50°C to +50°C. This performance was measured in a free-running mode using an input DC voltage supplied by a programmable power supply that can generate precise DC voltages.

For the higher-frequencies VCOs, bipolar transistors and GaAs FETs are being used. VCOs using cavity, ceramic, and coaxial resonators are commercially available from the low VHF frequencies to around 50 GHz, from a multitude of manufactures, making the job of a synthesizer designer much easier than in the past. Equally available are dielectric resonator oscillators (DROs) and yttrium iron garnet (YIG) oscillators.

A simplified schematic of a typical microwave DRO is shown in Figure 16.54. A DRO is a mechanically and/or electrically tunable fundamental oscillator that can be designed from 1 GHz to 100 GHz (typically up to 50 GHz). It includes a

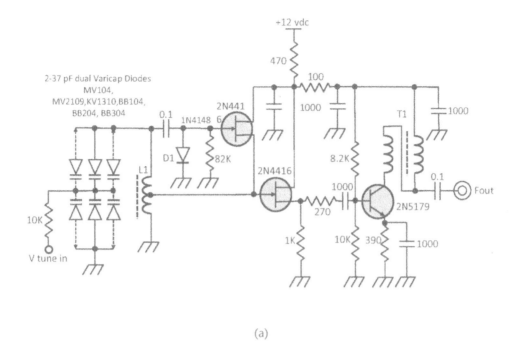

(a)

(b)

Figure 16.52   (a) Schematic diagram and actual prototype implementation of an actual VCO design using FETs. Frequency range is 110–165 MHz. Tuning range can be changed or extended using a different inductor and/or varactors. (b) Actual implementation of the prototype VCO.

custom-designed VCO, a monolithic microwave integrated circuit (MMIC), a dielectric resonator disk ("puck"), and two varactor coupling circuits, all laid out on a thick alumina substrate. The dielectric resonator disk (ceramic in nature) varies in diameter depending on the tuning frequency. The oscillator is mounted in a cavity metal housing. Frequency can be adjusted (or locked) typically over +/– 20 MHz.

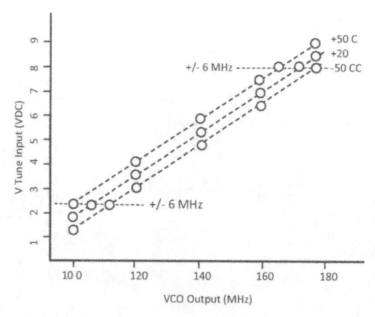

Figure 16.53   Linearity of the VHF VCO over the temperature range of −50°C to +50°C.

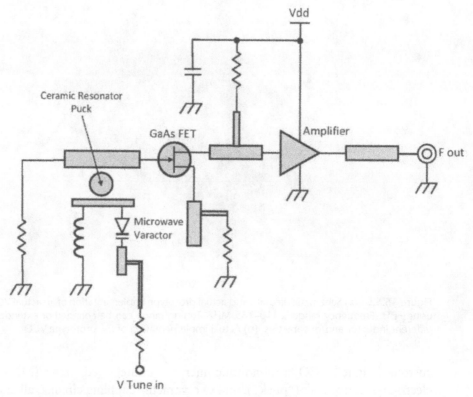

Figure 16.54   A simplified dielectric resonator VCO circuit.

The VCO/MMIC incorporates a negative-resistance oscillator amplifier along with a buffer amplifier. The resonator disk is coupled to a microstrip transmission line connected to the negative-resistance port of the VCO/MMIC. The two varactor coupling circuits include microstrip lines, laid out orthogonally to each other, for coupling with the resonator disk.

An example of a commercial DRO implementation is shown in Figure 16.55. It is the HMC-C200 from Hittite. It incorporates low phase noise technology and provides –122 dBc/Hz SSB phase noise at a 10-kHz offset at 8 GHz. It has an internal output buffer that provides 14.5 dBm of RF power. Internal temperature compensation allows this DRO to operate over a temperature range of –40°C to +85°C with a frequency drift of only 2 ppm/°C. The V tune port accepts an analog tuning voltage from +7V to +12V and provides a range of ±1 MHz from the center frequency.

(a)

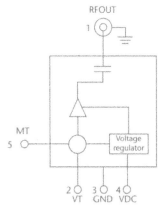

(b)

Figure 16.55   (a) A ceramic resonator VCO by Z-Comm. (Courtesy of Z-comm Corporation.) (b) A high-performance dielectric resonator oscillator from Hittite corporation, the HMC-C200. It operates at 8 GHz with a phase noise performance of –122 dBc/Hz at a 10-kHz offset. The DRO is tunable over +/– 1 MHz from the center frequency. The output is 14.5 dBm. (Courtesy of Hittite Microwave Corporation © 2000–2008 Hittite Microwave Corporation, All rights reserved.)

The YIG is yet another form of variable oscillator. YIG stands for yttrium iron garnet, a ferrite material that has a unique high Q frequency resonance property when exposed to a magnetic field. The resonator is a small sphere made from this composition, which is about 1 mm in diameter, and is placed between an electromagnet poles. Frequency tuning is possible by adjusting the current in the electromagnet, with the resonant frequency being directly proportional to the magnetic field strength or current applied. Changes in the permeability as a function of the magnetic field are the frequency-determined elements.

Often the frequency range requirement over which a VCO has to perform is very wide, with the result of it taking a long time for the loop to lock. This search would also be accompanied by extensive phase noise even after the loop has settled. This noise would be transferred into the receiver's circuits, accounting for serious intermodulation distortion problems.

Some designers use several limited-range VCOs that are automatically switched in from the synthesizer commands by complex digital circuitry. A block diagram example of this approach is shown in Figure 16.56.

Coarse steering of VCOs using D/A converters to bias varactors have also been used extensively by the industry. A less expensive method of achieving coarse steering and low-phase noise performance in a VCO is shown in Figure 16.57. In this example, limited frequency ranges are switched into the tank circuit of the VCO by placing the selected inductances in series with the main inductor. The switching is accomplished under microcomputer control with the use of PIN diodes utilized as current-controlled switches or by using microelectromechanical systems (MEMS). Similar results can be obtained by switching capacitors instead of inductors. Either technique can be implemented by building a strip line or microstrip printed-circuit board, which will contain a bank of tuned circuits together with their switching

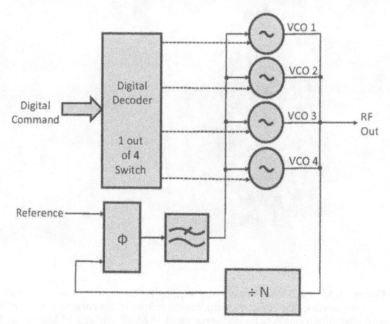

Figure 16.56 Switching VCOs in a PLL is a method of improving on the overall phase noise performance of a loop.

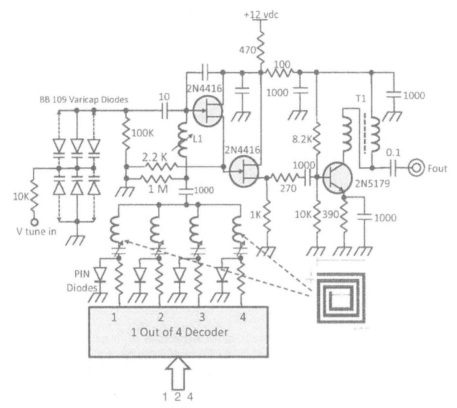

Figure 16.57 Extending VCO range by selecting additional inductance into the tank circuit of a VCO.

mechanisms, which could also include the digital decoder. The inductors could be etched directly on the board providing an economical means of production.

16.25 Modeling Phase Delays in Phase-Locked Loops

Many cookbook recipes deal with the design of phase-locked loops in the frequency domain. However, not many take into consideration the time-domain analysis or phase delays in a system. We will next look at this aspect of the PLL design.

We have seen that a PLL is a feedback system in which the VCO attempts to coherently follow the reference in frequency and phase.

However, due to phase detector acquisition delays, especially in mixer type phase detectors, the VCO phase can lead or lag the reference signal phase to the point that the loop can be out of the pull-in or capture range, causing false locking or not locking at all.

Although ways can be devised to steer the VCO until lock occurs, the problem remains a real one in many systems. This phenomenon is also known as transportation lag in PLLs [25, 26]. Consequently, a seemingly properly designed PLL becomes erratic or loses lock altogether. This problem can be difficult to troubleshoot.

We will next model a simple tracking type loop to better understand it. A simple tracking loop block diagram is shown in Figure 16.58. Let the reference input signal (A_R) be:

$$A_R \sin\left(\omega_R t + \Phi_i\right) \tag{16.20}$$

Let the VCO output signal (A_o) be:

$$A_o \sin\left(\omega_R t + \Phi_o\right) \tag{16.21}$$

where Φ_i is the input reference phase, Φ_o is the VCO phase referenced to the input reference frequency, and ω_R is the reference frequency in radians per second.

To find the phase relationship of the VCO (Φ_o) to the reference, we assume that ω_R is the nominal center of the phase detector correction. Then:

$$\Delta\omega = K_0 v_0 = \frac{d\phi_o}{d_t} \tag{16.22}$$

where K_o is the VCO tuning constant expressed in units of radians per second per volt, V_o is the control voltage applied to the VCO, and Φ_o is the VCO phase as referenced to the input reference phase.

To find Φ_o, we integrate using:

$$\Phi_0 \int_{-\infty}^{t} K_o v_o dt \tag{16.23}$$

or we can use the Laplace transform of (16.23):

$$\Theta_{o(s)} = \frac{K_o V_{o(s)}}{s} \tag{16.24}$$

From Figure 16.58, we use the phase detector model $K_d(\Theta_i - \Theta_o)$, where K_d is a constant expressed in volts/radian. Then:

$$K_d\left(\Theta_i - \Theta_o\right) \cdot F_s \cdot K_o/s = \Theta_o \tag{16.25}$$

The linear phase transfer function of the loop is then:

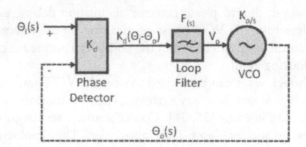

Figure 16.58 Model of a simple tracking type loop. The VCO becomes the phase error integrator for the entire loop.

$$\frac{\Theta_o}{\Theta_i} = \frac{K_o K_d F_{(s)/s}}{1 + K_o K_d F_{(s)/s}} = \frac{K_o K_d F_{(s)}}{s + K_o K_d F_{(s)}} \qquad (16.26)$$

We have seen that an important parameter of the dynamic behavior of a PLL is its pull-in range. This is usually affected by the transportation lag phenomenon inherently present in the loop. A model of a simple tracking loop has been used to investigate such effect on the tuning characteristic of the VCO. Our simple example looks at the loop filter transfer function and the effect of the transportation lag on the pull-in range. Consequently, the noise from the phase detector plus the noise from the loop filter will modulate the VCO output phase and can add to the VCO noise. Additional dividers in the loop will magnify this effect by the 20logN multiplying noise formula. By manipulating these elements, it is possible for a PLL to improve upon the phase noise performance of a VCO inside the loop filter bandwidth.

16.26   Designing a DDS-Driven PLL Synthesizer for the Upconvert, Double Conversion HPOI Receiver

We will now discuss a practical high-performance coherent DDS-driven PLL synthesizer. It uses a high-frequency DDS reference in conjunction with a microwave PLL as divided down for a better phase noise performance. It was developed for the HPOI transceiver that was previously discussed and achieves a high resolution due to the fractional nature of the direct digital synthesizer (DDS) used as a variable reference. It combines the merits of both the DDS and the PLL in a single design.

This synthesizer utilizes the 84-MHz MRU reference previously discussed. This high-quality signal is used to clock a DDS, which, in turn, is used as a high-resolution variable reference for a PLL in order to generate the first LO in the transceiver. As mentioned earlier, the MRU provides eight times the Nyquist requirement (only twice Nyquist reference is minimally required) to keep spurious problems out of the DDS output.

In this design, the PLL is used in a fixed integer divide-by-100 approach, as there is no need for the fractional N approach anymore. The DDS fulfills the high-resolution fractional requirements. The DDS output is consequently a fine-stepped, high-frequency signal in order to minimize reference feedthrough and help realize a wider loop filter in the PLL for faster switching.

In the block diagram from Figure 16.59, the first LO to the double conversion receiver has to be capable of producing a range of 77 MHz to 105 MHz continuously with a resolution of 10 Hz. It has to be able to switch frequencies in less than 10 ms and output a clean signal compatible with the receiver's MDS. In order to do this, the loop functions at 10 times the higher frequency than the required range, while the DDS operates in a fractional mode from 7.7 MHz to 10.5 MHz for an ultimate PLL resolution of 10 Hz over the 4-octave general coverage of the receiver. Consequently, the synthesizer operates at a range of 770–1,050 MHz in 1-Hz increments, and the output frequency is divided down by 10 using a high-performance static digital divider for better phase noise performance.

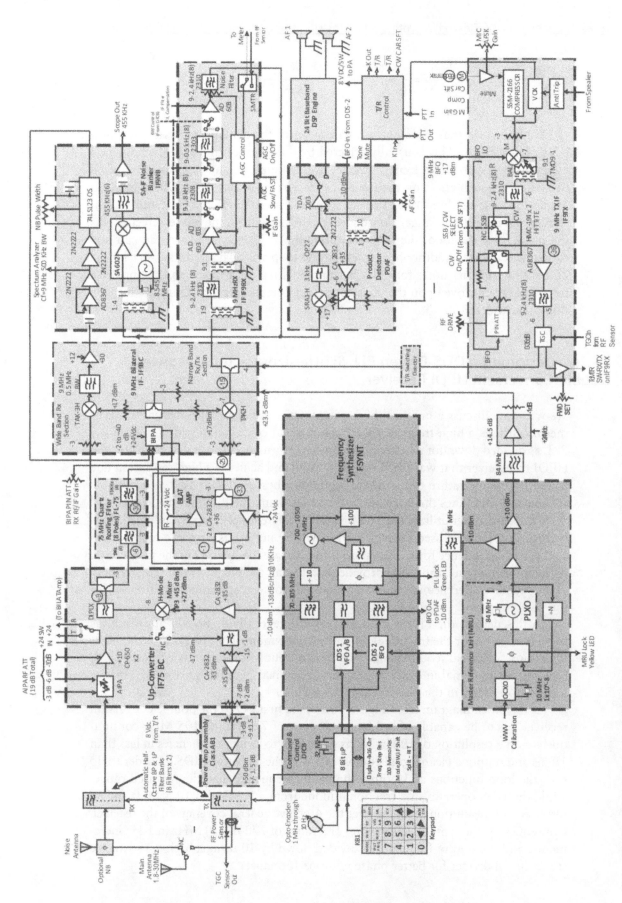

Figure 16.59 A block diagram of the HPOI transceiver showing the DDS-driven PLL section and other significant areas associated with it.

An interesting problem posed by this design was that theoretically the same phase noise should result from using a nonmultiplied DDS-PLL approach, operating from 77 MHz to 105 MHz. However, the phase noise performance of the multiplied DDS-PLL was better by 6 dB.

After calculations and measurements, it was decided that this improvement is because of the increased bandwidth of the loop at the higher frequencies, which provide an edge over the lower frequencies, and because the Q of the higher-frequency VCO is better than at the lower frequencies (percentage bandwidth advantage). This also justifies using a single VCO instead of several VCOs simplifying the design. If we assume that the phase noise performance of the higher-frequency VCO is only 1.58 times better than the lower-frequency VCO, this would account for the better performance of the microwave approach. Thus, we concluded that the phase noise performance improvement for using the microwave version DDS-driven PLL is due to a decreased percentage bandwidth at the multiplied frequency and/or by a 1.58 times improvement in the phase noise performance of the higher-frequency VCO. Both factors play a role in the phase noise performance of the synthesizer.

To maintain the same phase noise performance for the second IF of the receiver, the 84-MHz frequency from the MRU is being used as a fixed LO for the second conversion.

A second DDS in the synthesizer serves as the beat frequency oscillator (BFO) operating as slightly offset from 9 MHz to allow for various modes of operation. It uses the same 84-MHz reference filtered through a 9-MHz monolithic quartz filter for maintaining the same phase noise as the previous LOs. The entire system is coherent with the 84-MHz PLXO and therefore the 10-MHz WWV disciplined OCXO.

A schematic diagram of the DDS-driven PLL synthesizer is shown in Figure 16.60. Both DDSs are the AD-9850 from Analog Devices. The word clock information (WORD_CLK_1/2) and data (DATA_1,2) are communicated serially to the DDS 1 and DDS 2 from the command and control board via J1.

The data is validated via the FQ_UD lines. With more than 10,000 lines of code programmed in it, the command and control system manipulates the transceiver's over 2.8 million synthesized frequencies within 1×10^{-8} as impacted by various modes of operation in less than 10 milliseconds.

We will now discuss the DDS-driven PLL synthesizer design. In Figure 16.60, the output of a high-resolution DDS source (U1) goes through a lowpass filter of elliptic design. This filter is intended to reject any out-of-band spurious problems that may still exist despite the careful reference choice. The variable reference signal is further conditioned by Q1 and Q2 and U3 A and B to produce a fast-rise square wave signal to be presented to the reference input of the phase detector PLL, U4.

In this design, the divide-by-N function is done through the two divide-by-10 devices which have been cascaded in the loop for a total division of 100. A clean 77–105-MHz LO range is derived after the first divider. The synthesizer step resolution is therefore 10 Hz (for 1 Hz at the DDS) over the entire range.

In Figure 16.60, there are two things important to note. First, in implementing the DDS-driven PLL synthesizer, the variable reference produced by the DDS and presented to the PLL implies the use of a phase detector capable of accepting such reference frequencies. Many PLL chips have a minimum acceptable reference

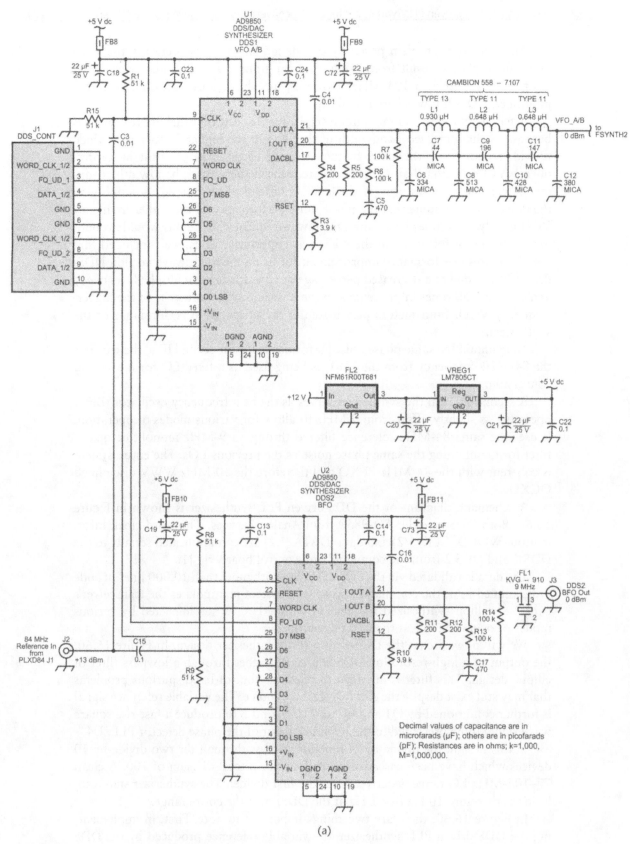

Figure 16.60 (a) Schematic diagram of the DDS-driven PLL. (*From:* [20]. Graphics courtesy of ARRL.) (b) Actual implementation of the DDS-driven PLL assembly.

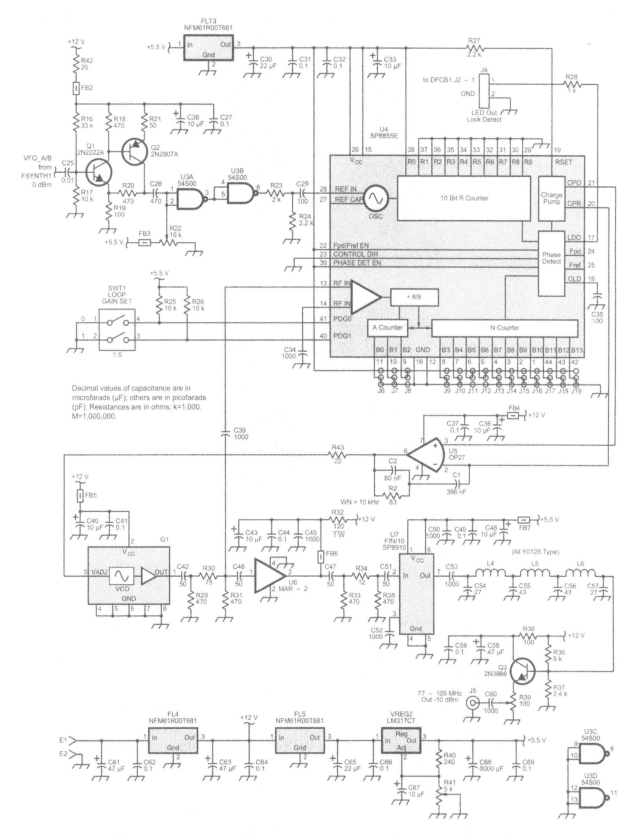

Figure 16.60 (continued)

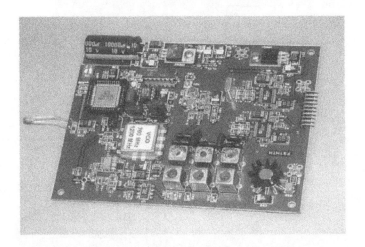

(b)

Figure 16.60   (continued)

frequency of around 10 MHz. This is the case in our design, and conditioning (squaring up the reference) is required as shown.

Second, most PLL chips on the market today contain a built-in reference programmable divider (called the R divider) that is intended to make it easy for the designer to tailor specific division ratios using the same reference. This programmability usually allows for division by binary numbers such as 2, 4, 8, 16, and so on. However, few PLL devices available on the market today allow for division by 1, which is what is needed in this DDS-driven PLL synthesizer. This is one of the reasons that dictated the choice of the PLL chip used. In addition, this chip allowed for parallel hard-wired programmability of the divide-by-N chain, which simplifies design.

The conditioned variable reference frequency is presented to one side of the phase detector. The charge pump output is presented to a third-order loop filter which was chosen over the more common second order type because the reference sidebands are reduced. The calculations for the loop filter elements are shown in

Figure 16.61. The loop bandwidth has been increased from an initial 5 kHz to about 20 kHz in the final design to reduce the phase noise "hump."

The VCO used covers a wide range of 700 MHz to 1,200 MHz. A fixed divide-by-$N = 100$ in the PLL chip completes the loop divisor to the second phase detector input.

An example of the math occurring in the entire system is shown in Figure 16.62. This simple example shows what is going on in the system for a received frequency of 14.24011 MHz. A DDS word is calculated in a separate example that is also part of the figure.

The command and control interface assembly schematic is shown in Figure 16.63.

$$f_D = \frac{(fLO - fIF)}{10}$$

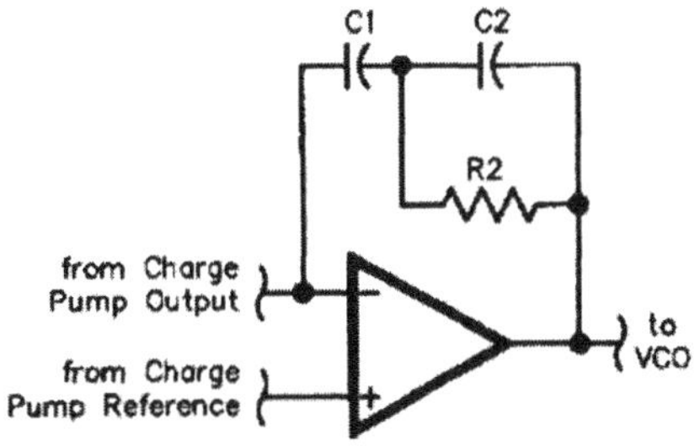

Third - Order loop filter with

$f = 5\text{ kHz};\ N = 100;$

$k_0 = 2\pi(10\text{ MHz}) = 62831853.1$

$\Phi = \dfrac{6.3\text{ mA}}{2\pi} = 1.0027\text{ mA/radian} = 0.001\text{ A/radian}$

$\omega_n = 2\pi f = 31415.927\text{ Hz} = 31.416\text{ kHz}$

$\Phi_0 = 45°$

$T_3 = \dfrac{-\tan\Phi_0 + \dfrac{1}{\cos\Phi_0}}{\omega_n} = 1.31848 \times 10^{-5}$

$T_2 = \dfrac{1}{\omega_n^2 T_3} = 7.6847 \times 10^{-5}$

$T_1 = \dfrac{k_0 k_\Phi}{N\omega_n^2}\left[\dfrac{1 + \omega_n^2 T_2^2}{1 + \omega_n^2 T_3^2}\right]^{\frac{1}{2}} = 1.54105 \times 10^{-6}$

$C1 = T_1 = 1.54105\ \mu\text{F}$

$\dfrac{T_2}{R2} - C1 = \dfrac{T_3}{R2}\ ;\ \dfrac{(T_2 - T_3)}{R2} = C1\ ;\ \dfrac{(T_2 - T_3)}{C1} = R2$

$R2 = 41.311\ \Omega$

$C2 = \dfrac{T_3}{R2} = 3.19162 \times 10^{-7} = 319.162\text{ nF}$

Figure 16.61   Calculations for the third-order loop filter used in the DDS-driven PLL synthesizer.

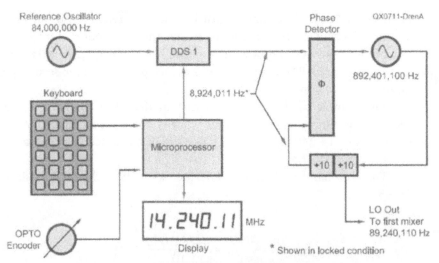

For a f_{IF} of 75,000,000 Hz (75 MHz) and a local oscillator output of 89,240,110 Hz, the control software will calculate the operating frequency that will be displayed.

$$f_D = \frac{(89,240,110 - 75,000,000)}{10} = 14,240,110$$

$$f_n = \left(\frac{2^{32}}{100 \times 10^6}\right) \cdot \left(\frac{75 \times 10^6 + \text{Display}}{10}\right)$$

$$f_n = \left(\frac{2^{32}}{100 \times 10^7}\right) \cdot 75 \times 10^6 + \left(\frac{2^{32}}{100 \times 10^7}\right) \cdot \text{Display}$$

where f_n is the tuning value.
Then:

$$f_n = 322122547.2 + 4.294967269 \cdot \text{Display}$$

Example: For a display value of 29.999990×10^6 yields:

$$f_n = 450,971,523$$

$$DDS_{out} = \frac{f\,\text{Desired} + fIF}{10}$$

and

$$DDS_{out} = \frac{DDSWord \cdot fREF}{2^{32}}$$

and

$$f_{kbd} = \frac{fDesired}{10}$$

$$\frac{DDSword \cdot f\,Ref}{2^{32}} = fkdb + \frac{f\,if}{10}$$

$$DDS_{word} = \frac{2^{32}}{f\,Ref} \cdot f\,kbd + \frac{2^{32}}{f\,Ref} \cdot \frac{f\,if}{10}$$

Let:
$$f_{ref} = 84 \times 10^6$$
$$f\,if = 75 \times 10^6$$

Then:

$$DDS_{word} = 51.13056304762 \times f_{kdb} + 383479222.8571$$

Figure 16.62 A mathematical example for calculating the local oscillator output of 89,240,110 Hz for an f_{IF} of 75 MHz. The control software calculates the DDS word and the actual receiver operating frequency to the display. (*From:* [20]. Graphics courtesy of ARRL.)

16.27   Performance of the DDS-Driven PLL

The phase noise performance of the microwave DDS-driven PLL is shown in Figure 16.64. This performance is true at any frequency within the range. The phase noise performance of the synthesizer is −138 dBc/Hz at a 10-kHz offset. As mentioned before, this is significant because it is harder to design good performance in high-resolution synthesizers when compared with brute-force synthesizers. The spurious performance of this synthesizer is better than −95 dBc, anywhere from 77 MHz to 105 MHz. The total sweep and lockup time is less than 10 milliseconds from one end of the 280-MHz range to the other, allowing for an unconditional split operation over a 28-MHz delta frequency at HF. The ultimate step resolution is 10 Hz anywhere in the entire range. There are 2.8 million available steps.

16.28   The Opto-Encoder and Its Application

In order to duplicate the tuning feel of the VFO, a synthesized receiver usually uses a digital optical encoder. It has been seen that a synthesizer responds to computer-like encoded information to step from one frequency to another. To input this information, it is easiest to use a detented thumb-wheel encoder or a keypad as shown in Figure 16.63. While such methods are fully adequate, fine-tuning a receiver with this method is rather impractical from an operator standpoint. A knob-like dial that will provide a feel similar to the familiar old VFO is desirable. This function is achieved with the help of the opto-encoder.

The opto-encoder is usually a film or metal disk with holes in certain locations. An infrared beam shines through the holes. Two photocells detect when the light is interrupted, as well as the direction of rotation (by means of who's on first). Information in the form of pulses is presented to a chain of up/down counters (or a microprocessor fulfilling this function) that are commanded to count in the direction of movement of the disk. This information is then presented to the synthesizer's dividers or the DDS data inputs, and an imitation VFO feel is realized by the operator. This process is illustrated in Figure 16.65.

16.29   Key Rules in Designing PLLs

We will next go through a PLL design process, but first let's look at a short list of design rules to follow when designing phase-locked loops. Following the list in Table 16.3 can keep us out of trouble most of the time.

16.30   Problems: Design a Synthesized Receiver System for the FM Broadcast Band

We will now turn to the system design of a simple superheterodyne receiver. The task is to design a coherent synthesized system for the FM broadcast band to compliment the receiver block diagram shown in Figure 16.66. To accommodate all FM broadcast frequencies in the United States, the FM band has been equally divided

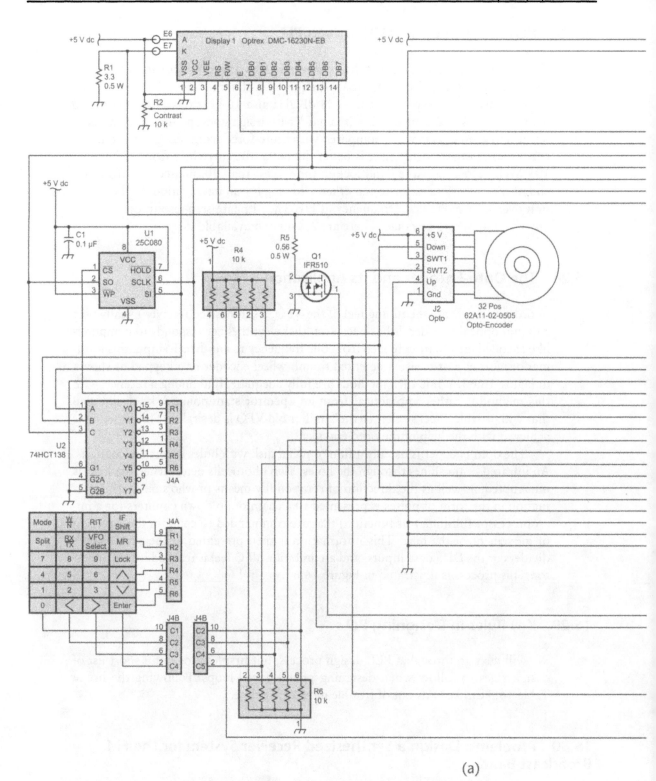

(a)

Figure 16.63   (a) The command and control assembly communicates with the DDS-driven PLL synthesizer via serial ports. (*From:* [20]. Graphics courtesy of ARRL.) (b) Actual implementation of the command and control assembly in the transceiver.

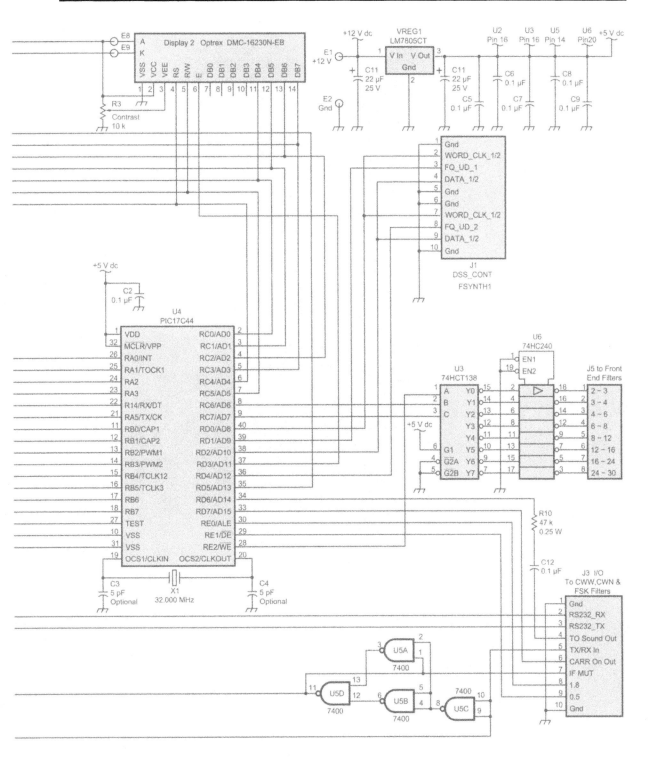

Figure 16.63   (continued)

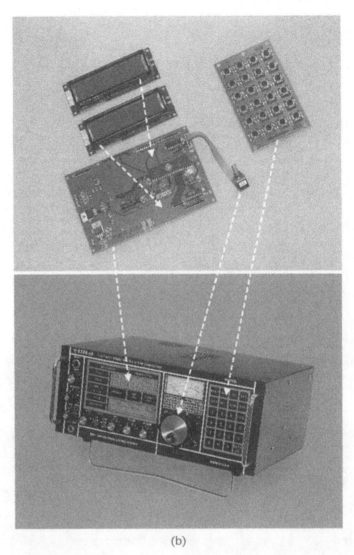

(b)

Figure 16.63 (continued)

into 100 channels, separated by 200 kHz each. Other channelization may be necessary in different parts of the world. The system requirements are as follows:

- The RF receiver band is 87.9 MHz to 107.9 MHz.
- The frequency step is 200 kHz (or less to accommodate other standards).
- Use an IF of 10.7 MHz to avoid images.
- The information bandwidth is 15 kHz.
- Design for a settling time of 10 ms at 10% overshoot with a maximum overshoot of 20%.

As can be seen from Figure 16.66, the required LO would have to tune from 98.6 to 118.6 MHz in 200-kHz steps (or less).

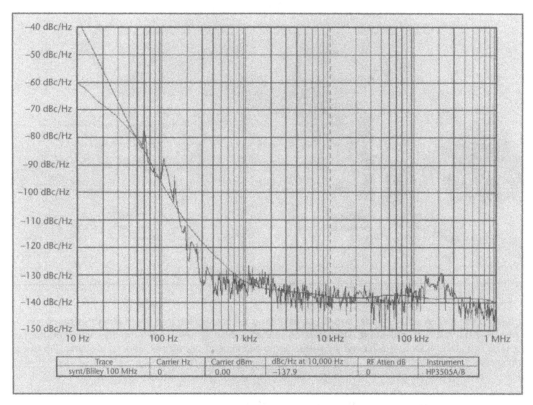

Trace	Carrier Hz	Carrier dBm	dBc/Hz at 10,000 Hz	RF Atten dB	Instrument
synt/Bliley 100 MHz	0	0.00	–137.9	0	HP3505A/B

Figure 16.64   Phase noise performance of the DDS-driven PLL as seen at the LO output. Phase noise performance is –138 dBc/Hz at a 10-kHz offset and has been found uniform (within 1 dB) over the entire frequency
coverage of 2.8 million steps.

There are many ways to accomplish the job, as we will soon find out. For simplicity, we chose one way to approach the problem by using an integer PLL, shown
in Figure 16.67. However, the following rules also apply to other PLL types including the dual modulus and the DDS-driven PLL.

As we can see from Figure 16.67, the phase detector reference was chosen on
purpose to be 10 kHz in order to adapt the system to other possible channelization
standards. The 10-kHz reference is derived from a 1-MHz quartz MRU used as the
reference oscillator. The phase detector is of the flip-flop 4π digital kind in order to
keep the capture range wide enough for the entire band. The VCO operates directly
at the required frequency band between 98.6 MHz and 118.6 MHz as offset by the
IF frequency. The loop uses a fixed prescaling divider (divide by 10) after the VCO
before the programmable integer divider, which can be set to divide between 986
through 1,186 to accommodate the entire FM band. Here is the process followed
in the design:

1. Calculate the range of the PLL to offset the IF. The PLL range is from 98.6
    to 118.6 MHz.
2. From (16.27), find $K = 10$. Using a 1-MHz crystal reference oscillator, de-
    rive a reference of 10 kHz (use divide by 100).

$$F_{ch} = KF_{ref} \tag{16.27}$$

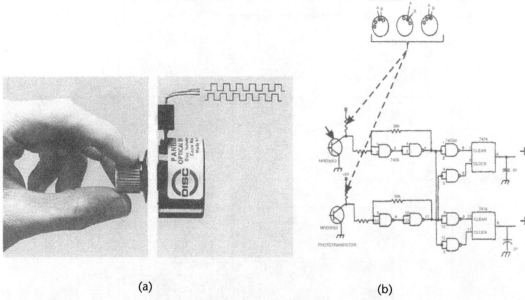

(a)   (b)

Figure 16.65 (a) An opto-encoder and (b) conditioning circuits. An infrared beam of light is shone through a film or metal disk with holes in it. The direction of the count is sensed by determining which photodetector is on first. Debouncing is achieved via simple logic circuits and signals are input to available I/O ports in a microprocessor.

3. From (16.28) and (16.29), find $N_{max} = 1,186$ and $N_{min} = 986$.

$$N_{max} = \frac{F_o(max)}{F_{ch}} \tag{16.28}$$

$$N_{min} = \frac{F_o(min)}{F_{ch}} \tag{16.29}$$

4. Determine the loop filter parameters as follows:
   a. Choose the desired channel spacing F_{ref} (the highest step resolution is 10 kHz).
   b. Calculate the range of the digital division from (16.28) and (16.29).
   c. Determine the VCO range from (16.30).

$$\left[2F_o(max) - F_{o(min)}\right] \leq F_{vco} \leq \left[2F_{o(min) - F_o(max)}\right] \tag{16.30}$$

   d. Choose a minimum value for the damping factor (ζ_{min}) depending on the type of filter chosen from the classic normalized transient response curves in Figure 16.49.
   e. Calculate the loop natural frequency ωn based on desired settling time t_s and the $\omega n \, t_s$ product obtained from the transient response curve.
   f. Calculate the minimum value of the capacitance for the loop filter, say, a simple first-order lowpass filter from Figure 16.47 using (16.31) and (16.32).

Table 16.3   Key Rules to Follow in Designing PLLs

1. Determine the frequency range.

2. Determine the channel spacing.

3. Determine the step resolution.

4. Calculate the digital division ratio.

5. Determine the VCO range.

6. Calculate the expected phase noise and spurious response.

7. Compare with the requirement for the receiver.

8. Back to the drawing board.

9. Determine the VCO range or ranges.

10. Choose the loop filter and type.

11. Choose the loop response as 1% of the reference.

12. Choose the minimum value for damping from the transient response curves.

13. Calculate the loop natural frequency based on the desired settling time and the product of the two from transient response curves.

14. Design the loop filter.

15. Determine the maximum damping factor.

16. Check the transient response for compliance with the initial requirement.

17. Breadboard and check the open loop performance.

18. Close the loop and check everything again.

19. Go back to the drawing board and redo.

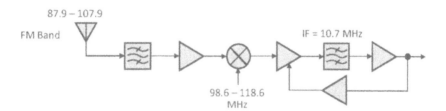

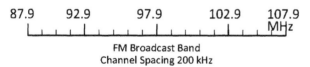

FM Broadcast Band
Channel Spacing 200 kHz

Figure 16.66   A block diagram of a typical FM receiver having an IF of 10.7 MHz. The task is to design a fully coherent synthesizer to match the chanelization requirements listed.

$$C_{min} = \frac{K_o K_p}{N_{max} R_{\omega n^2}} \tag{16.31}$$

$$C_{min} = \frac{K_o K_d}{N_{max} R_{1\omega n^2}} \tag{16.32}$$

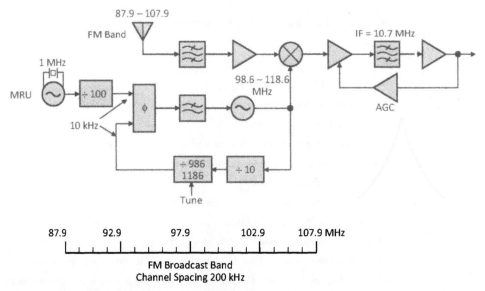

87.9   92.9   97.9   102.9   107.9 MHz

FM Broadcast Band
Channel Spacing 200 kHz

Figure 16.67 Using an integer PLL with a reference frequency of 10 kHz to tune the FM broadcast band.

g. Determine the maximum damping factor (ζ_{max}) from (16.33).

$$\zeta_{max} = \zeta \min \left(\frac{N_{max}}{N_{min}} \right)^{1/2} \tag{16.33}$$

h. Check the transient response for ζ_{max} to comply with the response requirement.

To actually do a design, we have to make some additional decisions on the loop filter type, the phase detector type, and the VCO tuning range.
Determine the VCO range from:

$$VCO\,F\min = 2fo(\min) - fo(\max) = 2(99) - 119 = 198 - 119 = 79\,\text{MHz}$$
$$VCO\,F\max = 2fo(\max) - fo(\min) = 2(119) - 99 = 238 - 99 = 139\,\text{MHz}$$

VCO range = ~79 MHz to 139 MHz for 98.6 MHz to 118.6 MHz +/– 20% overlap for both ends. Choose a VCO from the vendor catalog or design your own.
Determine the loop filter parameters. Tentatively use a first-order RC or an active lowpass. Use a normalized transient response curve from the classic table (Figure 16.49). We find that a damping factor of 0.8 should give an overshoot of 20%.

1. Find the damping and natural loop frequency. To reasonably pick values for ζ and ωn, one method is to design for a specified overshoot within a given settling time.
2. The overshoot is the maximum difference between transient and steady state values for a sudden change applied to the input of the PLL $(N +/– 1)$.

3. The settling time is (t_s) is time required for the transient response to reach and remain within the specified 10% of the steady state.
4. The damping factor (as a rule) is therefore chosen between 0.5 and 0.8. This leaves only the amount of overshoot and the settling time to decide upon. For example, choose a damping factor to be 0.5 so that the output is less than 10% of the steady state, and a value of 10 ms after a change in phase or frequency is applied. From the classic table, we see that the response reaches and remains within 10% of the steady state value at ωnt = 4.5 for a damping factor of 0.5. Since the settling time is 10 ms, the loop natural frequency is found from ωnt_s = 4.5.

$$\omega n = 4.5/10\,\text{ms} = 450\,/\text{sec} = 71.6\,\text{Hz}$$

Then, by picking a value for the filter's C and knowing the conversion gain of the VCO and the phase detector, we can determine R from:

$$\omega LPF = (K0 Ko \omega LPF)^{1/2}$$

Using a first-order active filter and the transient response curve from the classic table, we find that a damping factor of 0.8 will give an overshoot of less than 20%.

Also from the classic table, the transient response will be less than 10% at ωnts = 3.5, so for a settling time of 10 ms:

$$\omega n = \omega nts = |ts = 3.5|10\,\text{ms} = 350\,\text{rad/sec}\,(55.7\,\text{Hz})$$

Assume that we use a classic VCO with a $Ko = 11 \times 10$ rad/sec/V.
Assuming that we use a standard flip-flop phase detector with a $K = 0.12$ V/rad and picking R1 = 4.7 KΩ, then:

$$C\min = K\Theta\, Ko/N\max R1 \omega n^2$$
$$= (11\times10^6)\times(0.12)/(1,080)\times(4.7\,K)\times(350)^2 = 2.1\,\mu F\,(\text{use}\,2\,\mu F)$$

Then, by rearranging the dampening formula, we get:

$$R2 = 2\zeta\min/\omega n\,C\min = (2)(0.8)/(350)(2\,\mu F) = 2.285\,K\Omega\,(\text{use}\,2.2\,K\Omega)$$

Now, prove that for a dampening factor of 0.89 the transient response will have an overshoot of less than 10% within 10 ms:

$$\zeta\max = \zeta\min(N\max/N\min)^{1/2}$$
$$= (0.8)(1,187/987)^{1/2} = 0.87$$

Next, design and implement the loop filter circuit as connected with the chosen VCO and phase detector. Choose a divide-by-10 prescaler that will operate at the maximum VCO frequency of 139 MHz + 10%. Choose programmable up-down counters for the division ratio. Design the circuits. Brass board and test the open loop and then the closed loop. If necessary, redo the entire process. If there are no results, go back to the drawing board and review everything [repeating (16.27 through (16.33) here].

$$F_{ch} = KF_{ref} \tag{16.27}$$

$$N_{max} = \frac{F_o(\text{max})}{F_{ch}} \tag{16.28}$$

$$N_{min} = \frac{F_o(\text{min})}{F_{ch}} \tag{16.29}$$

$$\left[2F_o(\text{max}) - F_{o(\text{min})}\right] \le F_{vco} \le \left[2F_{o(\text{min})-F_o(\text{max})}\right] \tag{16.30}$$

$$C_{min} = \frac{K_o K_p}{N_{max} R_{\omega n^2}} \tag{16.31}$$

$$C_{min} = \frac{K_o K_d}{N_{max} R_{1\omega n^2}} \tag{16.32}$$

$$\zeta_{max} = \zeta \min\left(\frac{N_{max}}{N_{min}}\right)^{\frac{1}{2}} \tag{16.33}$$

There are several other ways to design a synthesizer for this application, shown in Figures 16.68 through 16.73. These are actual designs entered by students. Your job is to look at each implementation and decide which is the best implementation from a performance point of view and which is the most economical way to do the job.

The system in Figure 16.68 is what we have been discussing so far. However, there is a major drawback to this implementation. You should be able to find it quickly if you have read the previous chapters. The reference of 10 kHz causes the reference feedthrough to be almost in band with the LO signal producing a relatively noisy local oscillator, which, in turn, can impact the received signal. T-notch filters could be used in the loop to remedy this problem to some degree.

The system shown in Figure 16.69 uses a mixer PLL approach with a reference of 10.7 MHz feeding a fixed frequency DDS, which divides it down by a fractional number (53.5) to produce a reference of exactly 200 kHz. The 10.7-MHz quartz reference was chosen because of the easy availability of 10.7-MHz quartz crystals. This design is a perfect example of choosing LO frequencies that fall in band of the

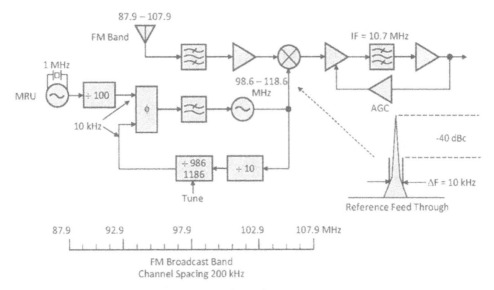

Figure 16.68   Simple integer PLL using a 10-kHz reference.

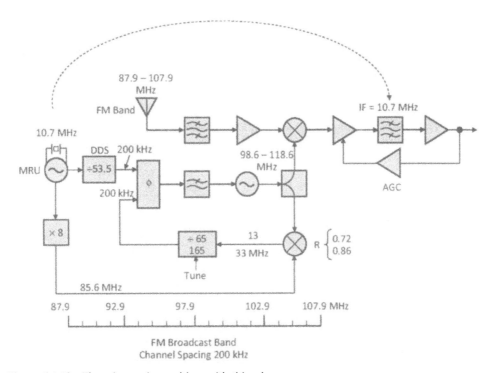

Figure 16.69   There is a major problem with this scheme.

receiver's IF. No matter how shielded the reference is, the 10.7-MHz signal from it will find its way into the IF and spoil the information to be received.

Although not as bad as the previous example, the 107-MHz reference from Figure 16.70 will spill into the RF band to be covered. This is not a major problem such as spilling into the IF, but this should also be avoided.

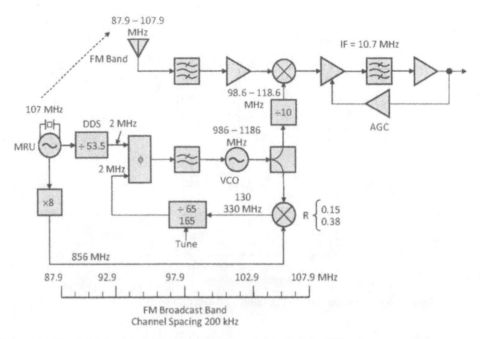

Figure 16.70   The reference frequency was chosen in band of the RF band.

In the example in Figure 16.71, the mixer PLL uses a ×8 multiplied reference that could impact overall performance. Phase noise performance analysis is in order here.

Figure 16.72 shows the proper way to implement the FM receiver synthesizer. The reference frequency is 1 MHz. This provides a 1-MHz step resolution from 986 MHz through 1,186 MHz. When divided by 10, this resolution becomes 100

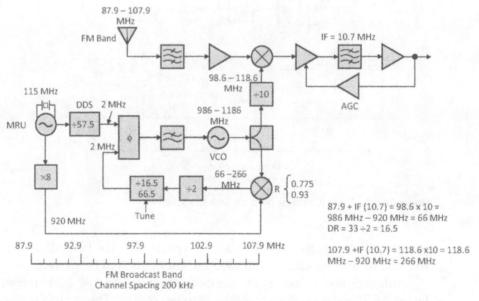

Figure 16.71   In this approach, there are no in-band spilling problems. However, the ×8 multiplier can spoil the phase noise performance of the mixer.

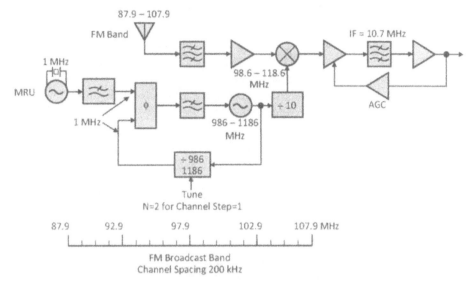

Figure 16.72   The right way to do the job. This is what we call in the RF world "the best of the worse" case. This method will provide the best performance over the previous methods.

kHz at 98.6–118.6 MHz. For $N = 2$ for each step, the final step resolution is 200 kHz. In addition, a 20log10 phase noise improvement (20 dB) occurs as a result of the divide-by-10 function. This is the right way to implement the system.

Figure 16.73 shows a very ingenious way of providing synthesis for the FM receiver. An 8-bit digital static up-down counter is ramped up or down and addresses a latched D/A, which has preset voltage values for all frequencies of interest over the frequency range of the VCO. Although perhaps not exactly on frequency due to the reduced number of bits, a frequency discriminator of the Foster Seeley kind in the receiver's IF steers the VCO exactly on frequency. Many inexpensive radios are implemented this way.

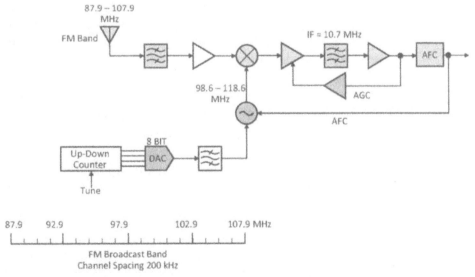

Figure 16.73   An ingenious and inexpensive way to synthesize the FM receiver.

16.31   Final Concluding Notes to Synthesizers

PLLs and DDS-driven PLLs can be designed with relative ease today using software programs such as Sim PLL from Analog Devices or other programs. These tools are good because one can set the op-amp parameters for an active loop filter and immediately see the effect on phase noise and reference spurs when adjusting the input bias and offset currents.

The art of designing PLLs lies in how well we can reject spurious problems while maintaining superior phase noise performance over wide bands. Several key observations result from designing PLLs and DDS-driven PLLs:

1. The output of a PLL will usually contain a myriad of spurs at the reference frequency and multiples of it. The higher the reference frequency, the better this situation will be.

2. These spurs are attenuated by the loop filter, but the loop dynamics are such that only so much attenuation can be achieved before impacting capture and lockup times.

3. The key to better PLL design is to reduce unnecessary pumping that is replacing charges lost to leakage between each phase detector cycle. With less leakage, the pulse width diminishes, which can be equated to lower amplitudes for the spurs.

4. When using a narrowband loop bandwidth in a PLL, it usually lends to a small hump in the overall phase noise profile at the intersection of the VCO curve meeting the PLL loop filter corner. This hump can be optimized (made smaller), by setting the loop bandwidth wider or by adjusting the loop noise floor and VCO noise profiles accordingly, or by designing a VCO with this point in mind. Even after such an optimization, there are still some residual interactions due to small time delay differences in the sampling.

5. Problems associated with the reference feedthrough and the hump can be improved upon if using very low leakage parts in the active loop filter and utilizing charge pumps with very good symmetry. A couple of very low noise operational amplifiers have been found to be superior. Although the OP-27 used in the above DDS-driven PLL design is a very good and inexpensive choice, the Analog Devices AD-829 or the Texas Instruments THS-4031 could improve on the performance. Here are a few additional rules to keep in mind when designing PLLs.

   - As a general rule, a PLL synthesizer can improve on the performance of a VCO inside the loop bandwidth provided that the phase noise contribution of the reference source is minimal.
   - Always keep in mind that the PLL is a multiplier. Its phase noise performance cannot be any better than the overall performance of its multiplied reference.
   - If a VCO has a phase noise performance worse than the PLL, then the performance inside the loop bandwidth can be improved.
   - If the VCO phase noise performance is better than the PLL's phase noise performance, then the overall performance is degraded by the PLL.

- The phase noise inside the loop bandwidth is dependent on the phase detector jitter, and the division number N (additive noise contribution). Smaller N numbers are obviously better.
- Each subsystem component can dominate the overall phase noise performance. The reference at low offsets blends inside the loop filter impacting phase noise performance as it cannot be filtered out. Corrective measures can minimize these problems.
- However, the long time between the corrective measures also has implications for pushing, pulling, and microphonics, which will also result in reference spurs. Frequency pushing is caused by the VCO's sensitivity to supply voltages. It is expressed in megahertz per volt. Frequency pulling is a measure of the frequency change due to a nonideal load. It is measured by noting the frequency change caused by a load having a nominal 2-dB return loss with all possible phases.
- Gaps in the range of divider ratios sometimes exist in manufactured PLL devices that use quadruple modulus designs. These have some limitations and cannot divide all integers, especially at the lower end of the range. Manufacturers usually state the minimum continuous divide ratio and the exceptions due to other internal setups, which result in illegal divide ratios.
- In extremely broadband multiple loop PLL synthesizers, it is sometimes impossible to be free of spurious problems. Depending on the application, a full spurious analysis should always be performed.

These discussions point out that the design of a complex synthesizer is not a simple task when one considers the mathematical and practical impacts of changing various elements in favor of others. Impractical mixing ratios or unrealizable prescaling requirements in PLLs, combined with fractional reference requirements in DDSs, can easily send a good engineer back to the drawing board several times before a good compromise is achieved. While loop behavior can be modeled accurately through computer programs, predicting phase-noise performance through analysis can be inexact and sometimes frustrating.

More often, as in the case of large projects, the designer views the synthesizer as a specialized piece of equipment for which there is no time to design. As various vendors of general-purpose synthesizers are approached, a choice is usually made based on knowledge of the requirements at the time. Understanding the techniques presented here can help in choosing the right products and eliminating costly mistakes.

Because of the vast meaning of the word *synthesizer*, there are many misunderstandings about what a synthesized receiver should be. It is clear that a synthesizer can take many forms and derives its stability from the reference oscillator whose stability and performance has a further impact on the total receiver's performance circuitry. Unless all other frequency sources used in the receiver are slaved to the same reference frequency, the receiver is considered as not fully synthesized.

In specialized radio communications, a radio receiver can be asked to do a variety of tasks, such as tune or scan several ranges of frequencies in a particular manner selecting different bandwidths. A radio receiver can also be programmed to be active at certain frequencies and/or times. Upon receiving a wanted signal, a second

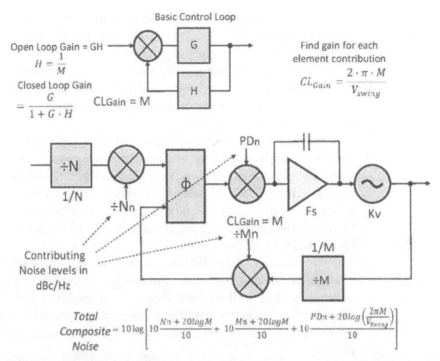

Figure 16.74 Accounting for noise contributions of the reference divider, the feedback divider, and the phase detector in PLL synthesizers.

receiver can be activated to analyze that frequency, while the first receiver continues its predetermined task. The importance of being on the right frequency as fast as possible is vital, especially in specialized data communications. If a receiver was not fully synthesized and inherently stable, errors would result at the output of the demodulator decreasing the probability of intercept in a system, which is already affected by multipath fading and noise phenomena.

16.32 Additive Noise in PLL Design

When designing a PLL, there are three key noise contributors to its performance: the reference divider, the feedback divider, and the phase detector. The designer should have an understanding of these contributions.

A simple linear control system equation of how to calculate the digital reference divider, the feedback divider, and the phase detector noise contributions to the loop is shown in Figure 16.74. These three factors often limit the noise performance within the loop bandwidth of a PLL circuit.

References

[1]   Edson, W. A., "Noise in Oscillators," *Proc. IRE*, August 1960.

[2]   Leeson, D. B., "A Simple Model of Feedback Oscillator Noise Spectrum," *Proc. IEEE*, February 1966.

[3]   Craninckx, J., and M. Steyaert, "Low-Noise Voltage-Controlled Oscillators Using En-
        hanced LC-Tanks," *IEEE Trans. on Circuits Syst. II*, Vol. 42, No. 12, December 1995.

[4]   Razavi, R., "A Study of Phase Noise in CMOS Oscillators," *IEEE Journal of Solid State
        Circuits*, Vol. 31, No. 3, March 1996.

[5]   Chenakin, A., "Building a Microwave Frequency Synthesizer," *High Frequency Electron-
        ics*, June 2008.

[6]   Brandon, D., and J. Cavey, "The Residual Phase Noise Measurement," Analog Devices
        AN-0982 Application Note, www.analog.com.

[7]   Zavrel, Jr., R. J., and M. Morini, "DSP for RF Designers," *Communication Design
        Conference*.

[8]   Scherer, D., "Design Principles and Test Methods for Low Phase Noise RF and Microwave
        Sources," *RF & Microwave Measurement Symposium*, Hewlett Packard.

[9]   Egan, W. F., *Frequency Synthesis by Phase Lock*, New York: John Wiley & Sons.

[10]   Drentea, C., "The Art of RF System Design," Three-day course.

[11]   Miller, R. L., "Fractional-Frequency Generators Utilizing Regenerative Modulation," *Pro-
        ceedings of the IRE*, Vol. 27, July 1939.

[12]   Mossammaparast, M., et al., "Phase Noise of X-Band Regenerative Frequency Dividers,"
        Poseidon Scientific Instruments, Fremantle, WA, Australia.

[13]   Noorfazila, K., et al., "A High Frequency Divider in 0.18 μm SiGe BiCMOS Technology,"
        The School of Electrical & Electronic Engineering, University of Adelaide, Adelaide, Aus-
        tralia. Institute for Telecommunications Research, University of South Australia, Macqua-
        rie University, Australia, The University of Sydney, Australia.

[14]   Khalil, A., "MMIC Cavity Oscillator," Hittite Microwave, http://sbir.nasa.gov/SBIR/ab-
        stracts/06/sbir/phase2/SBIR-06-2-S6.04-9389.html?solicitationId=SBIR_06_P2.

[15]   Shahriary, I., and K. M. McNab, "Apparatus and Method for Phase Noise and Post Tuning
        Drift Cancellation," U.S. Patent 4918748, 1990.

[16]   Gillette, G. C., "Digiphase Principle," *Frequency Technology*, August 1969.

[17]   Gillette, G. C., "Frequency Synthesizer," U.S. Patent 3582810, June 1, 1971.

[18]   Drentea, C., "Designing Frequency Synthesizers, Session B-1, Frequency Synthesis," *RF
        Technology Expo 1988*, Anaheim, CA, February 10–12, 1988.

[19]   Drentea, C., "Beyond Fractional-N, Part 1," *QEX*, March/April 2001.

[20]   Drentea, C., "Beyond Fractional-N, Part 2," *QEX*, May/June 2001.

[21]   Drentea, C., "The Star-10 Transceiver," QEX article series, January/February 2007, March/
        April and May/June 2008, ARRL.

[22]   Yao, X. S., and L. Maleki, "Opto-Electronic Oscillator Having a Positive Feedback with an
        Open Loop Gam," U.S. Patent 5,723,856, 1998.

[23]   Yao, X. S., and L. Maleki, "Multiloop Optoelectronic Oscillator," *IEEE Journal of Quan-
        tum Electronics*, Vol. 36, No. 1, January 2000.

[24]   OEwaves Promotional Letter, *2006 RF Symposium*, Dallas, TX, 2006.

[25]   Beam, W. C., and P. J. Rezin, "Design Considerations for Fast Switching PLL Synthesizers,"
        Watkins Johnson Tech-Notes, Vol. 18, No. 2, March 1991.

[26]   Wetenkamp, S. F., and K. J. Wong, "Transportation Lag in Phase-Locked Loops," *Watkins
        Johnson Tech-Notes*, Vol. 5. No. 3, May/June 1978.

Selected Bibliography

Agilent Technologies, Application Note 1314, "Testing and Troubleshooting Digital RF Communications Receiver Designs," cp.literature.agilent.com/litweb/pdf/5968-3579E.pdf.

Allan, D. W., "Clock Characterization Tutorial," *Proceedings of the 15th Annual Precise Time and Time Interval (PTTI) Applications and Planning Meeting*, 1983.

Allan, D. W., "The Impact of Precise Time in Our Lives: A Historical and Futuristic Perspective Surrounding GPS," *50th Anniversary Invited Talk at Institute of Navigation Annual Meeting*, Colorado Springs, CO, June 5–7, 1995.

Allan, D. W., "Statistics of Atomic Frequency Standard," *Proceedings of the IEEE*, Vol. 54, No. 2, 1966, p. 105.

Allan, D. W., "Millisecond Pulsar Rivals Best Atomic Clock Stability," *Proceedings of the 41st Annual Symposium on Frequency Control*, Philadelphia, PA, 1987.

Allan, D. W., "Should the Classical Variance Be Used as a Basic Measure in Standards Metrology," *IEEE Trans. on Instrumentation and Measurement*, Vol. IM-36, 1987.

Allan, D. W., "Time and Frequency (Time-Domain) Characterization, Estimation, and Prediction of Precision Clocks and Oscillators," *IEEE Trans. on Ultrasonics, Ferroelectrics, and Frequency Control*, Vol. UFFC-34, 1987.

Allan, D. W., and J. A. Barnes, "A Modified 'Allan Variance' with Increased Oscillator Characterization Ability," *Proceedings of the 35th Annual Frequency Control Symposium*, 1981.

Allan, D. W., and W. Dewey, "Time-Domain Spectrum of GPS SA," *Proceedings of 1993 Institute of Navigation ION GPS-93*, 1993.

Allan, D. W., M. A. Weiss, and T. K. Peppler, "In Search of the Best Clock," *IEEE Trans. on Instrumentation and Measurement*, Vol. 38, 1989.

Allan, D. W., et al., "Standard Terminology for Fundamental Frequency and Time Metrology," *Proceedings of the 42nd Annual Frequency Control Symposium*, Baltimore, MD, June 1–4, 1988.

Anderson, J. B., *Digital Transmission Engineering*, New York: IEEE Press, 1999.

Arora, H., et al., "Enhanced Phase Noise Modeling of Fractional-N Frequency Synthesizers," *IEEE Trans. on Circuits Syst. I, Regular Papers*, Vol. 52, No. 2.

Banerjee, D., "Fractional N Frequency Synthesis," Application Note 1879, National Semiconductors.

Banerjee, D., "PLL Performance, Simulation, and Design," National Semiconductors, 1998.

Bartlett, J., "Crystals Inside Out," QST, ARRL, January 1978.

Best, R. E., *Phase-Locked Loops: Design, Simulation, and Applications*, New York: McGraw-Hill, 1984.

Blanchard, A., *Phase-Locked Loops: Application to Coherent Receiver Design*, New York: John Wiley & Sons, 1976.

Braymer, N. B., "Frequency Synthesizer," U.S. Patent 3555446, January 12, 1971.

Brennan, P. V., *Phase-Locked Loops: Principles and Practice*, New York: Macmillan.

Brennan, P. V., et al., "A New Mechanism Producing Discrete Components in Fractional-N Frequency Synthesizers," IEEE 1549-8328, 2008.

Butterfield, D., and B. Sun, "Prediction of Fractional-N Spurs for UHF PLL Frequency Synthesizers," Qualcomm Inc., San Diego, CA, 1999.

Carson, R. S., *High-Frequency Amplifiers*, New York: John Wiley & Sons, 1975.

Carson, R. S., *Radio Communications Concepts*, New York: John Wiley & Sons, 1990.

Clairon, A., et al., *Symposium on Frequency Standards and Metrology*, Singapore, 1996, p. 49.

Coen, S., and M. Haelterman, "Continuous-Wave Ultrahigh-Repetition-Rate Pulse-Train Generation Through Modulational Instability in a Passive Fiber Cavity," *Opt. Lett.*, Vol. 26, 2001, pp. 39–41.

Coen, S., and M. Haelterman, "Modulational Instability Induced by Cavity Boundary Conditions in a Normally Dispersive Optical Fiber," *Phys. Rev. Lett.*, Vol. 79, 1997, pp. 4139–4142.

Conexant, "Fractional-N Synthesizers," White Paper.

Cox, T. G., "Frequency Synthesizer," U.S. Patent 3976945, August 24, 1976.

Crochiere, R. E., and L. R. Rabiner, *Multirate Digital Signal Processing*, Upper Saddle River, NJ: Prentice-Hall, 1983.

Daimon, M., and A. Masumura, "High-Accuracy Measurements of the Refractive Index and Its Temperature Coefficient of Calcium Fluoride in a Wide Wavelength Range from 138 to 2326 nm," *Appl. Opt.*, Vol. 41, 2002, pp. 5275–5281.

Del Haye, P., et al., "Optical Frequency Comb Generation from a Monolithic Microresonator," *Nature*, Vol. 450, 2007, pp. 1214–1217.

De Muer, B., and M. S. J. Steyaert, "On the Analysis of $\Delta\Sigma$ Fractional-N Frequency Synthesizers for High-Spectral Purity," *IEEE Trans. on Circuits Syst. II, Analog Dig. Signal Process.*, Vol. 50, No. 11, November 2003.

Derksen, R. H., H. -M. Rein, and K. Wörner, "Monolithic Integration of a 5.3 GHz Regenerative Frequency Divider Using a Standard Bipolar Technology," *Electronics Letters*, Vol. 21, October 1985.

Drentea, C., "Bragg-Cell Application to High Probability of Intercept Receiver," U.S. Patent 7,324,797.

Drentea, C., "High Stability Local Oscillators for Microwave and Other Applications," *Ham Radio*, November 1985.

Eliyahu, D., et al., "Improving Short and Long Term Frequency Stability of the Opto-Electronic Oscillator," *Proc. IEEE International Frequency Control Symposium*, 2002.

Federal Standard 1037C, *Glossary of Telecommunication Terms*, August 7, 1996.

Floyd, B. A., et al., "SiGe Bipolar Transceiver Circuits Operating at 60 GHz," *IEEE J. Solid-State Circuits*, Vol. 40, 2005, pp. 156–167.

Franca, J., and Y. Tsividis, (eds.), *Design of Analog-Digital VLSI Circuits for Telecommunications and Signal Processing*, Upper Saddle River, NJ: Prentice-Hall, 1993.

Gardner, F. M., *Phaselock Techniques*, New York: John Wiley & Sons, 1966.

Gibbs, J., and R. Temple, "Frequency Domain Yields Its Data to Phase-Locked Loop Synthesizer," *Electronics*, April 1978.

Golding, W. M., *Proceedings of the 1994 IEEE International Symposium on Frequency Control*, Piscataway, NJ, 1994, p. 724.

Gravel, J. F., and J. S. Wight, "On the Conception and Analysis of a 12-GHz Push-Push Phase-Locked DRO," *IEEE Trans. on Microwave Theory and Tech.*, Vol. 54, 2006, pp. 153–159.

Haelterman, M., S. Trillo, and S. Wabnitz, "Additive-Modulation-Instability Ring Laser in the Normal Dispersion Regime of a Fiber," *Opt. Lett.*, Vol. 17, 1992, pp. 745–747.

Hajimiri, A., and T. H. Lee, "General Theory of Phase Noise in Electrical Oscillators," *IEEE Journal on Solid-State Circuits*, Vol. 33, No. 2, February 1998.

Hardy, J., *High Frequency Circuit Design*, Reston, VA: Reston, 1979.

Harrison, R., "Theory of Regenerative Frequency Dividers Using Double-Balanced Mixer," *IEEE Microwave Theory and Techniques Society (MTT-S) Digest*, Long Beach, CA, 1989.

Hayden, M. E., M. D. Hurlimann, and W. N. Hardy, *Phys. Rev. A*, Vol. 53, 1996, p. 1589.

Hedayati, H., B. Bakkaloglu, and W. Khalil, "Closed Loop Nonlinear Modeling of Wideband Fractional-N Frequency Synthesizers," *IEEE Trans. on Microwave Theory and Tech.*, Vol. 54, No. 10, October 2006.

Helfrick, A. D., *Electrical Spectrum and Network Analyzers*, San Diego, CA: Academic Press.

Hewlett Packard, Microwave Design System, 1989.

Hewlett Packard, "Phase Noise Characterization of Microwave Oscillators," Product Note 11729B-1.

Hosoya, K., et al., "V-Band HJFET MMIC DROs with Low Phase Noise, High Power, and Excellent Temperature Stability," *IEEE Trans. on Microwave Theory and Tech.*, Vol. 51, 2003, pp. 2250–2258.

Ivanov, E. N., and M. E. Tobar, "Low Phase-Noise Microwave Oscillators with Interferometric Signal Processing," *IEEE Trans. on Microwave Theory and Tech.*, Vol. 54, 2006, pp. 3284–3294.

Ji, Y., X. S. Yao, and L. Maleki, "Compact Optoelectronic Oscillator with Ultralow Phase Noise Performance," *Electron. Lett.*, Vol. 35, 1999, pp. 1554–1555.

Kawakawa, K., et al., "A 15 GHz Single Stage GaAs Dual-Gate FET Monolithic Analog Frequency Divider with Reduced Input Threshold Power," *IEEE Trans. on Microwave Theory and Tech.*, Vol. MTT-36, December 1988.

Kenny, T. P., et al., "Design and Realization of a Digital $\Delta\Sigma$ Modulator for Fractional-N Frequency Synthesis," *IEEE Trans. on Vehicular Technology*, Vol. 48, March 1999.

Khalil, W., et al., "Analysis and Modeling of Noise Folding and Spurious Emission in Wideband Fractional-N Synthesizers," Intel Corporation, and A. Fulton School of Engineering, Arizona State University.

King, N. J. R., "Phase-Locked Loop Variable Frequency Generator," U.S. Patent 4204174, May 20, 1980.

Kingford, C. A., and C. A. Smith, "Device for Synthesizing Frequencies Which Are Rational Multiples of a Fundamental Frequency," U.S. Patent 3928813, December 23, 1973.

Kippenberg, T. J., S. M. Spillane, and K. J. Vahala, "Kerr-Nonlinearity Optical Parametric Oscillation in an Ultrahigh-Q Toroid Microcavity," *Phys. Rev. Lett.*, Vol. 93, 2004, p. 083904.

Klyshko, D. N., *Photons and Nonlinear Optics*, New York: Taylor and Francis, 1988.

Knapp, H., et al., "86 GHz Static and 110 GHz Dynamic Frequency Dividers in SiGe Bipolar Technology," *IEEE Microwave Theory and Techniques Society (MTT-S) Digest*, 2003, pp. 1067–1070.

Koelman, J. M. V. A., et al., *Phys. Rev. A*, Vol. 38, 1988, p. 3535.

Kouomou Chembo, Y., et al., "Determination of Phase Noise Spectra in Optoelectronic Microwave Oscillators: A Phase Diffusion Approach," arXiv:0805.3317 [physics.optics], May 2008.

Krauss, H. L., C. W. Bostian, and F. H. Raab, *Solid State Radio Engineering*, New York: John Wiley & Sons, 1980.

Kroupa, V. F., *Direct Digital Frequency Synthesis*, New York: IEEE Press, 1999.

Kucharski, D., and K. T. Kornegay, "A 40 GHz 2.1 V Static Frequency Divider in SiGe Using a Low-Voltage Latch Topology," *IEEE Radio Frequency Integrated Circuits Symposium*, Boston, MA, 2005.

Landén, L., C. Fager, H. Zirath, "Regenerative GaAs MMIC Frequency Dividers for 28 and 14 GHz," Chalmers University of Technology, Department of Microelectronics.

Lao, Z., et al., "35-GHz Static and 48-GHz Dynamic Frequency Divider IC's Using 0.2-μm AlGAs/GaAs-HEMT's," *IEEE J. Solid-State Circuits*, Vol. 32, October 1997.

Leeson, D. B., "A Simple Model of Feedback Oscillator Noise Spectrum," *Proc. IEEE*, Vol. 54, 1966, p. 329330.

MacLeod, J. R., et al., "Software Defined Radio Receiver Test-Bed," *IEEE VTS 54th Vehicular Technology Conference, VTC Fall 2001*, Vol. 3, 2001.

Majewski, J. J., "Digital Receiver System Design," MIT, Lincoln Laboratory, June 7, 2009.

Maleki, L., et al., "Whispering Gallery Mode Lithium Niobate Micro-Resonators for Photonics Applications," *Proc. SPIE*, Vol. 5104, 2003.

Manassewitsch, V., *Frequency Synthesizers, Theory and Design*, New York: John Wiley & Sons, 1980.

Matsko, A. B., et al., "High Frequency Photonic Microwave Oscillators Based on WGM Resonators," *2005 Digest of the LEOS Summer Topical Meetings*, 2005, pp. 113–114.

Matsko, A. B., et al., "Nonlinear Optics with WGM Crystalline Resonators: Advances and Puzzles," *2004 Digest of the LEOS Summer Topical Meeting WGM Microcavities*, 2004.

Matsko, A. B., et al., "Optical Hyper-Parametric Oscillations in a Whispering Gallery Mode Resonator: Threshold and Phase Diffusion," *Phys. Rev. A*, Vol. 71, 2005, p. 033804.

Matsko, A. B., et al., "Whispering-Gallery-Mode Resonators as Frequency References: I. Fundamental Limitations," *J. Opt. Soc. Am. B*, Vol. 24, 2007, pp. 1324–1335.

McFerran, J. J., et al., "Low-Noise Synthesis of Microwave Signals from an Optical Source," *Electron. Lett.*, Vol. 41, 2005.

Meninger, S. E., and M. H. Perrott, "A Fractional-N Frequency Synthesizer Architecture Utilizing a Mismatch Compensated PFD/DAC Structure for Reduced Quantization-Induced Phase Noise," *IEEE Trans. on Circuits Syst. II, Analog Digit. Signal Process*, Vol. 50, No. 11, November 2003.

Meninger, S., and M. Perrott, "Sigma-Delta Fractional-N Frequency Synthesis," Massachusetts Institute of Technology, June 2004.

Meyr, H., and G. Ascheid, *Synchronization in Digital Communications: Phase-Frequency-Locked Loops and Amplitude Control*, New York: John Wiley & Sons.

Mullrich, J., et al., "SiGe Regenerative Frequency Divider Operating Up to 63 GHz," *Electronic Letters*, Vol. 35, 1999.

Nanhin, P. J., *The Science of Radio*, New York: American Institute of Physics, 2001.

Nayfeh, A. H., *Perturbation Methods*, New York: John Wiley & Sons, 1973.

Ohira, T., "Mathematical Proof of Leeson's Oscillator Noise Spectrum Model," ATR Wave Engineering Laboratories, Kyoto, Japan.

Owen, D., "Fractional-N Synthesizers," Application Note, www.ifrsys.com.

Pace, P. E., *Detecting and Classifying Low Probability of Intercept Radar*, Norwood, MA: Artech House, 2004.

Pamarti, S., L. Jansson, and I. Galton, "A Wideband 2.4-GHz Delta-Sigma Fractional-N PLL with 1-Mb/s In-Loop Modulation," *IEEE J. Solid-State Circuits*, Vol. 38, No. 6, June 2003.

Perrott, M. H., M. D. Trott, and C. G. Sodini, "A Modeling Approach for ΣΔ Fractional-N Frequency Synthesizers Allowing Straightforward Noise Analysis," *IEEE J. Solid-State Circuits*, Vol. 37, No. 8, August 2002.

Rauscher, C., "Regenerative Frequency Division with a GaAS FET," *IEEE Trans. on Microwave Theory and Tech.*, Vol. 32, 1984.

Razavi, B., "Challenges in the Design of Frequency Synthesizers for Wireless Applications," Integrated Circuits and Systems Laboratory, University of California, Los Angeles.

Riely, T. A. D., M. A. Copeland, and T. A. Kwasniewski, "Sigma Delta Modulation in Fractional-N Frequency Synthesis," *IEEE Journal of Solid-State Circuits*, Vol. 28, May 1993.

Rowland, J. R., *Linear Control Systems: Modeling, Analysis and Design*, New York: John Wiley & Sons, 1986.

Rubiola, E., *Phase Noise and Frequency Stability in Oscillators*, New York: Cambridge University Press, 2008.

Rubiola, E., et al., "Phase and Frequency Noise Metrology," *Proc. 7th Symposium on Frequency Standards and Metrology*, October 2008.

Rylyakov, A., "A 51 GHz Master-Slave Latch and Static Frequency Divider in 0.18 μm SiGe BiCMOS," *IEEE Bipolar/BiCMOS Circuit and Technology Meeting (BCTM)*, Toulouse, France, 2003.

Sabin, W. E., and E. O. Schoenike, *Single-Sideband Systems and Circuits*, New York: McGraw-Hill, 1987.

Safarian, A. Q., and P. Heydari, "Design and Analysis of a Distributed Regenerative Frequency Divider Using Distributed Mixer," *International Symposium on Circuits and Systems*, Vancouver, Canada, 2004.

Savchenkov, A. A., et al., "All-Optical Photonic Oscillator with High-Q Whispering Gallery Mode Resonators," *Proc. IEEE International Topical Meeting on Microwave Photonics*, 2004, pp. 205–208.

Savchenkov, A. A., et al., "Low Threshold Optical Oscillations in a Whispering Gallery Mode CaF2 Resonator," *Phys. Rev. Lett.*, Vol. 93, 2004, p. 243905.

Savchenkov, A. A., et al., "Phase Noise of Whispering Gallery Photonic Hyper-Parametric Microwave Oscillators," OEwaves Inc.

Savchenkov, A. A., et al., "Optical Resonators with Ten Million Finesse," *Opt. Express*, Vol. 15, 2007, pp. 6768–6773.

Savchenkov, A. A., et al., "Whispering-Gallery-Mode Resonators as Frequency References. II. Stabilization," *J. Opt. Soc. Am. B*, Vol. 24, 2007, pp. 2988–2997.

Schawlow, A. L., and C. H. Townes, "Infrared and Optical Masers," *Phys. Rev.*, Vol. 112, 1958, pp. 1940–1949.

Scherer, D., "The Art of Phase Noise Measurement," *RF & Microwave Measurement Symposium*, March 1985.

Sirmans, D., and B. Urell, "Digital Receiver Test Results," WSR-88D, Commerce Defense Transportation, Next Generation Weather Radar Program, www.roc.noaa.gov/eng/docs/digital%20 receiver%20test%20results.pdf.

Sklar, B., *Digital Communications: Fundamentals and Applications (with System View)*, Upper Saddle River, NJ: Prentice-Hall, 1988.

Skolnik, M., *Radar Handbook*, 2nd ed., New York: McGraw-Hill, 1990.

Stadler, M., "Calculation and Reduction of Low Frequency Noise in LC Oscillators," Electromagnetic Fields and Microwave Electronics Laboratory (IFH), Zurich, Switzerland, *International Conference on Signals and Electronic Systems (ICSES)*, Poznan, Poland, September 2004.

Stremler, F. G., *Introduction to Communication Systems*, Reading, MA: Addison-Wesley, 1977.

Sullivan, D. B., et al., "Characterization of Clocks and Oscillators," NIST Tech Note 1337, BIN: 868, 1990.

Temporiti, E., et al., "A 700-KHz Bandwidth $\Sigma\Delta$ Fractional Synthesizer with Spurs Compensation and Linearization Techniques for WCDMA Applications," *IEEE Solid-State Circuits*, Vol. 39, No. 9, September 2004.

Tjoelker, R. L., J. D. Prestage, and L. Maleki, *Symposium on Frequency Standards and Metrology*, Singapore, 1996, p. 33.

Underwood, M. J., "Frequency Synthesizer," U.K. Patent 1447418, August 25, 1976.

Vanier, J., and C. Audoin, *The Quantum Physics of Atomic Frequency Standards*, Philadelphia, PA: Adam Hilger, 1989.

Varnum, F., et al., "Comparison of Time Scales Generated with the NBS Ensembling Algorithm," *Proceedings of the 19th Precise Time and Time Interval (PTTI) Meeting*, 1987.

Vassiliev, V. V., et al., "Narrowline-Width Diode Laser with a High-Q Microsphere Resonator," *Opt. Commun.*, Vol. 158, 1998, p. 305312.

Vessot, R. F. C., et al., *Phys. Rev. Lett.*, Vol. 45, 1980, p. 2081.

Vig, J. R., "IEEE Standard Definitions of Physical Quantities for Fundamental Frequency and Time Metrology–Random Instabilities" (IEEE Standard 1139-1999), IEEE 1999.

Volyanskiy, K., et al., "DFB Laser Contribution to Phase Noise in an Optoelectronic Microwave Oscillator," *Proc. Laser Optics Conference*, St. Petersburg, Russia, June 23–28, 2008.

Wang, L., et al., "Low Power Frequency Dividers in SiGe: BiCMOS Technology," *Proc. Silicon Monolithic Integrated Circuits in RF Systems (SiRFIC)*, Greece, 2006.

Wells, J. N., "Frequency Synthesizers," European Patent EP 125790, November 21, 1984.

Wolaver, D. H., *Phase-Locked Loop Circuit Design*, Upper Saddle River, NJ: Prentice-Hall, 1991.

Wurzer, M., et al., "71.8 GHz Static Frequency Divider in a SiGe Bipolar Technology," *IEEE Bipolar/BiCMOS Circuit and Technology Meeting (BCTM)*, Monterey, CA, 2002.

Yu, N., E. Salik, and L. Maleki, "Ultralow-Noise Mode-Locked Laser with Coupled Optoelectronic Oscillator Configuration," *Opt. Lett.*, Vol. 30, 2005.

Zirath, H., and I. Angelov, "A 10 GHz to 5 GHz Regenerative Frequency Divider," *Proc. 24th European Microwave Conference*, Vol. 2, 1994.

Intermediate Frequency (IF) Receivers

Throughout this book, we have discussed the role of the IF in a superheterodyne receiver. IF filter design and technology have been presented in detail, and a quartz crystal filter design for a 9-MHz IF has been investigated in Chapter 4. The role of double-conversion receivers has been explained in Chapter 5. We have seen that the front end is very important to the overall receiver performance along with equally important follow-up conversions. The noise figure of a system was presented as being key to the performance of an actual receiver. We then investigated the meaning of low intermodulation distortion in the front end and the first mixer of a high probability of intercept (HPOI) receiver. These topics and more were presented in Chapters 12, 13, and 14. We finally discussed the importance of synthesizers and their phase noise performance affecting an entire receiver's behavior in Chapter 16.

We will discuss next the last conversion IF in a communications receiver and how to achieve various ultimate bandwidths using switched or cascaded IF filters.

17.1 Switched and Cascaded IF Filters

The last IF section of a superheterodyne receiver is called the *IF receiver* and is usually treated as a separate design subject. This is where the ultimate bandwidth is achieved in receiver design.

Typically this stage has a very high gain. A major part of the receiver's automatic gain control (AGC) is applied here, as we will discuss in Chapter 18.

In a double- or multiple-conversion superheterodyne receiver, the last IF stage usually contains several narrowband filters that can be selected via an OR function switching mechanism. In an upconvert, downconvert receiver, these filters are usually at a lower IF frequency than the first conversion because high Q filters can be achieved better at these lower frequencies. They are usually of a ladder quartz crystal design such as the 9-MHz filter designed in Chapter 4. They serve as the ultimate bandwidth elements of the receiver prior to detection or to provide preconditioning for the following digital signal processing (DSP) sections, which, in turn, can use bandpass tunable delta-sigma (Δ-Σ) modulators and/or FIR filters.

The switching of different ultimate bandwidth electrical IF filters in a receiver is typically accomplished with silicon (PN) or PIN diodes in a way similar to how preselector filters are switched in the front end. Some receivers use FET switches for better isolation. However, intermodulation distortion caused by these diodes becomes an important factor in the performance of a receiver, and some of the most modern radios today use high reliability RF relays or microelectromechanical systems (MEMS) to perform this switching function. We will discuss next the use of diodes in these switching mechanisms.

The use of diodes in IF filter selectors isn't a new technique. In the past, the standard method of switching filters in an IF used economical silicon (PN) junction diodes such as the IN914, IN4148, or Schottky (hot carrier) diodes to perform this function. However, the use of these diodes has been actually the wrong thing to do in these applications. While these diodes may appear to improve the low-end signal handling of receivers (creating more noise), their overall performance as basic RF switches is marginal when multiple strong signals are present in the RF path. Simply put, silicon or Schottky diodes act as mixers as opposed to near perfect mechanical RF switches, such as relays or MEMS. More often, PIN diodes are being used to provide this switching functionality. To understand why PIN diodes are better RF switches than silicon or Schottky diodes, let's look closer at the various types of diodes available to the receiver designer.

Simple PN junction or Schottky diodes are intended to generate products when presented with different frequency signals. This nonlinear behavior is a desirable feature, which is usually exploited in controlled mixer applications. However, this property is undesirable when these diodes are used to switch bandpass filters in IF circuits or at the input of a radio receiver. Ideally, such function requires hard, mechanical-type switching, such as provided by high-quality RF relays. This kind of switching would be preferable from an electrical point of view, because relays generate very little intermodulation distortion (IMD). On the other hand, using several relays in a complex front end or an IF could prove a cumbersome and expensive proposition. In addition, it may be totally prohibited in certain airborne applications.

PIN diodes, because of their physics, act closer to a perfect distortionless mechanical switch defined in this application as one that will not generate mixing products. If PIN diodes are used to switch in bandpass filters, the receiver performance maintains a certain integrity in the third-order intercept point when compared with either PN junction diodes or Schottky diodes. This translates into crisper, more intelligible weak signal handling in the presence of strong ones (this is not to say that PIN diodes are as good as relays from an IMD point of view).

Before delving further into the physics of PIN diodes, let's look at how diodes are configured as filter switches in an IF. Looking at Figure 17.1, we can see how filter selection takes place in a bank of switched IF filters using diodes.

Looking at Figure 17.1(a), one of several control lines selects a particular filter by turning on the NPN drive transistor Q1 or Q2. DC currents then flow through the Ql or Q2 transistors, and through both ends of the selected filter circuit into the diodes, as shown. The DC path is completed to ground through the current limiting resistors Rl and R2. A similar process is shown in Figure 17.1(b). The difference here is that all diodes have voltage applied constantly to their anode sides with current limiting resistors R3 and R4 placed in series. The circuit is completed to ground through the open collector transistors Q1 or Q2. When properly biased, a filter is selected in much the same manner as shown in Figure 17.1(a). In either case, just one filter is selected at the time leaving the other filters out of the picture. This means that only the pair of diodes corresponding to the selected filter is being biased at any given time. The diodes are said to be forward biased, a state that al-

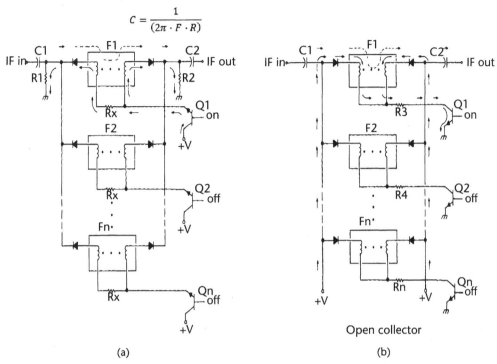

Formula for calculating value of coupling capacitors (C1 and C2) is:

Where F is minimum 3-dB cutoff frequency and R is the value of load resistance

$$C = \frac{1}{(2\pi \cdot F \cdot R)}$$

Figure 17.1 Typical implementation of IF filter selection in a receiver. Silicon (PN) switching diodes or PIN diodes are used to provide the function. (a) Standard forward bias. (b) Forward switching bias with open collector. Note the coupling capacitors basic formula for the frequency of interest. This formula can be used in any capacitor coupling design case from audio to microwave frequencies.

lows RF at the IF frequency to flow through the selected filter and perform the IF filtering function.

Now, let's discuss the occurrence of IMD in these example circuits when using three kinds of diodes, silicon PN junction, Schottky and PIN types.

When using PN junction or Schottky switching diodes, the RF signals coming from the previous stages cross the I-V characteristics of the activated diodes, creating additional unwanted signal products that will show up in the receiver's output. These products translate into distortion and an increased noise floor, which can be confused with high sensitivity by an unskilled user. This can be further aggravated by an inferior synthesizer phase noise performance.

When unmatched PN junction pairs of diodes are used at the input and output of a filter, the various signals going through the biased filter circuit cross the uneven I-V characteristics of the input and output diodes at different points on their curves. This, in turn, creates even more intermodulation products problems. Schottky diodes eliminate some of these problems by providing guaranteed matched I-V curves between the diodes. However, as we previously discussed, Schottky diodes have been designed to generate products, which are useful in mixer applications, not in the static switching of RF filters. The behavior of a switching diode is shown in Figure 17.2.

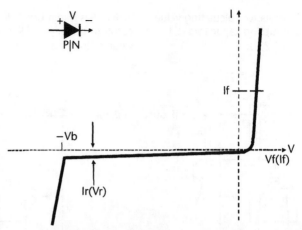

Figure 17.2 Current versus voltage (I-V) characteristic of a typical PN junction signal diode. In standard unmatched diodes, this curve will vary from part to part. In-band RF signals will cross the curves of input and output diodes at different points producing unpredictable distortion. In Schottky diodes, the position of these I-V curves is guaranteed to be the same for all parts over a wide frequency range. Better signal handling results from this, especially at the lower end on the curves, giving an impression of a more sensitive receiver. PIN diodes, on the other hand, minimize these problems by acting as current-controlled RF resistors or switches, without exhibiting the intermodulation distortion problem of the PN junction or Schottky diodes. Although PIN diodes are not perfect switches either, they are recommended for filter switching in IFs and the front-end preselector filter banks.

In contrast with PN junction or Schottky diodes, the switching characteristics of PIN diodes work on a totally different principle. PINs act as variable RF resistors whose resistance depends on DC current excitation. That is, RF energy will pass through these diodes when a small amount of DC current flows through them. In a switched IF filter bank or a front-end switched preselector bank using PIN diodes, RF signals are superimposed over the DC current.

Their incident energy causes an additional rectified current to be generated in the diodes. This, in turn, further lowers the diode resistance in such a way that the RF will flow undisturbed through them. When properly biased, only the fundamental frequencies of the signals pass through the filters, much as with mechanical switches. That is not to say that PIN diodes are as good as RF relays from an intermodulation point of view. They do exhibit a certain amount of intermodulation distortion (IMD).

A closer look at the physics of a PIN diode reveals that it is constructed very differently than a PN junction or a Schottky diode. This is shown in Figure 17.3. While signal diodes are manufactured on the principle of a metal deposited on a P or N semiconductor, the PIN diode is constructed with a thick high resistance nearly intrinsic silicon layer (I) sandwiched between the P and the N semiconducting materials, hence the name PIN diode. The resistance of the I layer with no DC applied to it is on the order of about 10 kΩ. This is known as the isolation property of a PIN diode. The PIN diode's depletion capacitance is also reduced due to the thick intrinsic layer between the P and the N layers, allowing the diode to operate as a good attenuator or switch at much higher frequencies than the regular PN junction diode.

When a PIN diode is forward-biased using the proper current, holes and electrons are injected into the I region from the P and N regions. These holes and

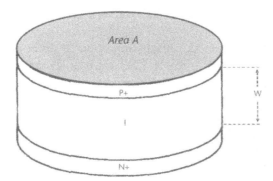

Figure 17.3 Construction of a PIN diode. Unlike a signal diode, the PIN diode uses a thick near-intrinsic silicon material (W-thick), which is sandwiched between the P and the N materials. When properly biased with DC current, RF energy will propagate through the diode with a minimum of internally generated distortion.

electrons don't immediately recombine in the I region as may be expected, but rather coexist for a period of time. This is called latency. The average time before recombination is called the carrier lifetime. The mean distance over which a charge travels before recombination is the diffusion length, which in turn, is related to the carrier lifetime. This injection recombination process is continuous when the diode is forward biased. This results in a steady-state charge in the I layer, which depends on the forward current and the carrier lifetime. The final effect is to lower the resistance of the I region of the diode, slowly making it a better RF conductor. Looking at it another way, it can be said that PIN diodes use their dynamic intrinsic properties to make RF signals remain in an on state longer than PN junction or Schottky diodes, much like relays do, and without the distortion-rich crossings of the I-V curves characteristic of the other two types of diodes. The carrier lifetime dictates the low frequency limit where the PIN diodes operate effectively as current controlled resistors; below this frequency limit, they operate as normal PN diodes.

There are also epitaxial PIN diodes, which are more suited as low current controlled switches, but they are not as linear. These are often called band switch diodes. For IF filter switching, conventional, "thick I section" PIN diodes are usually used to avoid IMD issues. For switching sources or VCO sections, the "thin I section" epitaxial PIN diodes can be used.

We have seen that the RF conductance of PIN diodes depends on the DC current pumped through them. Within reason, the more current is injected the better RF switches PIN diodes make, and the better their intermodulation distortion performance is. Overall improvements of 8 to 10 dB in IP3SFDR have been realized when using PIN diodes over the PN types.

PINs have been classified as power-hungry devices when used in RF switching circuits. In general, current specifications for PIN diodes vary upwards to about 200 mA for a hard turned on diode. Low-current PIN diodes have also been manufactured. For example, the classic HP 5082-3080 PIN diode requires only 10 mA for an on resistance of 8Ω, while the HP 5082-3081 exhibits 10Ω for the same forward current. Other PIN diodes are designed for less than 1Ω for the same current.

17.2   Implementing a High-Performance IF in the Star-10 Receiver

We will now return to our discussion about IF filter switching. Looking at Figure 17.1, it can be seen that the IF filter selector shown is inefficient. That is the OR function approach used to select filters takes advantage of only one filter at the time while leaving the other filters unused. If the filters could be cascaded, the shape factor of the entire system could be improved.

A more effective way of using all IF filters in an IF filter bank is shown in the novel IF approach used in the Star-10 transceiver example. This method is shown in Figure 17.4.

We will next discuss the actual implementation of the cascaded IF as designed and used in the Star-10 example transceiver used throughout the book. The block diagram of the transceiver system showing how the second 9-MHz IF of the receiver fits in the entire system is shown in Figure 17.5.

It can be seen that in this IF, the first and last quartz crystal filters are permanently in the circuit. They have the same bandwidth of 2.4 kHz, which is the maximum required bandwidth for the SSB modulation mode used in this receiver. While the first filter sets the bandwidth, the last filter is intended to clean up the IMD noise generated by the previous gain stages, a common technique used in receiver design. The two AGC IF amplifiers following the first filter provide the main gain of about 80 dB of linear-logarithmic amplification. The two IF filters in the center are inserted in the cascaded circuit by the command and control microprocessor depending on the modulation selected. If not used, they are shorted out. The last IF amplifier is used to compensate for the additional insertion loss of the two filters when inserted. The switching is accomplished with miniature RF relays for best intermodulation performance.

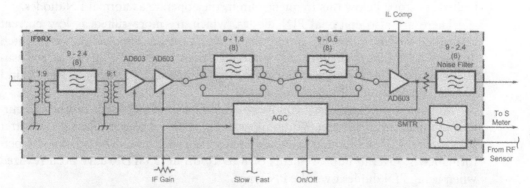

Figure 17.4 A novel approach of selecting IF filters in a cascaded IF. This approach has been used in the Star-10 example receiver. The first and last quartz crystal filters are always in the circuit. They have the same bandwidth, the maximum required as defined by the widest modulation mode required. The last filter is intended to clean up the noise generated by the previous gain stages, a common technique used in receiver design. The two AGC IF amplifiers following the first filter provide the main gain, about 80 dB of linear-logarithmic amplification. The middle two IF filters are inserted in the circuit by the command and control microprocessor depending on the mode selected. If not used, they are shorted out. The last IF amplifier is used to compensate for the additional insertion loss from the various combinations of filters. When compared with the selector method from Figure 17.1 (which usually offers only up to 8 poles of selectivity regardless of mode used) this IF chain switching method offers a minimum of 16 poles of cascaded quartz filtering and as many as 32 poles. (*From:* [1–3]. Courtesy of ARRL.)

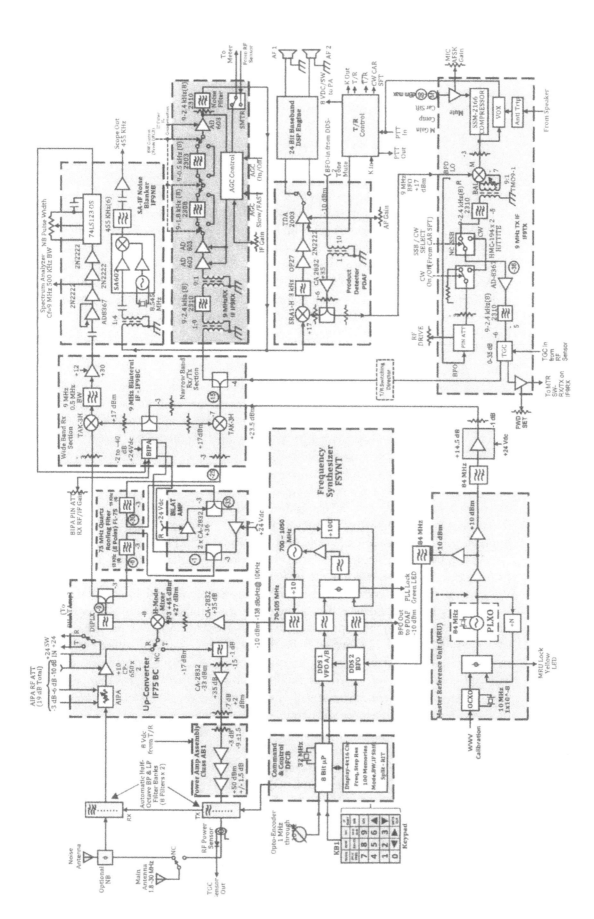

Figure 17.5 Block diagram of the transceiver showing the receiver's 9-MHz cascaded IF assembly. The section is shown in darker gray.

It can be seen that this IF offers a minimum of 16 poles of cascaded quartz filtering and as many as 32 poles when all filters are engaged. This design is much more efficient than the OR function switching of individual filters from Figure 17.1.

We will now discuss the circuit design of the cascaded filter IF. The gain blocks in this IF are the Analog Devices AD 603. This is a voltage-controlled amplifier for use in RF and IF AGC systems. This IC was chosen on purpose because of its logarithmic-linear response, low noise, and high dynamic range. It provides accurate, pin selectable gains of −11 dB to +31 dB with a bandwidth of 90 MHz. Any gain range may be chosen using a single external resistor. Power consumption is 125 mW. The gain-control attack time is less than 1 ms for a 40-dB change. The gain-control interface allows the use of either differential or single-ended positive or negative control voltages. The AD 603 comprises a fixed-gain amplifier, preceded by a broadband passive attenuator of 0 dB to 42 dB, having a gain-control scaling factor of 40 dB per volt. A typical application of the AD 603 is shown in Figure 17.6.

The schematic diagram of the cascaded 9-MHz IF is shown in Figure 17.7. The IF9RX board provides 100 dB of gain (80-dB AGC) using three AD 603 high dynamic range logarithmic/linear IF blocks. The first two amplifiers are AGCed, while the third is used to compensate for different filter combinations insertion loss as we previously discussed in the block diagram from Figures 17.4 and 17.5. The IF bandwidth selection is provided by four cascaded 8-pole crystal filters that were specially designed for the example transceiver.

Instead of selecting individual filters like in conventional IF designs, the Star-10 IF9RX filter assemblies are combined in a cascaded AND function for a total of 32 poles as discussed above. This architecture makes the IF9RX a unique design that works in tandem with the system's command and control software. The actual implementation of the cascaded IF assembly is shown in Figure 17.8.

As can be seen in Figure 17.7, the two 8-pole crystal filters with a bandwidth of 2.4 kHz are always used at the beginning and the end of the 9-MHz IF chain for good noise management. Then, additional 8-pole crystal filters of narrower bandwidths are inserted or removed between the gain stages (for a maximum of 32 poles in CW narrow) depending on the mode selection. The proper selection combination of filters is achieved from the microprocessor controlled unit, which selects the proper combination based on a designated mode priority as impacted by additional operator-selected bandwidth. Automatic insertion loss compensation control is then achieved automatically depending on the diverse filters configurations so there is no difference in signal amplitude when changing filters and bandwidths. Because of the 100-dB gain provided by this board, and because of its limited size of 4.5 × 5.5 inches, this board had to be specially laid out to prevent self oscillation. A special effort was made to provide a layout using many plated through ground stitches in the double-sided ground planes, to make the system as quiet as technically possible. A method of compensating for the filters insertion loss is shown in Figure 17.9.

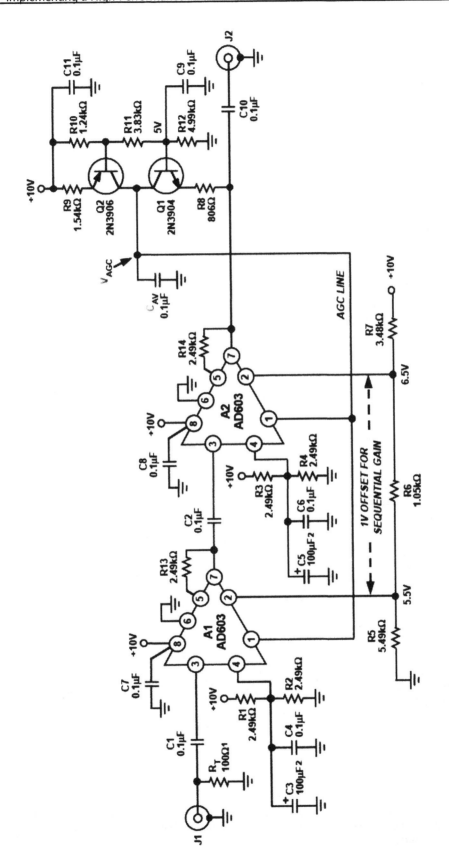

Figure 17.6   Typical application of the AD 603 in an AGCed IF. (Adapted from Analog Devices, Inc.)

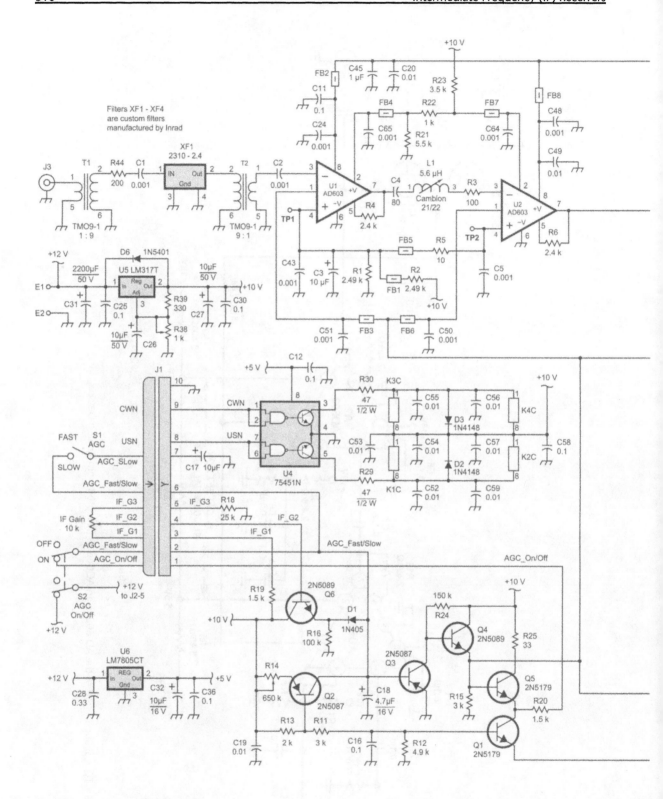

Figure 17.7 Schematic diagram of the cascaded 9-MHz IF assembly. Instead of selecting individual filters like conventional IF designs, the Star-10 IF9RX filter assemblies are combined in a cascaded AND function (rather than an OR function) for a total of 32 poles. (*From:* [1–3]. Courtesy of ARRL.)

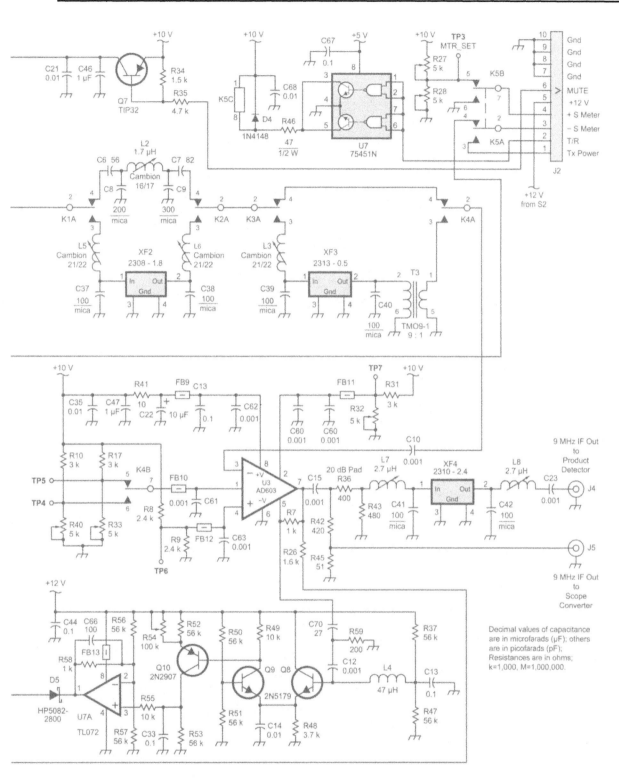

Figure 17.7   (continued)

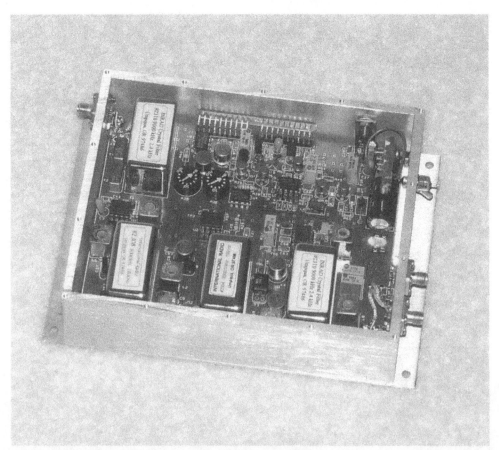

Figure 17.8 Actual implementation of the cascaded 9-MHz IF assembly. Two 8-pole crystal filters with a bandwidth of 2.4 kHz are always used at the beginning and the end of the 9-MHz IF chain for good noise management. Additional 8-pole crystal filters of narrower bandwidths are inserted or removed between the gain stages (for a maximum of 32 poles) depending on the mode selection. Because of the 100-dB gain and the limited board size of only 4.4 × 5.5 inches, the layout of this double-sided plated through board uses many ground stitches between the top and bottom ground planes, to make the system as quiet as possible.

17.3   Logarithmic IFs

Logarithmic amplifiers are used as measurement devices in many applications. Among them are direct RF detectors operating from near-DC to over 15 GHz. They are also used in wideband crystal video and TRF receivers as we will discuss in Chapter 26. In addition, they are used as IFs in radar and spectrum analyzer receivers. Other applications include but are not limited to receivers intended for extracting data such as amplitude-shift keying (ASK), or frequency-shift keying (FSK) in low-cost integrated circuit superheterodyne receivers.

Their purpose is to compress dynamic range (and eliminate the need for AGC) and provide a close logarithmic approximation of the input RF voltage amplitude directly in decibels (dB). Logarithmic amplifiers are also used frequently as video amplifiers requiring a large IF dynamic range to be adapted to a limited input dynamic range of A/D converters.

There are several kinds of logarithmic amplifiers. Among them is the "linear-in-dB" gain type such as the AD 603 discussed in the previous pages. The most

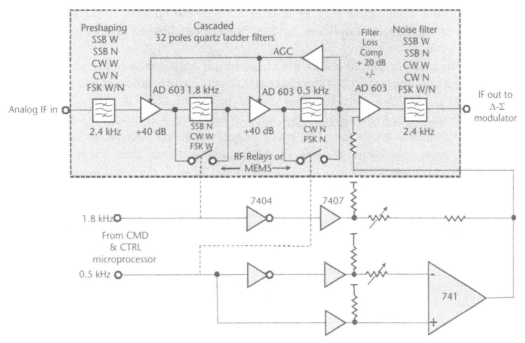

Figure 17.9   Method of insertion loss compensation for a communications receiver in a cascaded filter IF approach using three AD 603 IF amplifiers from Analog Devices.

common logarithmic amplifier is the multistage progressive compression limiting type. This type of logarithmic amplifier uses a rectifier (a detector) for each of the amplifier stages (typically five to ten are used) as shown in Figure 17.10. The outputs are summed together to produce an average filtered voltage, which is scaled in decibels and is sometimes referred to as a received-signal-strength indicator (RSSI) because of its logarithmic nature. This signal can be used directly to detect data or drive signal strength meter indicators in receivers. We will discuss how the

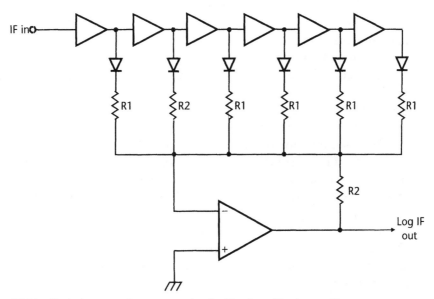

Figure 17.10   Typical progressive compression limiting logarithmic amplifier.

logarithmic RSSI can be used as an IF in the next paragraphs. Additional applications of logarithmic amplifiers include, but are not limited to precision AGC or TGC systems.

The concept behind the limiting logarithmic amplifier is based on the idea of progressive compression of the input signal as it passes through several amplifier-limiter stages. Each cascaded stage usually provides voltage amplification of about 10 dB and a rectified output proportional to the input. The outputs of all stages are then summed together for a composite output logarithmically related to the input.

Logarithmic amplifiers are fundamentally different then linear amplifiers. While linear amplifiers provide gain proportional with the input signal, the logarithmic amplifiers base their performance on the progressive compression properties of the sequential stages. They exhibit a gain inversely proportional to the input voltage magnitude. Their transfer function is a nonlinear relationship between the input and the output as shown in

$$V_{out} = V_y \log \frac{V_{in}}{V_x} \qquad (17.1)$$

where V_{in} is the instantaneous input voltage, V_{out} is the instantaneous output voltage, V_y is the scaling voltage, and V_x is the intercept voltage.

17.4 Using Logarithmic Amplifiers in Low-Cost High-Performance ASK Data Receivers

We will discuss next a receiver design for an ASK wireless data communication application using a logarithmic amplifier integrated in a low-cost RF receiver IC, the inexpensive MC 3372.

This is a low-power RF receiver circuit with an operating RF frequency of up to 60 MHz. Its low-voltage design provides low power drain, excellent sensitivity, and good image rejection in narrowband FM voice and data link applications. This part combines a mixer with a logarithmic IF using a signal strength indicator (RSSI) output. In addition, there is a quadrature detector, an active filter, and a squelch trigger circuit. The oscillator is an internally biased Colpitts type with the collector, base, and emitter connections at pins 4, 1, and 2, respectively. This oscillator can run under crystal control as shown in Figure 17.11. The mixer is doubly balanced to reduce intermodulation distortion. Conversion gain is typically 20 dB.

Although initially intended for low cost FM receivers, the MC 3372 can take advantage of its RSSI logarithmic IF to constitute the base for a reliable data communications receiver using amplitude-shift keying (ASK). Such a receiver can use many remote control applications (e.g., model aircraft) operating at distances of up to a half a mile (0.8 km) or more if used in combination with an automatic threshold detector inspired by the U.S. Patent 2,999,925, and a 1-watt data transmitter. The variable threshold detector circuit improves the probability of intercept considerably in multipath situations. As much as a 10-dB improvement over regular circuits has been realized using this circuit.

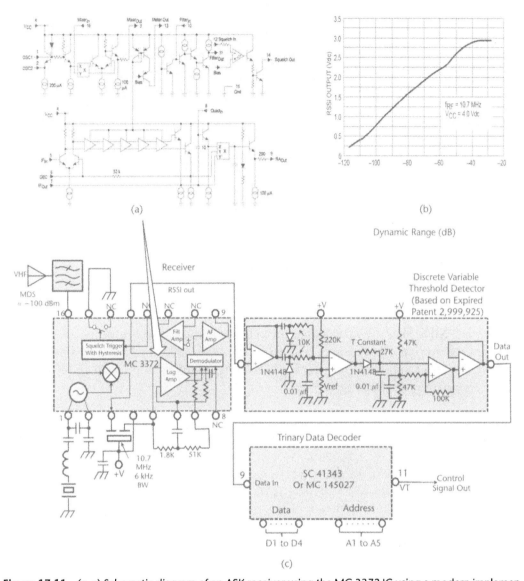

Figure 17.11   (a–c) Schematic diagram of an ASK receiver using the MC 3372 IC using a modern implementation of the variable threshold detector (Based on patent 2,999,925) and a trinary data decoder that can provide up to 243 commands. The design takes advantage of the progressive logarithmic amplifier in the MC 3372 receiver, which is intended as the RSSI. The signal is then threshold detected automatically from an MDS of about −100 dBm and up. The receiver design can be used reliably over a distance of a half a mile (0.8 km) when equipped with a miniature antenna at VHF [4, 5].

The low-cost ASK receiver design is shown in Figure 17.11(c). It is made of three basic circuits, the MC 3372 receiver, the automatic threshold detector mentioned above, and a trinary data detector usually intended for decoding infrared commands on TV remotes. The FM section of the MC 3372 is not used in this application.

The receiver and its logarithmic IF are shown in Figure 17.11(a), while the dynamic range response of the receiver is shown in Figure 17.11(b).

The mixer amplifier converts a VHF amplitude-shift keying (ASK) input signal to a 10.7-MHz IF. This IF is in turn filtered via a ceramic or quartz IF filter with a 3-dB bandwidth of 6 kHz. Because of the AM approach, the ASK modulation is very forgiving from a frequency stability point of view and such a bandwidth can easily accommodate up to 2.4-kbps data rates over a wide temperature range. Although ASK is considered to be inferior to other modulation methods from a bit error rate point of view, the addition of the variable threshold detector to this circuit makes it comparable with more sophisticated modulation schemes. This, in turn, makes it an excellent choice for low-cost RF data communications systems.

As can be seen from the figure, the IF signal output is fed through a logarithmic limiting amplifier and a detection circuit. Because of the logarithmic nature of the receiver, there is no need for AGC. The receiver operates over the dynamic range of the logarithmic IF amplifier with an MDS of approximately −100 dBm, which includes the mixer conversion gain, and when using a well-tuned VHF miniature antenna.

17.5   Variable Passband Filters and Analog IFs

Variable bandwidth can be implemented in a receiver as shown in Figure 17.12. In this example, the bandwidth of a 455-kHz filter is slid through the bandwidth of the 9-MHz filter (by varying the 455-kHz BFO) and only the information contained in the overlapped portion of the two filters is passed to the baseband amplifier. Although this approach to variable passband is still being used, more modern receivers perform this function in DSP.

17.6   Noise Blankers

The IF amplifier is the place where noise blankers are usually implemented. The ideal function of a noise blanker is to discriminate against all forms of electrical noise, which could interfere with the desired information. The degree of discrimination is limited by the nature of the noise and the type of noise blanker used.

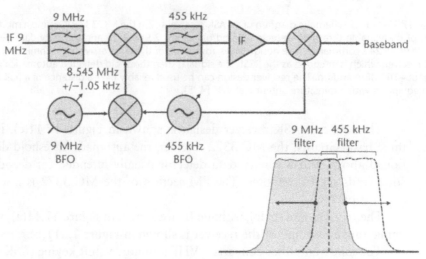

Figure 17.12 Concept of achieving variable IF bandwidth uses two IFs and two filters to provide continuous bandwidth adjustment.

Consequently, noise blankers take many forms and are not perfect in their performance. Shown in Figure 17.13 are simple forms of noise blankers: the shunt and series diode limiter. These methods of reducing pulsating noise can be implemented in the IF or baseband circuits of a receiver. Looking at Figure 17.13(b), we can see a resemblance with the filter switching mechanism from Figure 17.1. This circuit operating as a limiter works as follows.

If a sufficiently positive DC voltage is applied to the anodes of the diodes through the 6.8 kΩ resistor, the diodes will conduct. As long as the diodes remain conducting, an AC signal applied to the cathodes of the two diodes will appear clipped if the two diodes are well balanced. If the positive peak(s) of the signal are high enough, the diode on the left will be momentarily reverse biased and the signal peak(s) will be truncated. Likewise, if the signal peak(s) go negative enough, the right diode will be reverse biased and the negative peak(s) will be truncated. Therefore, we have a variable threshold signal peak limiter—the threshold being set by the 10-kΩ potentiometer. A simple 0.01 μF is used as a simple filter to help "smooth out the rough edges" of the clipped signal.

This low-cost method provides relatively poor results compared to the more sophisticated approaches. However, it can improve on signal pumping AGCs and certain kinds of pulsating noise.

Shown in Figure 17.14 is another popular method of implementing a noise blanker circuit. The IF amplifier chain uses dual MOSFETS and is sampled for pulsating noise prior to the crystal filter over a wide bandwidth, allowing for amplification and detection of fast rise-time pulses, which are then amplified by another MOSFET, a DC amplifier, which in turn will make the 2N2222 transistor provide the ground path for the IF signal during the duration of a pulse. The net result of this effect is that the receiver will be blanked during the pulse, obscuring the noise, and providing improved copy of the desired signal. The disadvantage of this noise blanker is that it responds only to a limited category of pulses, ignoring others.

17.7   The Variable Pulse Noise Blanker and the Star-10 Receiver Noise Blanker

There are several other noise blanker designs today. Among them, the variable pulse noise blanker using one-shots. This is the method used in the Star-10 receiver as shown in Figure 17.15.

The phase cancellation method introduced by Collins Radio uses a separate receiver, which is tuned to a frequency where there are no intelligent signals expected. The pulsating noise is processed through a superheterodyne or a TRF approach and the detected signals are used to shut off the main receiver much the same as previously described.

While the methods presented in this chapter are effective in most noise situations, sometimes they do not perform as well depending on the nature of the pulsed signals received. New and complex adaptive noise blankers using a combination of methods including DSP cancellation filters process the information received and adapt to the nature and length of the signals, providing selective cancellation of certain types of pulses.

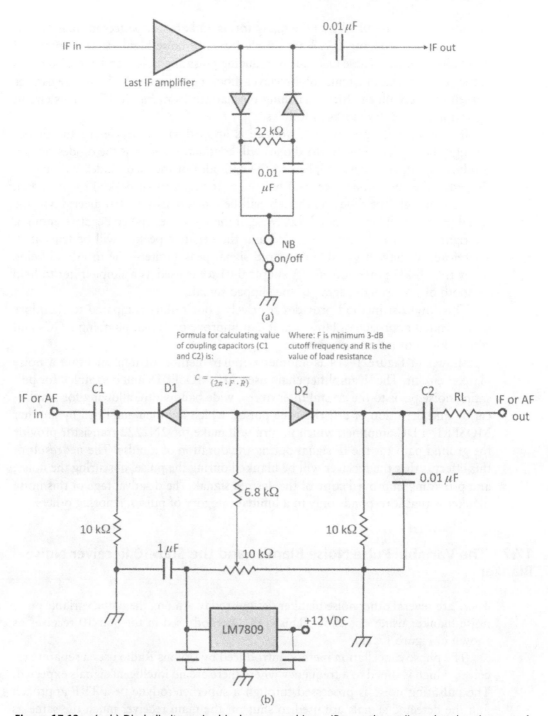

Figure 17.13   (a, b) Diode limiter noise blankers as used in an IF or at the audio or baseband stages of a communications receiver.

17.8   The Notch Filter and the Bandpass Tuning Mechanism

The notch filter is a very useful function in a communications receiver. Its purpose is to allow the operator to cancel one or more interfering signals, which fall within

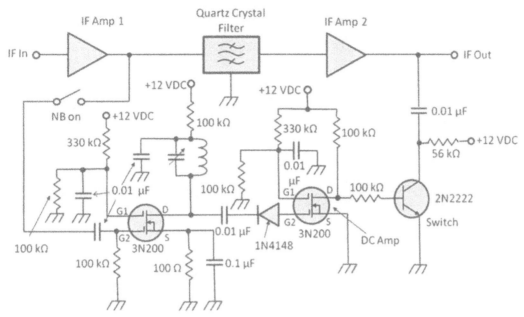

Figure 17.14   Implementation of an IF noise blanker circuit.

the receiver's IF passband without impacting the dynamic range. This is usually accomplished at baseband with the help of a tunable audio rejection filter of very high Q. There are also front-end cancellation filters and DSP notch filters.

Figure 17.16 shows the schematic diagram for a variable Q, wide-range notch filter, which can be used in a communications receiver and can provide 40 dB of attenuation within its passband. The design is a twin "T" network filter using feedback. The frequency control is performed by R2 and R3, which are two potentiometers coupled together on the same shaft.

This example shows that the notch filter is a powerful tool in combating unwanted interference. A more efficient way of implementing notch filters is to design them directly at IF frequencies prior to AGC.

Figure 17.17 shows a means of pulling a 9-MHz crystal in an IF providing a variable notch over the bandpass of the IF where a DC voltage is used with a wide-range varactor in a series-resonant crystal configuration. Better than 50 dB of rejection can be achieved with this method.

These examples show that implementing bandpass tuning in a communications receiver is a complex matter. High-level mixers should be used to minimize intermodulation distortion, which could be created quite easily in multiconversion systems.

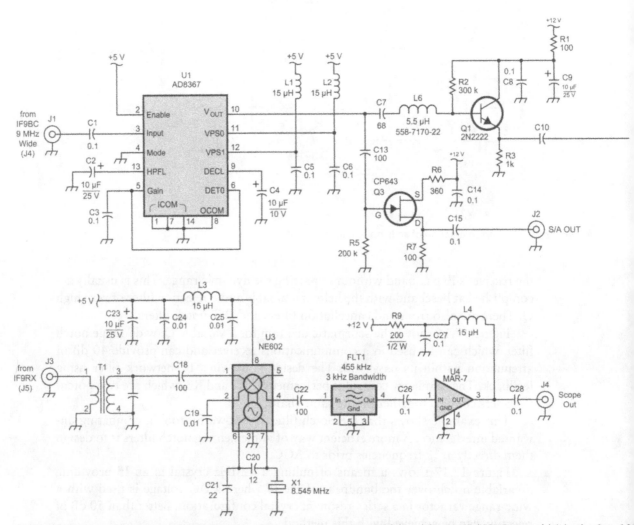

Figure 17.15 A more complex noise blanker implemented in a separate wideband 9-MHz IF (IF9NB assembly) in the Star-10 receiver. The wideband 9-MHz IF is derived separately (not from the narrow band IF9RX). Response time is adjustable via the one-shot (74HCT123) to the required noise pulse width. The output of the one-shot drives the PIN diode attenuator at BIPA (shown Figure 18.5) located in the second converter assembly (IF9BC). Additional functions such as wideband spectrum analyzer and oscilloscope processing are also provided by IF9NB assembly. (*From:* [1–3]. Figure courtesy of ARRL).

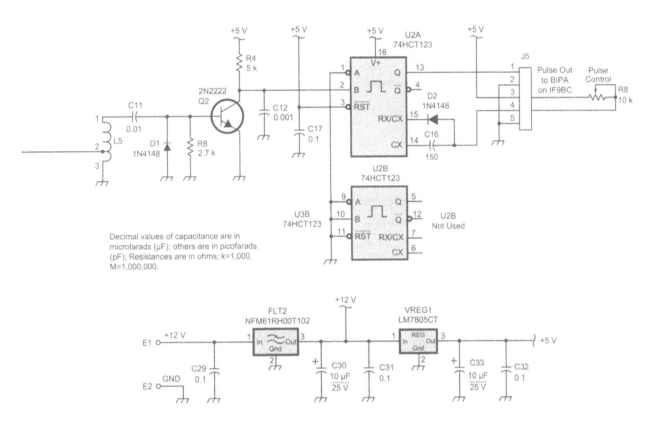

Decimal values of capacitance are in
microfarads (μF); others are in picofarads
(pF); Resistances are in ohms; k=1,000,
M=1,000,000.

Figure 17.15   (continued)

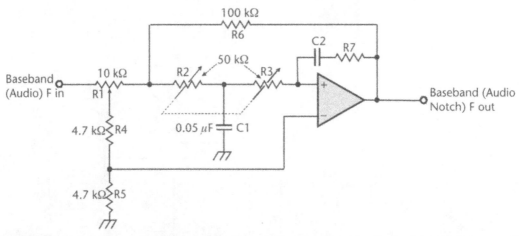

Figure 17.16 A simple and effective tunable T-notch filter can provide 40 dB of attenuation at audio frequencies.

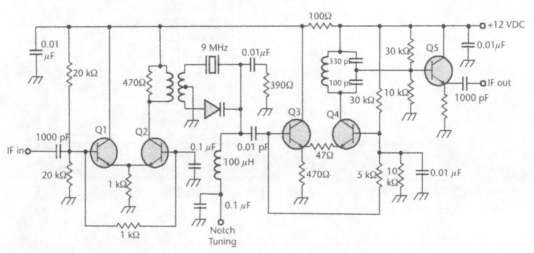

Figure 17.17 Implementation of an IF notch filter.

References

[1]   Drentea, C., "The Star-10 Transceiver, Part 1," *QEX—ARRL*, November/December 2007.

[2]   Drentea, C., "The Star-10 Transceiver, Part 2," *QEX—ARRL*, March/April 2008.

[3]   Drentea, C., "The Star-10 Transceiver, Part 3," *QEX—ARRL*, May/June 2008.

[4]   Thomas, E., "Variable Decision Threshold Computer," U.S. Patent 2,999,925.

[5]   Rousos W.N., and R.B. Denny, "Threshold Correction System in FSK Transmissions," U.S. Patent 3,947,769.

Selected Bibliography

Bullock, S. R., *Transceiver and System Design for Digital Communications*, Atlanta, GA: Noble Publishing, 2000.

Bullock, S. R., "Wireless Data Links & Spread Spectrum Communication, Fundamentals for Communication and Guidance Control of Missiles, Bullock Engineering Research."

Carson, R. S., *High-Frequency Amplifiers*, New York: John Wiley & Sons, 1975.

Carson, R. S., *Radio Communications Concepts*, New York: John Wiley & Sons, 1989.

Drentea, C., "The Art of RF System Design," RAD047, Raytheon.

Drentea, C., "Designing a Modern Receiver," *Ham Radio*, November 1983.

Drentea, C., "Improving Receiver Performance in Modern Transceivers," *Communications Quarterly Journal*, Fall 1991.

Drentea, C., *Radio Communications Receivers*, New York: McGraw-Hill, 1982.

Edde, B., *Radar, Principles, Technology, Applications*, Upper Saddle River, NJ: Prentice-Hall, 1993.

Hardy, J., *High Frequency Circuit Design*, Reston, VA: Reston Publishing, 1979.

Hayward, W., *Introduction to Radio Frequency Design*, Upper Saddle River, NJ: Prentice-Hall, 1982.

Helfrick, A. D., *Electrical Spectrum and Network Analyzers*, San Diego, CA: Academic Press, 1991.

Hickman, I., *Practical RF Handbook*, Boston, MA: Newnes, 1997.

Inglis, A. F., *Electronic Communications Handbook*, New York: McGraw-Hill, 1988.

Krauss, H. L., C. Bostian, and F. H. Raab, *Solid State Radio Engineering*, New York: John Wiley & Sons, 1980.

Sabin, W. E., and E. O. Schoenike, (eds.), *Single Sideband Systems and Circuits*, New York: McGraw-Hill, 1987.

Shrader, R. L., *Electronic Communication*, New York: McGraw-Hill, 1959.

Skolnik, M., *Radar Handbook*, New York: McGraw-Hill, 1970.

Stinson, G. W., *Introduction to Airborne Radar*, New York: IEEE Press.

Stremler, F. G., *Introduction to Communication Systems*, Reading, MA: Addison-Wesley, 1977.

Vizmuler, P., *RF Design Guide*, Norwood, MA: Artech House, 1995.

CHAPTER 18

Automatic Gain Control (AGC)

18.1 Introduction

Automatic gain control (AGC) is the electronic equivalent of a natural control loop such as the one used by the human eye when entering a dark room from a bright light environment. Upon entering the room, the iris opens up to let more light in. Conversely, when going from a dark room to a bright environment, the iris closes down to limit the amount of light falling on the eye's retina. While this seems like a simple concept, the AGC used in receivers can be a more complex problem.

18.2 Linear Control Systems—Feedback Systems and Their Significance in Receivers

Just as in the eye example, AGC is necessary in receivers to compensate for large slow variations in signal strength at the antenna terminal due to propagation phenomena such as ionospheric multipath fading.

However, in addition to this, an AGC system also has to compensate for fast selective fading caused by either the ionospheric multipath, in the case of HF communications, or by urban roaming multipath in VHF/UHF and microwave communications. In airborne radar, this becomes even more complicated as large signal reflections, known as clutter, caused by mountains or water bodies can obscure small signal reflections. This, in turn, can cause early compression of the system's gain stages, resulting in intermodulation distortion and loss of information.

The slow fading variations can sometimes exceed 100 dB, while the fast fading variations can exceed 30 dB. They can modulate a receiver's AGC at more than a 100-Hz rate. A proper AGC system has to compensate for all these problems at the same time without distortion.

18.3 Achieving High Dynamic Range with AGC: The Concept of Composite Dynamic Range

The major function of an AGC is to extend the limited linear dynamic range of a receiver system in the face of the above conditions. Without AGC, the linear dynamic range of a receiver would be limited by the compression points of its stages.

This is true in analog radios as well as in digital signal processing (DSP) receivers using direct sampling techniques, where the processing gain and compression point equivalents are dictated by the number of bits. Consequently, a receiver without AGC has a very limited linear dynamic range.

The AGC is intended to extend the composite linear dynamic range (discussed in Chapter 13) of a given receiving system such that any signal, from the MDS and up to a large RF input signal level can always be reproduced with high fidelity at the output of the receiver, combating the system's limited compression point as well as the propagation phenomenon slow and fast variations. The more stages requiring AGC in a receiver system, the larger the composite linear dynamic range is as previously explained in Chapter 13.

The AGC can also serve to ensure the delivery of a proper signal level to a limited dynamic range A/D (meaning a limited number of bits), in the last stages of an otherwise analog receiver. In this case, the signal presented to an A/D would be always funneled to approximately one third of the way down from the compression point equivalent of the A/D's dynamic range regardless of how small or how

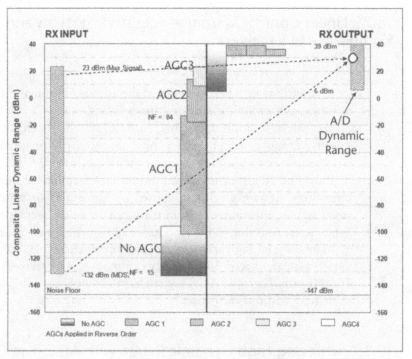

Figure 18.1 The effect of a triple AGC on the linear composite range of a high dynamic range receiver. This plot is the computer modeling for the actual performance of the Star-10 receiver. The left column shows the ramping of the RF input signal in actual dBm readings from the MDS to the compression point of the system using the total AGC action. The composite linear dynamic range in decibels is obtained from subtracting the MDS value in dBm from the system's compression point in dBm. The result is the linear composite dynamic range expressed in decibels. The middle bands show the compression points of the various AGC sections, indicating where the next AGC action overlaps and starts. The band on the far right shows the funneled uncompressed signal range after all AGC actions have been performed. If an A/D converter is used at this point, the signals from the large RF input range funnel through undisturbed about a third of the way down from the top of the A/D dynamic range equivalent as shown.

large the input signal to the receiver is within its composite dynamic range. This is shown in Figure 18.1.

18.4 Deriving and Applying AGC in Receivers

The AGC mechanism is intended to follow RF input variations rapidly and properly react to them, so that the output will remain nearly constant over time variable RF input amplitude ranges. The AGC process can become even more complicated by accommodating for different modulation schemes and transmission data rates, and suppressing the effects of adjacent channel interference and noise.

An AGC system usually behaves like a controlled negative feedback loop. In a multiple stage/multiple conversion receiver, the AGC control is always applied to various stages in reverse order, from the back end of the receiver toward the front end as shown in Figure 18.2. It can be seen that the impact of AGC action on the total noise figure of a receiver is due to inserting a loss between its stages or reducing gain in variable gain amplifiers (VGA). As previously discussed, loss means increased noise figure for the individual stages as well as for the entire system. Thus, the noise figure of such a receiver is directly proportional with the amount of AGC applied, with the front end being by definition the most affected stage.

That is why, as the RF input signal increases, the AGC controls are applied progressively from the last stages toward the front stages of the receiver, but only as needed such that the total noise figure of the receiver at any point in time does not exceed the RF level of the received signal. As the input RF signal increases, the AGC action switches closer toward the front-end stages. As soon as each AGC stage nears compression level, the next AGC stage toward the antenna is activated. The higher the input RF signal is, the more AGC action is applied closer to the front end (and consequently the higher the total system noise figure becomes). However, the noise figure of the entire system remains always below the RF signal level so best SNR is guaranteed.

We have seen how the highest RF input signal activates the last AGC stage, which is the front end of the receiver where the noise figure and the RF input signal are the highest; consequently, the RF is always higher than the noise figure. The result is good signal-to-noise (SNR) ratio at any RF level over the entire linear composite dynamic range of the receiver, from the MDS to the highest RF level allowed.

AGC performance can be analyzed using variations of the universal feedback or feed-forward control system theory. The methodology is similar to that used in negative feedback systems or the phase-locked loop (PLL) control theory. The difference between PLL control theory and AGC control theory is that instead of a phase or frequency-correcting mechanism it uses a gain-correcting mechanism. Most negative feedback loops can be analyzed and explained as simple second-order systems. However, they can get a bit more complicated due to the linear in dB gain response and the need for linearization, which is usually determined experimentally.

An AGC system is considered a nonlinear system. Conventional feedback loop theory, such as negative feedback or phase-locked loop, deals with subtraction. However, in AGC feedback theory, we deal with multiplication. Many theoretical models have been presented in the literature. Several are listed in the References

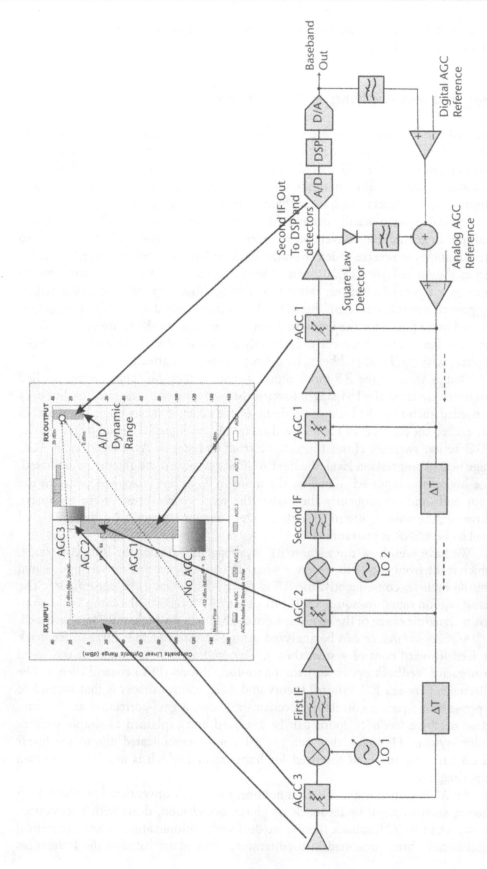

Figure 18.2 Complex AGC system as applied to a double-conversion receiver using DSP. The AGC is applied from the back end of the receiver progressively toward the front end. As soon as the AGC stage approaches compression, the next AGC stage to the left starts being controlled. The process continues until the last stage (the front end) is AGCed. Noise figure is progressively increased as the front end is approched. However, the SNR is always maintained because the signal level is always higher than the noise.

and Selected Bibliography at the end of this chapter. We will use [1] and [2] for our discussion.

An AGC system model is shown in Figure 18.3. Looking at Figure 18.3 AGC model, we can see that the relative change in output voltage V_o is a nonlinear function of V_{in} and V_{ref}.

The fluctuating RF/IF input signal V_{in} is amplified by the variable gain amplifier (VGA), which is controlled by the correction signal $x_{(t)}$. This results in the output signal V_1, which is further amplified by the fixed gain amplifier A intended to produce a proper output V_o for the following stages as well as having enough signal to perform an envelope detection process. The signal is sampled by the diode detector D and is further presented to the lowpass filter F, which strips all modulation components information or transients from the detected signal leaving only the variable carrier amplitude level information. The V_{ref} allows the system to be triggered based on a threshold level and keeps the output signal at some maximum value until the next RF/IF burst happens.

The dynamic response of the resulting AGC signal is generally slower when compared with the RF/IF frequency sampled. Consequently, the bandwidth of the AGC loop has to be limited to a value lower than the lowest modulating frequency component. On the other hand, the response has to be fast enough that the AGC system can produce a rapid control voltage, which must be applied to the VGA at the beginning of each burst message element received. The transfer functions for the AGC model in Figure 18.3 are as follows. The output of the VGA is given by:

$$V_1 = V_{in} \cdot P_{(x)} \tag{18.1}$$

Looking at Figure 18.3, the equivalent gain of the VGA is then given by the operating point p:

$$p = \frac{dV_1}{dx} = V_{in} \cdot \frac{dP_{(x)}}{dx} \tag{18.2}$$

The loop gain (L_g) is then expressed by

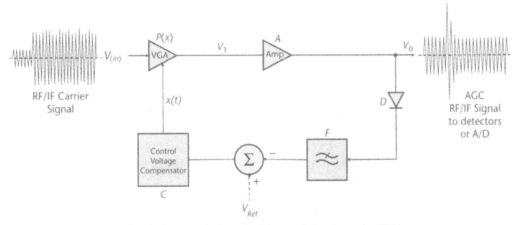

Figure 18.3 AGC model block diagram. (Adapted with permission from the IEEE.)

$$L_g = C \cdot p \cdot A \cdot F \cdot D \tag{18.3}$$

Equations (18.2) and (18.3) show that the loop gain is usually dependent on the carrier amplitude V_{in}. This is apparent when a linear multiplier type control amplifier is used. In this case, the gain control of a linear multiplier is:

$$P_{(x)} = V_{in} \cdot k \cdot x \tag{18.4}$$

where k is a constant. Equation (18.3) then becomes:

$$L_g = C \cdot k \cdot V_{in} \cdot A \cdot F \cdot D \tag{18.5}$$

Consequently, the loop gain L_g is directly proportional to the carrier V_{in} amplitude.

The purpose of the above discussion is to derive an automatic gain control law, which will maintain a constant amplitude output using a constant loop gain which is independent of the V_{in} amplitude. We will call this output amplitude $V_o{}^\circ$. Then:

$$V_o^0 = V_{in} \cdot P_{(x)} \cdot A \Big|_{o=V_o^0} = A \cdot V_1^0 \tag{18.6}$$

For a constant p, we can then obtain a constant loop gain expression:

$$p = \frac{dV_1}{dx}\Big|_{V_1=V_1^0} = \frac{V_1^0}{P_{(x)}} \cdot \frac{dP_{(x)}}{dx} = K_1 \tag{18.7}$$

where K_1 is an arbitrary constant. Then P_x can be obtained from (18.6) and (18.7):

$$P_{(x)} = K_2 \cdot \exp\left(\frac{K_1 \cdot A}{V_o^0} \cdot x\right) \tag{18.8}$$

where K_1 and K_2 are arbitrary constants. Consequently, by forcing a loop constant gain, which is independent of the carrier amplitude, the control law of an ideal controlled gain amplifier results in:

$$L_g = C \cdot K_1 \cdot A \cdot F \cdot D \tag{18.9}$$

An AGC system using this control law will react rapidly to the RF/IF input amplitude, but the control signal will be independent of the modulation frequency. This, in turn, guarantees a system with an exponential control effect, which ensures a fast and accurate AGC action over a wide signal range lending to using transconductance amplifiers as the VGA.

Practically speaking, we can use analog circuit simulators to model AGCs and other feedback systems by replacing various system elements with op amps as amplifiers or integrators. Most AGC system designs can be satisfied with simple second-order feedback loop designs.

The first thing in designing a receiver AGC system is to determine the total amount of controlled gain needed for a given linear composite dynamic range. This was discussed in Chapter 13 and at the beginning of this chapter. Based on the calculated MDS and the composite linear dynamic range, and using a cascaded linear gain stacking tool, we can determine the amount of AGC gain required and where in a system this gain is placed.

By using a cascaded gain computer program capable of computing noise figure and compression points, the designer will have to thoroughly investigate different system configurations as impacted by real parts until a minimum of AGC stages is modeled with the widest possible linear composite dynamic range. This process can take a long time in order to balance the various portions of the AGC and prove that no compression occurs in any of the stages over the entire RF input linear dynamic range. This is an important part of the receiver system design. Only then can we proceed further with the actual design.

Another important step in designing an AGC system is determining the amount of gain (or loss) needed for the individual AGC stages. The required amount of gain variation has to be matched with the variable gain amplifiers chosen along with the voltage-controlled ranges. As gain does not usually follow a linear voltage curve, linearization of the curves has to be fully understood and corrected; otherwise distortion will occur. Simple linearizer circuits can be used to correct gain variations in the AGC slopes in order to avoid variations in loop gain and noise bandwidth. Linearizers can be designed using diodes and op amps to achieve slope corrections approaching the required linear curves. For better results, independent logarithmic or linear logarithmic amplifiers can be used for detection and generation of feedback signals.

Most important in the analysis is determining the lowpass filter's bandwidth. The bandwidth's corner frequency should be at least ten times lower than the minimum modulation frequency to be passed through the IF chain, so no modulation distortion is reflected in the AGC action.

Excessive loop bandwidth can also cause gain pumping despite theoretical analysis. This can deteriorate signal fidelity especially under rapid fading situations. Oftentimes, the loop dynamics and bandwidth of an AGC are determined experimentally (after the theoretical work has been done) by actually observing the response of a brassboard on an oscilloscope under a proper modulation simulation in order to determine the exact characteristics of an AGC. The results are then compared with the theoretical analysis in order to make the right decisions.

18.5   Understanding and Using Logarithmic Detectors

Today, logarithmic amplifiers, such as those discussed in Chapter 17, can be used successfully to provide logarithmic in dB linear AGC feedback. These devices can be easily applied to an AGC for a feedback or feed-forward control system.

This type of detector produces an output proportional to the logarithm of the RF input voltage as previously discussed. Because this behavior is complementary to that of the linear-in-dB VGA in the loop, the resulting loop dynamics are those of a linear system, assuming that signal level fluctuations during transients remain within the measurement range of the logarithmic detector. Subject to that assumption, the AGC loop's response to large and abrupt changes in RF/IF input level will not be impacted.

18.6 Square-Law Detectors

Square-law detectors are often used in AGC systems. They exhibit an instantaneous output that is proportional to the square of the RF/IF input voltage. Thus, the output is directly proportional to the input power. When used with a wide-bandwidth AGC loop, the square-law detector provides an average output independent of the modulation. Just as in the case of a simple envelope detector, the rectified output can only be positively affected by the lowpass filter incorporated into the loop. This results in the loop having a limited instantaneous response to fast transitions in the RF signal amplitude.

18.7 True-RMS Detectors

This kind of detector uses a standard square-law detector and a lowpass filter followed by a square-root function. The lowpass filter function performs the "mean" operation associated with the root-mean-square (RMS) function, and it should have a sufficiently long time constant to smooth the output variations of the squaring detector that would otherwise arise from the modulation of the signal. Because of the square root element in this detector, the average output is proportional to signal voltage, not the power. Therefore the loop's response to small, abrupt decreases or increases of signal level are basically the same as those of an envelope detector, provided that the added filter pole within the RMS detector is compensated correctly elsewhere in the loop. The fact that the added pole is located in a region of the signal path that behaves by the square law actually improves on the large-step response, making the transition smoother than in the case of the simple envelope detector. Note that the true RMS detector has a slightly slower recovery time from large downwards amplitude steps than the standard envelope detector, but a slightly faster recovery time (with a small overshoot) from a step up in the RF/IF input amplitude.

As in the square-law detector case, the true RMS detector will result in a more stable loop equilibrium, which is independent of the RF signal waveform.

The long time constant of a lowpass filter in such a detector impacts the loop dynamics of the AGC loop. This filter will then provide a dominant role in the AGC design. The loop bandwidth must then be well coordinated with the remainder of the entire AGC for good stability.

18.8 Attack and Release Time, Hanged AGC, and the Star-10 AGC System

In benign general receiver applications, AGCs use 50% duty cycle for attack and decay times. However, specific modulations such as SSB or burst data communications require fast attack and various degrees of slow release times.

In a typical AGC system, the attack time usually depends on natural delays in circuits and filters. The release time depends on the discharge of a capacitor. Although these are desired functions for SSB or KCW reception, the false alarm caused by noise bursts can deactivate the receiver for an unnecessary amount of time, thus the dual action and "hanged" AGCs were created. The "hanged" AGC is designed to change the time constants and compensate for noise bursts, while still maintaining the characteristics of the fast attack slow-release AGC for SSB and KCW signals. This type of AGC can also be viewed as a noise discriminator. Such a design usually involves IF as well as audio- or baseband-derived AGC systems. For receiving a single-sideband signal, a 2-millisecond rise time is usually sufficient to overcome any thumping tendency at the beginning of a word. The release requirement is much longer, 2 to 3 seconds in the case of voice communication using SSB. The fast attack, slow release concept has been used extensively by the manufacturers, but the delay timing requirements have been disputed by operators. As a generality, typical circuits having attack times ranging from 2 to 200 milliseconds and release times ranging from 0.5 to 4 seconds have been used extensively in communications receivers.

The Star-10 IF and AGC example from Figure 17.7 uses a combination of fast attack and slow release IF-derived AGC by utilizing several application points of the AGC signal as was explained in Figure 18.1 and is summarized in Figure 18.4.

The IF-derived AGC obtains its information at the output of the last IF stage, before the last noise filter, usually through a simple Schottky diode detector. The averaged DC voltage obtained here is further amplified and applied to all the gain stages in the IF amplifier, reducing the gain as necessary. The delayed release is

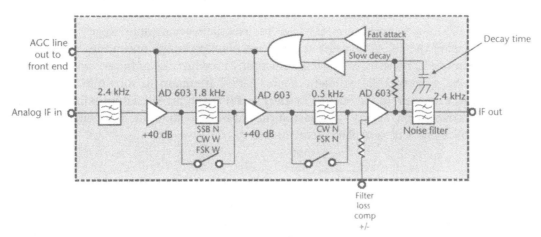

Figure 18.4 Block diagram of the fast attack, slow release AGC system in the Star-10 IF receiver. Circuit details can be found in the IF schematic diagram from Figure 17.7.

applied separately as shown. Most communications receivers today use IF-derived AGC. In some cases, linearization requires additional tapping of signals from several points in the IF.

18.9 Audio-Derived AGC

Another type of AGC is audio derived. In this approach the correction voltage necessary to control the gain in the IF stages is obtained from the audio section of the receiver. The rectified voltage is further amplified and applied to the gain stages of the IF amplifier much the same way as in the IF-derived AGC.

This method of AGC is not very popular in communications receivers because of its relatively poor performance. The narrow bandwidth of the signal at this point in a receiver makes for relatively long delays, which, in turn, do not allow for millisecond rise times and produce AGC gain pumping.

18.10 The PIN Diode Attenuator Used for AGC

As we discussed before, it is possible to apply AGC to the front end of a communications receiver (as a last resort), providing attenuation directly at the antenna. This makes sense as AGC is applied progressively from the back toward the front for judicious noise figure management. This technique is used successfully to prevent front end overload, considerably reducing intermodulation distortion in this stage and beyond. This is usually accomplished by using a PIN diode attenuator, which is used in series with the antenna terminal. The PIN diode will conduct RF only if biased properly as we previously discussed.

A ladder network of several PIN diodes could be substituted for the single diode, providing a greater range of attenuation. This approach can provide as much as 80 dB of attenuation, but greater attenuation cannot usually be achieved because of the physical layout, which makes the "wrap around" leakage phenomenon (or isolation) the limiting factor (100 dB of isolation has been achieved with extreme care in physical design).

PIN diode attenuators can also be used between various stages of IF chains replacing the VGA, in order to improve upon existing conventional AGCs. The technique is, however, somewhat reluctantly used because of the possible intermodulation distortion that it might create. A PIN attenuator is shown in Figure 18.5. It has been used in the Star-10 receiver as BIPA and provides approximately 30 dB of variable gain. This is only part of the entire AGC loop, the rest being applied to the IF amplifiers as well as the front end.

18.11 Digital AGCs

Another AGC control method, which is relatively unknown, is the digital counter AGC. An implementation is shown in Figure 18.6. In this approach, the AGC output is a function of an 8-bit digital number, which is present at the binary inputs of a digital-to-analog (D/A) converter. This number is continuously changing, and

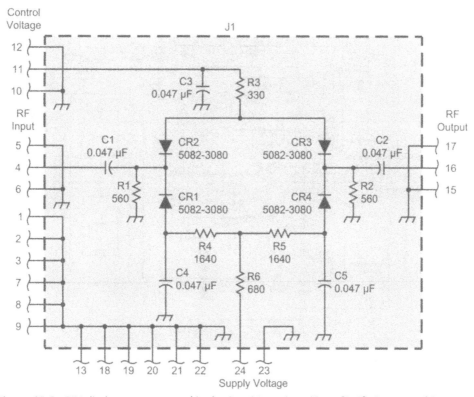

Figure 18.5 PIN diode attenuator used in the Star-10 receiver. (*From:* [3–5]. Courtesy of ARRL.)

reflects the counting state of two cascaded up/down counters (74C193) at any given time. A clock is fed to the counters and when the count-up command is activated, the AGC voltage at the output of the D/A converter will increase with every count until reaching the maximum voltage allowed by the number of bits. It is evident that the attack time of this voltage is directly proportional to the frequency of the clock. The higher this frequency is, the faster the maximum digital number (and therefore the maximum AGC voltage) is reached. When the countdown line is activated on the up/down counters, and the clock is slowed down, a reduction in the AGC voltage will be accomplished, but this time at a slower rate than when it went up. The attack and release times can then be programmed at will by merely changing between two clock speeds when selecting between the count-up and the countdown modes. For a voice SSB application, the two frequencies are usually 40 Hz for the attack time (count up) and 4 Hz for the release time (countdown). Signal level information from two points of interest in the receiver (usually the IF and the audio amplifier) control the gating of the properly chosen clock frequency for the up/down counter, as well as the up or down commands.

This system can be imagined as a continuously changing counter, which is going up fast and coming down slow, with the AGC voltage following this pattern. Because of the complex switching involved in this type of AGC, low-current logic should be used in the design. Our example was implemented with CMOS logic for this reason. This relatively expensive AGC control system has the advantage of being extremely reliable. The precise programmability of the attack and release times also makes it an ideal choice for a communications receiver.

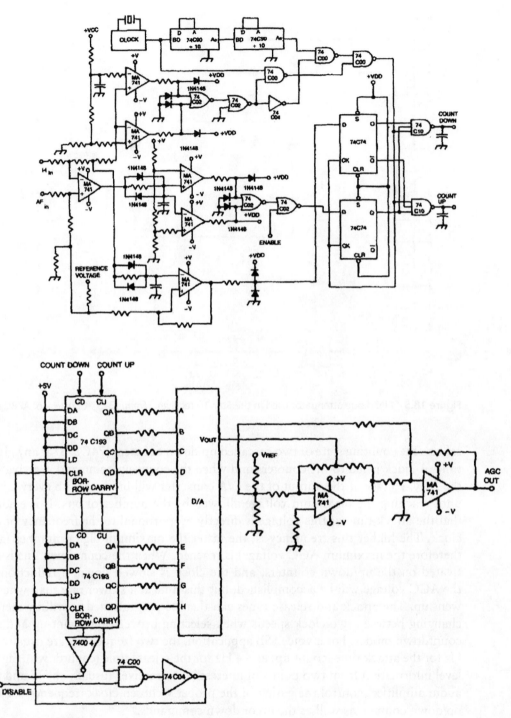

Figure 18.6   Implementation of a digital counter AGC.

18.12   Other Considerations for AGC Detectors

Practical AGCs are sometimes much more complex than the examples shown above. They have to cope with additional timing complications imposed by data rate and burst restrictions presented by multiple modulation schemes such as encountered in data communications over RF. For DSP AGCs, it is the A/D pipeline latency and the digital filter delays that make the job more complex. For analog AGCs, the important part of an analysis is the response timing, where to tap the loop inputs from, and where to apply the controls using step attenuators or variable gain amplifiers as we discussed at the beginning of this chapter.

Various software can be used in performing AGC analysis. Of particular interest is the System View software by Eagleware Elanix, which is a good tool for simulation along with MATLAB. The analysis shown in this book have been implemented with a proprietary software developed by the author entitled Victoria Falls.

Various manufacturers have been helping with new devices, which can make the job of designing AGCs easier. A simple and modern AGC system using a precision RMS detector chip available from Analog Devices is shown in Figure 18.7. The AD8361 is a TruPwr detection RFIC, offering RMS-responding power detection in integrated form. The device is capable of converting a complex modulated RF signal, from up to 2.5 GHz, into a DC voltage representing the RMS level of the signal. The device is highly linear and temperature stable. It is useful for detection of CDMA, QAM, and other complex modulation schemes.

The system in Figure 18.7 closely resembles our theoretical model from Figure 18.3. It uses the AD8367 linear-in-dB VGA amplifier from Analog Devices. This is a high-performance variable gain amplifier designed to be used at IF frequencies up to 500 MHz. An externally applied analog gain control voltage of 0 to 1 volt, scales a 45 dB of gain control range to provide 20 mV/dB output.

In the past, dual gate MOSFETs have also been used frequently in low-cost AGC systems. Although they have disappeared in recent years, they are making a comeback in VHF/UHF television receivers. Shown in Figure 18.8 is a modern dual MOSFET AGC part from Infineon, the BF5030.

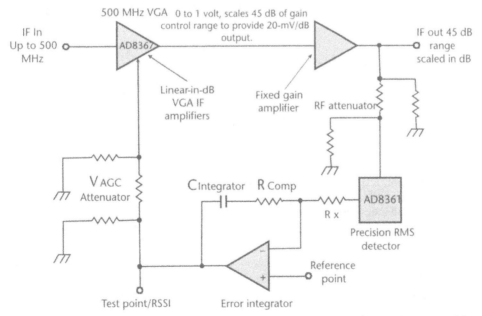

Figure 18.7   The AD8361 precision RMS detector from Analog Devices is used in this 500-MHz AGC system to provide 45-dB linear-in-dB AGC.

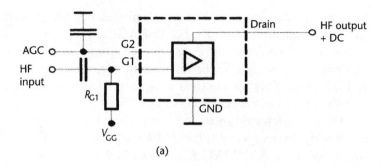

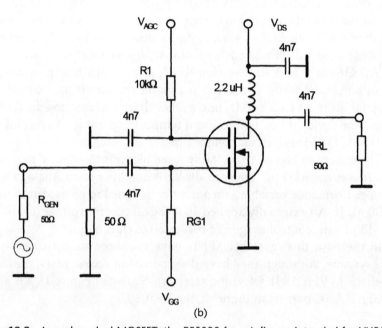

Figure 18.8 A modern dual MOSFET, the BF5030 from Infineon intended for VHF/UHF television receivers.

Much more material exists regarding AGC applications. For additional information on this topic, the reader is directed to the comprehensive Selected Bibliography at the end of this chapter.

References

[1] Tacconi, E. J., and C. F. Christiansen, "A Wide Range and High Speed Automatic Cain Control," Superconducting Super Collider Laboratory, Dallas, TX, IEEE 0-7803-1203-1/93503.00, 1993.

[2] Martinez, G., "Automatic Gain Control (AGC) Circuits Theory and Design," ECE1352, University of Toronto.

[3] Drentea, C., "The Star-10 Transceiver, Part 1," *QEX—ARRL*, November/December 2007.

[4] Drentea, C., "The Star-10 Transceiver, Part 2," *QEX—ARRL*, March/April 2008.

[5] Drentea, C., "The Star-10 Transceiver, Part 3," *QEX—ARRL*, May/June 2008.

Selected Bibliography

Abidi, A. et al., "The Future of CMOS Wireless Transceivers," *Int. Solid-State Circuits Conf.*, San Francisco, 1997.

Agilent Technologies, "Applications of PIN Diodes," Application Note 922.

Agilent Technologies, "A Low-Cost Surface Mount PIN Diode Attenuator," Application Note 1,048.

Analog Devices, *Nonlinear Circuits Handbook: Designing with Analog Function Modules and ICs*, Analog Devices.

Banta, E. D., "Analysis of an Automatic Gain Control (AGC)," *IEEE Trans. on Automatic Control*, Vol. AC-9, 1964.

Brown, B., "AGC Using the Diamond Transistor OPA660," Application Bulletin, Burr Brown/Texas Instruments.

Bullock, S., "Simple Techniques Yields Errorless AGC Systems," *Microwaves & RF*, August 1989.

Bullock, S. R., *Transceiver and System Design for Digital Communications*, 2nd ed., Tucker, GA: Noble Publishing, 2000.

Cavalli, M., and W. A. Serdijn, *The Design of a Translinear AGC Technical Report*, Ubicom Technical Report, Ubiquitous Communications, Delft University of Technology, June 1998,

Couch II, L. W., *Digital Analog Communication Systems*, New York: Macmillan.

"Designing Automatic Gain Control of Systems, Part 1 Design Parameters," *IEEE*, December 1964, pp. 43–47.

"Designing Automatic Gain Control of Systems, Part 2 Circuit Design," *IEEE*, January 1965, pp. 53–57.

Filanovsky, I. M., and V. A. Piskarev, "Automatic Gain Control by Differential Pair Current Splitting," *Journal Electronics*, Vol. 62, No. 2, 1987.

Franklin, G. F., D. Powell, and A. Emami-Naeini, *Feedback Control of Dynamic Systems*, Upper Saddle Ribver, NJ: Prentice-Hall, 2006.

Gardner, F. M., *Phaselock Techniques*, New York: John Wiley & Sons, 1966.

Gill, W. J., and W. K. S. Leong, "Response of an AGC Amplifier to Two Narrow-Band Input Signals," *IEEE Trans. on Comm. Tech.*, 1966.

Goyal, P., "Automatic Gain Control in Burst Communications Systems," *RF Design Magazine*, February 2000.

Gray, P. R., R. G. Meyer, and S. H. Lewis, *Analysis and Design of Analog Integrated Circuits*, 4th ed., New York: John Wiley & Sons, 2001.

Halford, P., and E. Nash, Analog Devices Inc., "Integrated VGA Aids Precise Gain Control," *Microwaves & RF*, March 2002.

Haque, T., L. Kazakevich, and A. Demir, "The Analysis and Design of a User Equipment Grade, All Digital Gain Control, Direct Conversion Radio," *Microwave Journal*, September 2004.

Henning, C. W., "The AGC Module," *VHF Communications Magazine*, January 2008.

Hugh, R. S., *Automatic Gain Control: A Practical Approach to Its Analysis and Design with Applications to Radar*, 2nd ed., Wexford Press.

Hughes, R. S., *Analog Automatic Control Loops in Radar and EW*, Norwood, MA: Artech House, 1988.

Hughes, R. S., "Design Automatic Gain Control Loops the Easy Way," *EDN*, October 1978.

Imcgeehan, J. P., and D. F. Burrows, "Large Signal Performance of Feedback Automatic Gain Control Systems," *IEEE Proc.*, 1981, p. 128.

Khourly, J. M., "Fixed Time Constant AGC Circuits," *IEEE International Symposium on Circuits and Systems*, Hong Kong, June 1997.

Licqurish, C., M. J. Howes, and C. M. Snowden, "Dual Gate FET Modelling," *IEEE Colloquium on Microwave Devices Fundamentals and Applications*, 1988.

Meijer, G., *Implementatiekunde*, Internal report of Electronics Research Laboratory, Delft University of Technology, 1993.

Mercy, D.V., "A Review of Automatic Gain Control Theory," *The Radio and Electronic Engineer*, Vol. 51, No. 1 l/12, 1981.

Meyr, H., and G. Ascheid, *Synchronization in Digital Communications*, New York: John Wiley & Sons, 1990.

Moskowitz, "Linear Feedback AGC with Input Level-Invariant Response Times," *Journal of Acoustical Society of America*, Vol. 62, No. 6, December 1977, pp. 1149–1456.

Mulder, J., "Static and Dynamic Translinear Circuits," Ph.D. Thesis, Delft University of Technology, October 1998.

Mulder, J., et al., "Dynamic Translinear Circuits—An Overview," *Proceeding of the IEEE*, Kopenhagen, April 28–30, 1998.

National Semiconductor Corporation, "CLCXXX Variable Gain Amplifiers," Product Datasheet.

Neamen, D. A., *Electronic Circuit Analysis and Design*, 2nd ed., New York: McGraw-Hill, 2001.

Newman, E. J., "X-Amp, A New 45-dB, 500-MHz Variable-Gain Amplifier (VGA) Simplifies Adaptive Receiver Designs," *Analog Dialogue*, 2002.

Nordholt, E. H., *Design of High-Performance Negative-Feedback Amplifiers*, New York: Elsevier, 1983.

Ogata, K., *Modem Control Engineering*, Upper Saddle River, NJ: Prentice-Hall.

Ohlson, J., "Exact Dynamics of Automatic Gain Control," *IEEE Trans. on Communication*, Vol. 22, January 1974, pp. 72–75.

Oliver, M., "Automatic Volume Control as a Feedback Problem," *Proceedings of the IRE*, Vol. 36, April 1948, pp. 466–473.

Pappenfus, E. W., W. D. Bruene, and E. O. Shoenike, *Single Sideband Principles and Circuits*, New York: McGraw-Hill, 1964.

Piazza, F., et al., "A 2 mA/3 V 71 MHz IF Amplifier in 0.4 μm CMOS Programmable over 80 dB Range," *Intl. Solid-State Circuits Conf.*, San Francisco, CA, 1997.

Plotkin, S., "On Nonlinear AGC," *Proceedings of the IEEE*, Vol. 51, February 1963.

Porter, J., *AGC Loop Control Design Using Control System Theory, RF Design*, Intertec Publishing.

Recommendation 118-2, "Hearing Aids with Automatic Gain Control Circuits," 1983.

Ricker, D. W., "Nonlinear Feedback System for the Normalization of Active Sonar Returns," *Journal Acoust. Soc. Am.*, 1976, p. 59.

Rijns, J. J. F., "CMOS Low-Distortion High-Frequency Variable-Gain Amplifier," *IEEE J. of Solid-State Circuits*, Vol. 31, No. 7, 1996.

Robertson, D., "Selecting Mixed Signal Components for Digital Communication Systems–III: Sharing the Channel," *Analog Dialogue*, Vol. 31, No. 1, 1997.

Rosu, I., "Automatic Gain Control (AGC) in Receivers," http://www.qsl.net/va3iul/.

Rowland, J. R., *Linear Control Systems, Modeling Analysis and Design*, New York: John Wiley & Sons, 1986.

Sabin, W. E., and E. O. Schoenike, (eds.), *Single Sideband Systems and Circuits*, New York: McGraw-Hill, 1987.

Sandell, M., S. K. Wilson, and P. O. Börjesson, "Performance Analysis of Coded OFDM on Fading Channels with Non-Linear Interleaving and Channel Knowledge," *Proceedings of Vehicular Technology Conference*, Phoenix, AZ, 1997.

Sayre, C. W., *Complete Wireless Design*, New York: McGraw-Hill, 2001.

Schachter, H., and L. Bergstein, "Noise Analysis of an Automatic Gain Control System," *IEEE Trans. on Automatic Control*, Vol. AC-9, July 1964.

Schilling, D. L., *Electronic Circuits: Discrete and Integrated*, New York: McGraw-Hill, 1968.

Serdijn, W. A., et al., "Low-Voltage Low-Power Fully Integrable Automatic Gain Controls," *Analog Integrated Circuits and Signal Processing*, Vol. 8, 1995.

Serdijn, W. A., C. J. M.Verhoeven, and A. H. M. van Roermund, *Analog IC Techniques for Low-Voltage Low-Power Electronics*, The Netherlands: Delft University Press.

Simpson, R. S., and W. H. Tranter, "Baseband AGC in an AM-FM Telemetry System," *IEEE Trans. on Comm. Tech.*, Vol. 18, 1970.

Smith, J. R., *Modern Communication Circuits,* 2nd ed., New York: McGraw-Hill, 1998.

Tacconi, E. J., "Amplificateur a Large Gamme Dynamique, Fiable Bruit et Commande Automatique de Gain pour le Systeme de Controle RF du Faisceau PSB," *CERN PS/BR*, No. 85-3, 1985.

Tadjpour, S., F. Behbahani, and A. A. Abidi, "A CMOS Variable Gain Amplifier for a Wideband Wireless Receiver," Electrical Engineering Department, University of California, Los Angeles, CA.

Tedeschi, F. P., *How to Design, Build & Use Electronic Control Systems*, Blue Ridge Summit, PA: Tab Books, 1981.

Trout, B., "A High Gain Integrated Circuit RF-IF Amplifier with Wide Range AGC," Application Note AN-513, Motorola.

Valten, T., "Digital Signal Processing in Short Wave Reception Technology," *Amateur Radio Conference*, Munich, Germany, 2001.

van der Plas, J., "Synchronous Detection in Monolithically Integrated AM Upconversion Receivers," Ph.D. thesis, Delft University of Technology, November 1990.

van Lieshout, P. J. G., and R. J. van de Plassche, "A Power-Efficient, Low-Distortion Variable Gain Amplifier Consisting of Coupled Differential Pairs," *IEEE J. of Solid-State Circuits*, Vol. 32, No. 12, 1997.

van Staveren, A., "Sructured Electronic Design of High-Performance Low-Voltage Low-Power References," Ph.D. thesis, Delft University of Technology.

Victor, W. K., and M. H. Brockinan, "The Application of Linear Servo Theory to the Design of AGC Loops," *Proc. IRE*, 1960.

Waldhauer, F. D., *Feedback*, New York: John Wiley & Sons, 1982.

Whitlow, D., *Design and Operation of Automatic Gain Control Loops for Receivers in Modern Communications Systems*, Analog Devices.

Yamawaki, T., et al., "A 2.7-V GSM RF Transceiver IC," *IEEE J. of Solid-State Circuits*, Vol. 32, No. 12, 1997.

Product Detectors and Beat Frequency Oscillators (BFO)

The last stage of a communications receiver contains one or more detectors. Among them are amplitude and frequency (AM and FM) detectors along with suppressed carrier or single-sideband (SSB) demodulators.

While AM and FM demodulation is implemented using simple detectors, of significant importance to our discussion is the suppressed carrier SSB demodulation and as a function of it, the CW demodulation.

In single-sideband (SSB) communications, the carrier is removed at the transmitter to be reinserted at the receiver. Only one sideband is usually transmitted and consequently received. This reduces the transmission bandwidth by a factor of two, while at the same time eliminating to a great degree the carrier power, thus economizing on power consumption and doubling the utilization of the spectrum. Independent sideband is also possible utilizing each sideband for different information.

A double sideband amplitude-modulated spectrum is shown in Figure 19.1 where each term contributes a positive and a negative frequency component according to the Euler identities [1–3]. This is shown in (19.1) and (19.2).

$$\cos\alpha = \frac{1}{2}\left(e^{j\alpha} + e^{-j\alpha}\right) \tag{19.1}$$

$$\sin\alpha = \frac{1}{j2}\left(e^{j\alpha} - e^{-j\alpha}\right) \tag{19.2}$$

Figure 19.1 shows that on each side of the main carrier frequency ωc we get two sidebands, $\omega_c - \omega_x$ for the lower sideband, and $\omega_c + \omega_x$ for the upper sideband. These sidebands can contain reduced carrier components, and the information can be demodulated at the receiving point after the last conversion in a superheterodyne receiver equipped with a double-balanced mixer called a product detector and a carrier reinsertion local oscillator called the beat frequency oscillator (BFO). The BFO is offset against the last IF filter to tailor the received baseband frequency response according to the upper or lower sideband modes of operation.

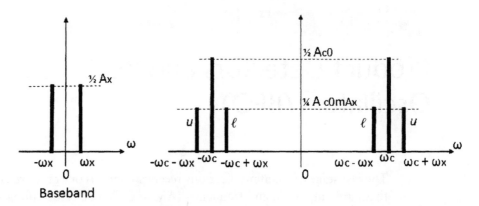

Figure 19.1 Double-sideband amplitude-modulated spectrum showing the upper and lower sideband components.

19.1 I and Q Demodulation Process: The Concept of Demodulation

Shown in Figure 19.2 is a model for generating and receiving SSB signals using in-phase and quadrature components of a desired baseband signal. It can be seen that the transmitter side takes advantage of a 90° phase shift between the baseband components of the in-phase and quadrature branches ($xt = A_x \cos \omega xt$), which, in turn, determine the sign of the shift, and consequently the upper or lower sideband choice. At the receiver end, the process is reversed.

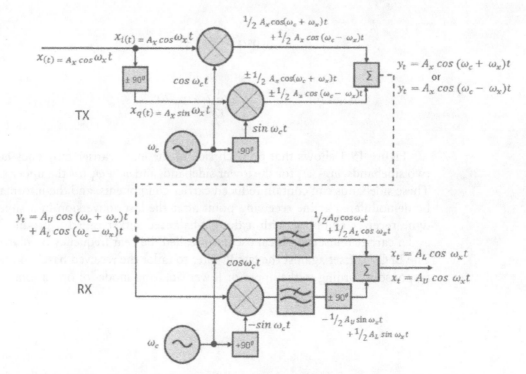

Figure 19.2 Phasing method of a SSB modulator and demodulator scheme. The sign of the 90° phase shifter determines the upper (U) or lower (L) demodulation.

After filtering out the high-frequency component based around $\cos(2\omega t)$ and the DC component C, the original message will be recovered.

19.2   Other Demodulation Techniques

There are many other kinds of product detectors as well. In digital signal processing (DSP), one may experiment with a wide range of techniques. For instance, it is possible to multiply the incoming signal by the carrier, times the square of another carrier, 90° out of phase. This will produce a copy of the original message, and another AM signal at the fourth harmonic, by means of the trigonometric identity:

$$\sin^2 \theta \cos^2 \theta = \frac{1 - \cos^4 \theta}{8} \tag{19.3}$$

The high-frequency component can again be filtered out, leaving the original signal. If $m(t)$ is the original message, the AM signal can be shown as:

$$x_{(t) = (c + m_{(t)})}\cos_{(\omega t)} \tag{19.4}$$

Multiplying the AM signal by the new set of frequencies yields:

$$\begin{aligned}
y_{(t)} &= \left(C + m_{(t)}\right)\sin^2\left(\omega t\right)\cos^2\left(\omega t\right) \\
&= \left(C + m_{(t)}\right)\frac{1 - \cos 4\omega t}{8} \\
&= \frac{\left(C + m_{(t)}\right)}{8} - \frac{\left(C + m_{(t)}\right)\cos 4\omega t}{8}
\end{aligned} \tag{19.5}$$

After filtering out the component based around $\cos(4\omega t)$ and the DC component C, the original message will be recovered.

In superheterodyne receivers, a product detector is usually implemented with a mixer, which is used in conjunction with the beat frequency oscillator (BFO) to produce an audible tone when a SSB, KCW, or FSK signals are received. A lowpass filter is used at the output of the detector in order to reject high-order products generated by the mixer.

19.3   The Star-10 Receiver Product Detector

Shown in Figure 19.3 is the schematic diagram of the product detector (PDAF assembly) in the example receiver we have used throughout this book, the Star-10 receiver. In the context of the system, the product detector board (PDAF) follows the 9-MHz IF previously discussed in Chapter 17. It provides final receiver conversion

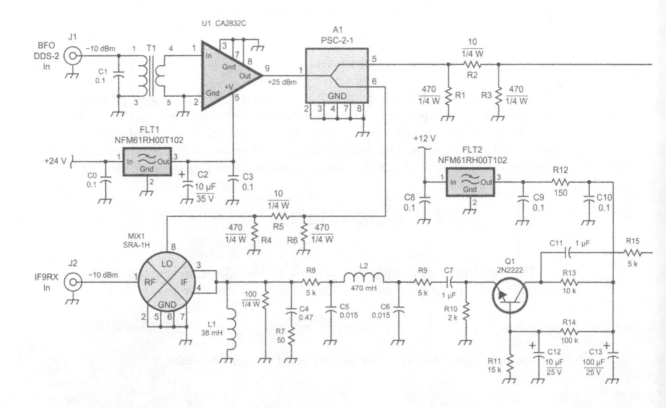

Figure 19.3 Product detector, BFO, and audio amplifier implementations in the Star-10 receiver. (*From:* [4–6]. Figure courtesy of ARRL).

to baseband audio frequencies of the detected signals after they have been filtered and conditioned (by the AGC) for final detection in all modes of operation.

In this example, the product detector is a high-level class II passive double-balanced mixer driven by a class A amplifier (U1), which obtains its BFO carrier from the coherent synthesizer discussed earlier in Chapter 16. The 9-MHz BFO signal is produced by a simple DDS in the synthesizer assembly and is filtered via a 9-MHz, 15-kHz-wide quartz monolithic filter to keep spurious products to a minimum. Because the BFO will be offset only slightly from the ultimate bandwidth filters in the IF assembly (discussed in Chapter 17), a DDS-only synthesizer was deemed sufficient to provide a pure coherent signal to the product detector.

Looking at Figure 19.3, a properly shifted BFO signal coming from FSYNTH enters the PDAF assembly at J1. This signal is slightly offset against the 9-MHz IF

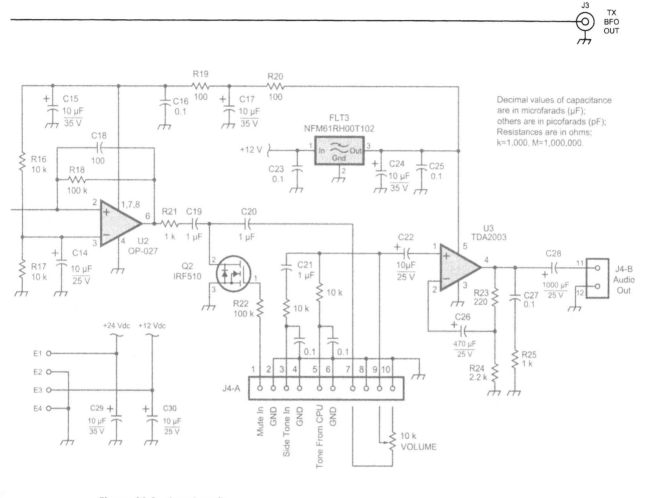

Figure 19.3   (continued)

filters center frequency by the synthesizer depending on the mode being selected by the operator through the command and control (DFCB) assembly. This provides the required baseband response dictated by the mode of operation required.

The BFO signal, which is amplified by the class A amplifier U1, is further split by A1, which is a Mini-Circuits PSC2-1 part. Half the signal is passed on to J3, which distributes it to the transmitter mixer on IF9TX. The other half goes through an attenuator pad made of R4, R5, and R6 and is input to MIX 1, the SRA-1H mixer that serves as the product detector. This mixer was purposely selected for this function, as it is suitable for baseband frequency response output compatible with audio frequencies, and has a reported IP3 of +28 dBm.

It is important that the last mixer in a superheterodyne receiver (the product detector) should be as good as the previous mixers in order to retain the fidelity of the system's performance.

The conditioned 9-MHz IF receiver signal coming from IF9RX is input to the product detector mixer at J2. The mixed-down audio product is matched and filtered via L1, C4, R7, R8, C5, L2, and C6, and is further processed by Q1 and U2 to be finally presented to U3, a TDA2003 audio block and output at J4-B. The L2, C5, and C6 make up the audio lowpass filter, which is intended to suppress noise beyond 3 kHz. Tones from the CW side tone generator along with various feedback tones coming from the DFCB command and control assembly are audio mixed and injected in this circuit via the J4-A connector as shown. In addition, volume control wires from the front panel audio control are input via this connector along with the mute signal from the T/R assembly, which silences the receiver via Q2 when transmitting.

The actual PDAF assembly implementation is shown in Figure 19.4. Shown in Figure 19.5 is a mathematical representation of how the BFO is offset against the 9-MHz IF filter to produce an audio baseband of 300 Hz to 2,700 Hz for upper and lower SSB communications. Other BFO shifts are shown for FSK and CW operation. It is important to notice that in any of the cases, the center frequency of the IF filters remains the same, while baseband response is shifted about the

Figure 19.4 Actual implementation of the product detector assembly in the Star-10 receiver.

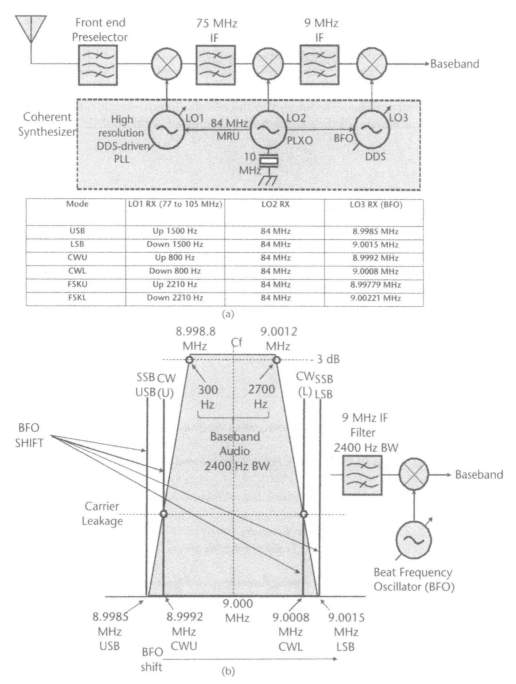

Figure 19.5 The mathematics behind SSB demodulation in the Star-10 receiver. Shifting the BFO offsets the audio baseband against the 9-MHz IF filters response from the previous IF extracting the baseband in the upper and lower SSB cases. The figure shows a 300-Hz to 2,700-Hz audio bandwidth, which is the case for a 2.4-kHz filter centered at 9 MHz. For simplicity, the other filters bandwidths have not been shown. The concept applies to any product detector and IF design.

9-MHz center frequency according to the mode requirements. The Command and Control software of the Star-10 under operator control addresses the various LOs in the system to position them such as to represent the correct frequency and display. It would also be possible to implement a sideband selector similar to this by using different switched filters and maintain a fixed BFO. However, this would at least triple the amount of quartz filters in the IF, which would be more expensive. Figure 19.5 also shows how the BFO signal crosses the skirts of the last IF filter. This undesirable phenomenon is also known as carrier leakage and depends on the shape factor of the 9-MHz IF filters. The carrier leakage should be as far down on the filters skirts as possible but no less than 40 dB. In our example, it is better than 60-dB down because of the cascaded filters composite of the 9-MHz IF.

References

[1]   Nahin, P. J., *Dr. Euler's Fabulous Formula: Cures Many Mathematical Ills*, Princeton, NJ: Princeton University Press, 2006.

[2]   Nahin, P. J., *The Science of Radio*, New York: American Institute of Physics, 2001.

[3]   Vidkjær, J., "Modulation, Transmission, and Demodulation, 31415 RF-Communication Circuits," http://www.inatel.br/docentes/dayan/TP504/Antes2009/ClassNotes/ModulationTransmissionDemodulation.pdf.

[4]   Drentea, C., "The Star-10 Transceiver, Part 1," *QEX—ARRL*, November/December 2007.

[5]   Drentea, C., "The Star-10 Transceiver, Part 2," *QEX—ARRL*, March/April 2008.

[6]   Drentea, C., "The Star-10 Transceiver, Part 3," *QEX—ARRL*, May/June 2008.

Selected Bibliography

Abidi, A. A., P. R. Gray, and R. G. Meyer, (eds.), *Integrated Circuits for Wireless Communications*, New York: IEEE Press, 1999.

Abromowitz, M., and I. A. Stegun, *Handbook of Mathematical Functions*, New York: Dover, 1965.

Benedetto, S., E. Biglieri, and V. Castellani, *Digital Transmission Theory*, Upper Saddle River, NJ: Prentice-Hall, 1987.

Drentea, C., *Radio Communications Receivers*, New York: McGraw-Hill, 1982.

Gradshteyn, I. S., and I. W. Ryzhik, *Tables of Integrals Series and Products*, San Diego, CA: Academic Press, 1965.

Lathi, B. P., *Modern Digital and Analog Communication Systems*, 3rd ed., New York: Oxford University Press, 1998.

Proakis, J. G., *Digital Communications*, 3rd ed., New York: McGraw-Hill, 1995.

Proakis, J. G., and M. Salehi, *Communication Systems Engineering*, Upper Saddle River, NJ: Prentice-Hall, 1994.

Simon, M. K., S. M. Hinedi, and W. C. Lindsey, *Digital Communication Techniques*, Upper Saddle River, NJ: Prentice-Hall, 1995.

Sklar, B., *Digital Communications Fundamentals and Applications*, Upper Saddle River, NJ: Prentice-Hall, 1988.

Stüber, G. L., *Principles of Mobile Communication*, Boston, MA: Kluwer, 1996.

CHAPTER 20

Audio and Baseband Amplifier Design Considerations

The audio amplifier in a receiver is just as important as the previous stages. Low distortion is required especially in systems involving digital signal processing (DSP) where reconstruction of the baseband signal in the D/A converter is the result of a rather limited number of bits. This, in turn, results in a staircase-like response, which can exhibit square waveform edges that can ring and create additional distortion.

In addition, out of the IF bandpass noise generated in the last IF amplifier stages has to be controlled in the audio amplifier to match the IF filters bandwidth. Otherwise, noisy audio signals result.

Consequently, using low-distortion audio amplifiers in combination with additional audio filtering is imperative in the design of a high-quality communications receiver.

As with any amplifiers, audio power amplifiers are classified in accordance with their input-biasing configurations in various classes of operation such as A, B, AB and so forth. These classes of operation can provide performance ranging from a near linear output (but with low efficiency—e.g., Class A ~40%) to non-linear output performance (with a higher efficiency—e.g., Class AB ~70%). More efficient amplifiers such as class C are not used in high-fidelity audio applications because of the increased distortion.

Too many times, designers of otherwise well-engineered communications receivers fall into the trap of using inexpensive, readily available audio integrated circuits without considering the total harmonic distortion (THD) performance of some of these devices, which can sometimes exceed 20%.

A well-designed audio amplifier capable of continuous power output of 2 to 3 watts (Institute of High Fidelity, IHF) and exhibiting a total harmonic distortion of no more than 1.5% at full output power is highly recommended here. The designer has the opportunity to either choose carefully from a supply of readily available integrated circuit blocks or to design discrete audio amplifiers in which noise can be properly tailored along with the total harmonic distortion (THD).

The total harmonic distortion and noise (THD+N) in an audio amplifier is a measure of the harmonics created by a single tone in a device such as an audio amplifier, as previously discussed at the beginning of Chapter 15. The total harmonic distortion and noise (THD+N) of an audio amplifier is expressed as a percentage for a certain gain, when a single frequency is input to the device (for example, 1 kHz) at almost full scale as shown in Figure 20.1.

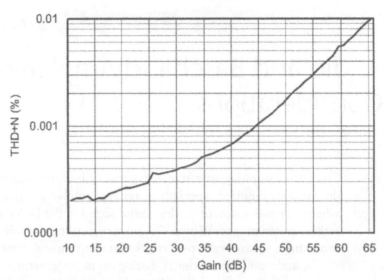

Figure 20.1 Example of a THD+N in a high-fidelity audio amplifier.

Tailoring the frequency response to match that of the ultimate IF bandwidth in the receiver is also highly recommended. Usually this is achieved by installing a lowpass audio filter network following the product detector mixer and prior to the audio amplifier as shown in the Star-10—PDAF assembly schematic from Figure 19.3. This figure also shows the schematic diagram of a 10-watt audio amplifier, the TDA-2003, which exhibits a typical harmonic distortion performance of 0.15% at 7.5 watts. Two-tone intermodulation distortion was measured for the TDA-2003 and was found to be better than 70-dB down at 4 watts output using 400 Hz and 2-kHz tones mixed 4:1.

The intermodulation distortion performance of the audio amplifier is just as important as the intermodulation distortion of the rest of the receiver if performance is to be maintained throughout. This statement is also true for the transducer, speaker, or headphones used.

Additional selectivity can be obtained in the audio portion of a communications receiver by using additional audio circuits. This is accomplished through the use of active filters using operational amplifiers usually in unity-gain configurations. Such designs provide selective bandpass and peak-tuning for KCW and SSB signals over the baseband frequency range as shown in Figure 20.2.

Shown in Figure 20.3(a) is a high performance 3-watt discrete audio amplifier design suitable for a high probability of intercept communications receiver. The amplifier features a complementary symmetry emitter-follower output circuit, which provides a typical harmonic distortion of no more than 1%. The amplifier is preceded by a bandpass audio filter, which is tailored to match an IF filter frequency response in order to reduce noise produced by previous IF amplifiers. Shown in Figure 20.3(b) is the actual measured composite audio frequency response of the amplifier against a typical IF filter, with the audio filter switched in and out.

More information about designing high-fidelity audio amplifiers can be found in the Selected Bibliography.

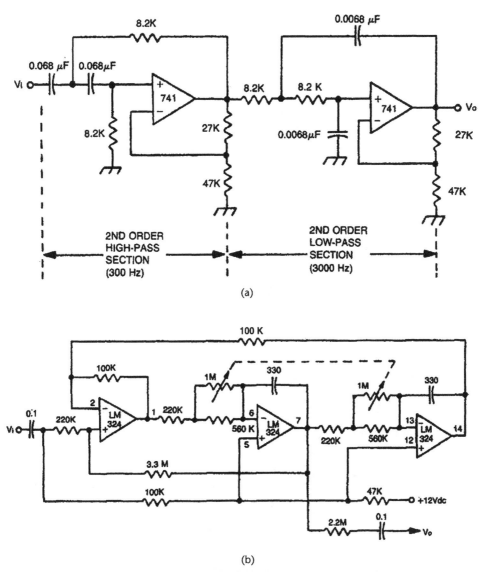

Figure 20.2 (a, b) Tailoring frequency response of an audio amplifier in a communications receiver. Additional selectivity can be obtained in the audio portion of a communications receiver by using additional audio circuits. This is accomplished through the use of active filters using operational amplifiers usually in unity-gain configurations. Such designs provide selective bandpass and peak-tuning for KCW and SSB signals over the baseband frequency range.

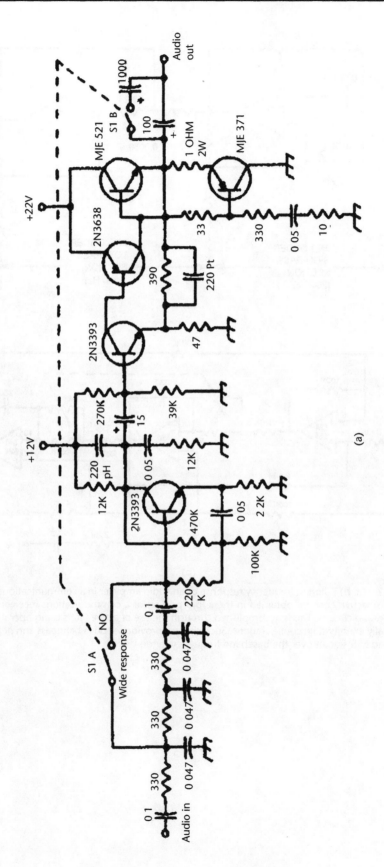

Figure 20.3 (a) Design of a practical discrete low-distortion audio amplifier for a communications receiver. This design is tailored so that the frequencies above 3 kHz are attenuated (S1 A and B open) providing effective cancellation of noise generated in the gain stages following the IF filters. A wider response can be achieved by closing the S1 A and B switch. (b) Actual measured frequency response of the amplifier with the filter switched in. Area of canceled IF noise is visible.

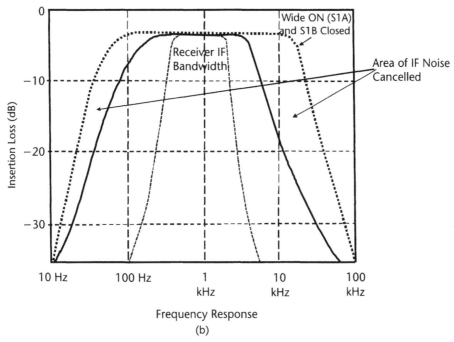

Figure 20.3   (continued)

Selected Bibliography

Drentea, C., *Radio Communications Receivers*, New York: McGraw-Hill.

Green, S., "Audio ADC Buffer Design Secrets," Cirus Logic, http://www.audiodesignline.com/ho wto/210605596;jsessionid=KDFYD4IBL4NRFQE1GHPSKH4ATMY32JVN?pgno=4.

Horowitz, P., and W. Hill, *The Art of Electronics*, New York: Cambridge University Press.

"Introduction to Amplifier Design," http://my.integritynet.com.au/purdic/rf_amp.htm.

"Introduction to Amplifiers," http://www.electronics-tutorials.ws/amplifier/amp_1.html.

Jung, W., *Op Amp Applications Handbook*, Analog Devices.

Lancaster, D., *Active Filter Cookbook, Second Edition*, Newnes.

Malvino, A. P., and D. J. Bates, *Electronic Principles*, New York: McGraw-Hill.

Pease, R., *Analog Circuits (World Class Designs)*, Newnes.

Slone, G. R., *High-Power Audio Amplifier Construction Manual*, New York: McGraw-Hill.

Williams, T., *The Circuit Designer's Companion*, 2nd ed., EDN.

CHAPTER 21

The Power Supply

The power supply should not be neglected when designing a high-performance communications receiver. When using linear power supplies, high-efficiency full-wave rectification should be used coupled with double regulation and adequate filtering. It is not uncommon to find two series regulators feeding each other in a high-performance linear power supply as shown in Figure 21.1. This type of power management is evident in the Star-10 schematic diagrams presented throughout this book where power for the various sections is filtered through special PC board-mounted tubular lowpass filters and is regulated on each of the boards.

Switching power supplies have been generally avoided in communications receivers because their fast rise time-switching waveforms generate wideband RF noise that is guaranteed to severely pollute the receiver's sensitive circuitry.

Switching power supply engineering is a fascinating topic dealing with many design areas such as power conversion through DC-to-ultrasonic frequency energy conversion using special property magnetics. Consequently, the design of EMI clean switching power supplies is considered an art in itself. "Noiseless" switching power supplies are generally an order of magnitude more complex to design than analog linear supplies.

The modern switched-mode power supply (SMPS) is an electronic supply based on a transformer-coupled switching regulator operating at ultrasonic frequencies to provide the required output voltage at high efficiency. It converts power from a source such as a battery or the electrical power grid to a load such as a radio receiver. The basic function of the converter is to provide a clean and stable output voltage at a different level than the input voltage. The smaller size and lower weight resulting from the elimination of bulky 50/60 Hz transformers and/or linear regulators translate into lower heat, higher efficiency, and higher power density, all desirable characteristics.

While a linear regulated power supply regulates the output voltage by dissipating excess power and heat in pass transistors and resistive elements, a switched-mode power supply regulates either the output voltage or the current by switching near ideal energy storage elements, such as inductors and capacitors, into and out of different electrical configurations. The SMPS switching frequency is typically ~500 times higher than a conventional ~60 Hz supply and as such, it uses magnetic devices that are a fraction of the volume of the 60-Hz transformers.

Near-ideal switching elements such as transistors operated in "hard" switching digital modes have minimum resistance when "closed" and carry zero current when "open," so the converters can theoretically operate near their ideal 100% efficiency.

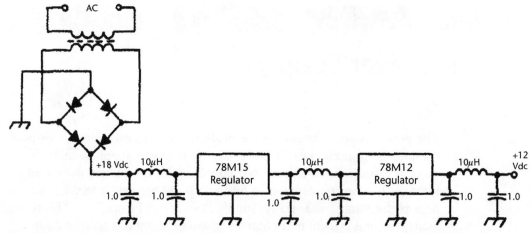

Figure 21.1 Doubly regulated linear power supply is used to ensure superior performance in a communications receiver.

Unfortunately, in an SMPS, the output current flow depends on the switching waveform used to drive the switching elements. The spectral density of these switching waveforms generate a myriad of high frequencies as discussed in Chapter 15. The resulting switching transients create ample EMI, which is usually not treated in common designs but can seriously degrade receiver performance.

Consequently, switching power supplies can emit EMI energy either directly from their active elements, through the grounds as well as chassis, and especially back into the power lines through the input AC lines.

The solution to properly designing noiseless switching power supplies lies in using multiple section lowpass filters at the AC input and the DC outputs. A triple (or quadruple) common mode PI section, lowpass filter (Figure 21.2) should be used on both the AC input and DC outputs.

The greatest amount of EMI corruption is emitted back out through the AC power line. It is caused by the pulsating input current to the switching regulator.

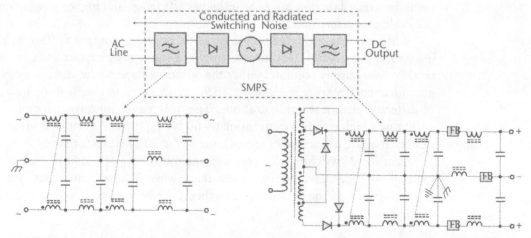

Figure 21.2 Simplified switching power supply block diagram shows EMI reduction circuits using common mode lowpass filters at the AC input as well as the DC output in order to prevent RF switching noise from being radiated outside of the power supply.

The DC outputs are less of a problem since the currents in this part of the converter topology are triangular in nature and continuous, and as such, have far less wideband EMI spectral energy.

As a rule, the corner frequency of the common mode lowpass filters should be at least a decade lower than the fundamental switching frequency in the power supply. In addition, properly referencing these filters to the chassis—a seemly minor point—is absolutely crucial to EMI mitigation. Most common SMPS designs allow the chassis to float electrically, which accounts for inferior EMI rejection by some 30 dB.

Many low-cost communications receivers power their audio amplifiers (and their oscillators) from the same power supply. Because of poor regulation in the power supply (i.e., it does not have low enough output impedance), the output voltage to critical voltage-sensitive circuits will be modulated by the peak current pulled by the audio power amp. The result is pulling variable tuning oscillators or phase-locked loops, which in turn can frequency modulate the output baseband of the receiver. When designing a receiver, use separate power regulators for the front end and IF amplifiers.

The Star-10 front end uses 24-volt DC for high dynamic range. The baseband amplifier should also have its own power supply and a separate ground if possible to avoid ground loops, which are also a consideration.

When designing multiple power supplies, there is no good way of predicting ground loops in a communications receiver, and they can be the final obstacle in achieving high performance. They manifest themselves by introducing severe amounts of hum in the output of the receiver. Good grounding techniques may or may not solve the problem. Although theoretically understood, ground loops will almost invariably impact the performance of a newly designed receiver. Removing them has always proved to be a time-consuming trial and error process.

More information on designing linear and switching power supplies can be found in the Selected Bibliography.

Selected Bibliography

Ang, S. S., *Power-Switching Converters*, New York: Marcel Dekker.

Brown, M., *Power Supply Cookbook*, Newnes.

Brown, M., *Practical Switching Power Supply Design*, San Diego, CA: Academic Press.

Lenk, J. D., *Simplified Design of Switching Power Supplies*, Boston, MA: Butterworth-Heinemann.

Maniktala, S., *Switching Power Supply Design & Optimization*, New York: McGraw-Hill.

Pressman, A. I., K. Billings, and T. Morey, *Switching Power Supply Design*, New York: McGraw-Hill.

Lee, Y. S., *Computer-Aided Analysis and Design of Switch-Mode Power Supplies*, New York: Marcel Dekker.

CHAPTER 22

Putting It All Together

Designing and building a high-performance communications receiver is probably one of the hardest tasks an individual or a company can undertake. Although it is relatively easy to put together building blocks and combine them to achieve electrical performance, the implementation of these blocks in the total ensemble of the package can make the difference between failure and success. There are many considerations.

The actual design of a communications receiver or transceiver involves many sciences from mechanical to digital. Figure 22.1 shows the actual implementation of a high-performance communications receiver, which I designed and built. Electrically the system is a double-conversion approach with a first IF at 75 MHz and the second IF at 455 kHz. A high-level mixer is used in the front end allowing for an intercept point of +22 dBm. A classic open loop VFO is used in conjunction with crystal oscillators providing a resolution of 2-kHz per revolution of the main knob. This, combined with narrow switchable filters, makes for an extremely sharp communications receiver. To provide maximum isolation, Collins mechanical filters have been used in the second IF and are switched with relays. Frequency is read through a digital counter display with a resolution of 100 Hz. The layout of the receiver shows a compartmented approach.

All circuits are built on 3.5 × 5-inch PC boards and gold-plated connectors are used throughout. The advantage of this design is mainly that circuit modifications and improvements are possible without altering the actual receiver.

Over the years, many modifications have been implemented in a continuous search for technical excellence. Mechanically, the receiver is made of anodized black aluminum. All parts have been extensively machined and there are 186 screws in the system. The gear drive exhibits no backlash because of its different design.

Light magnesium gears are used throughout. The variable frequency oscillator is driven by this gear train and is located in a closed compartment in the middle row close to the variable tuning oscillator (board), which can also be unplugged. The receiver has three tops, a T-like top, which covers the mechanics, and two hinged side covers, which cover the electronics. Upon closing the covers, each circuit board is fully isolated in its own housing because of the walls between the compartments. The electrical interconnections between the boards are provided through two motherboards located on the bottom of the receiver. All interconnect circuits are either shielded or etched into the motherboards. The bottom cover is only one-quarter of an inch from the circuits, providing extra shielding. There are two extender boards provided, should any troubleshooting be necessary. When not

(a)

(b)

Figure 22.1 (a, b) Actual implementation of a communications receiver. All circuits are built on 3.5 × 5-inch PC boards and gold-plated connectors are used throughout. The advantage of this design is mainly that circuit modifications and improvements are possible without altering the actual receiver.

in use, they are packed in a plastic bag and housed in the transformer compartment. The receiver weighs 15 pounds and is fully portable.

Shown in Figure 22.2 is the actual implementation of the Star-10 transceiver. Because the Star-10 is a fully synthesizer radio, the precision mechanical machining used in the previous examples is not necessary. Instead, the Star-10 uses individual assemblies as shown in Figure 22.3. The chassis mechanics are shown in Figure 22.4.

The accent is put more on electrical isolation between the assemblies as well as the interconnect harnesses.

Designing and building a high-performance communications receiver or transceiver is a good test of technical maturity for any individual who dares to attempt it. It is my opinion that such a multiscience test should be of significant value in asserting the engineering level of new college graduates, who will be faced with signal processing problems much similar to those encountered in a communications receiver in their future careers.

Figure 22.2   Actual implementation of the Star-10 transceiver.

Figure 22.3   The actual implementation of some of the Star-10 subassemblies already discussed in this book.

22.1   Packaging and Mechanical Considerations

Today, the electrical performance of a communications receiver can be mathematically predicted with the help of computers. As a result, the amount of circuit design required by a manufacturer is minimal. Of much more significance is the amount of performance that can be achieved for the price in view of short life cycles of parts. The question remains how to build a very high-performance, low-cost receiver. The answer lies in how the receiver is packaged. Designing a communications receiver

Figure 22.4 Actual implementation of the Star-10 chassis and final mechanical assemblies.

with stringent cost and labor boundaries automatically leads to the idea of using as much circuit integration as possible and using plug-in modules, as previously discussed.

These modules would functionally resemble the block diagram of the receiver as pointed out throughout this book. They should be designed such as to be individually manufactured and tested before integration into the receiver's frame. Fault isolation is automatically achieved with the use of diagnostic computers, thus reducing the manufacturing cost and allowing for more performance to be incorporated in the design for the same price. Because of the performance goals required in a communications receiver, the problem of interconnecting the modules economically and efficiently becomes a major cost element of the design. Demanding shock, vibration, and temperature requirements, as well as electromagnetic interference (EMI) standards, require new and economical ways of packaging. The problem is further aggravated by the highly digital nature of today's communications receiver. As many as a hundred control lines are sometimes needed to allow functional access to the various modules from the main microprocessor as well as from the commands and interfaces located on the front and back panels. The digital noise generated by these lines can greatly interfere with the performance of the receiver. Introducing lowpass filters in each one of the lines would prove too excessive for the cost of the receiver, while the use of ceramic capacitors to bypass each digital circuit would only provide suppression at the self-resonant frequency of the capacitors.

The economical solution to the problem is the use of a special flexible flat cable, which is wrapped in metal foil. This provides a cost-effective solution. It has been found that the filtering provided by this type of interconnection technique costs five times less than conventional line filtering. The method has been successfully applied in production by various manufacturers in an attempt to manufacture low-cost, high-performance communications receivers.

Additional manufacturing cost reductions have been obtained by the elimination of time-consuming milling and tapping of the chassis, while substituting punching and bending of sheet metal parts. This has become very feasible in synthesized receivers, which do not require the mechanical rigidity of past products. The use of nuts and washers has been reduced to a minimum, while emphasis was put on efficient ways of interconnecting the functional blocks. Extensive use of automated board assembly as well as automatic test equipment (ATE) makes the manufacturing of today's receiver one of the most automated production processes of its kind, allowing for a true low-cost, high-performance communications receiver at a fraction of the cost of other comparable receivers.

In conclusion, receivers can be built today with exceptional performance, but the manufacturing cost can make them prohibitively expensive.

The greatest task remains how to get the most performance for the lowest cost. While electrical design remains important in this regard, the solution lies in using more ingenious packaging techniques, as well as the automation of production lines and testing techniques.

CHAPTER 23

Radio Astronomy and the Search for Extraterrestrial Intelligence (SETI) Receivers

This book spans about 100 years of radio receiver technology. While this is a relatively short period of time, the advent of communications receivers has had its technological impact. One may ask: Where are we today? While the superheterodyne is still the most popular approach, with IFs as high in frequency as 30 GHz, the trend is toward even higher IFs. While selectivity is still achieved at lower frequencies compatible with filter technologies and percentage bandwidths, new technological discoveries will impact the high probability of intercept receivers of tomorrow.

The development of low-cost microprocessor-controlled synthesizers, such as the one used in the Star-10 transceiver, will allow for more sophisticated communications receivers at lower costs in the near future. Direct sampling receivers using digital signal processing are becoming the norm at HF frequencies in software-defined radios. Large-scale integrated circuits (LSIC) are being developed to replace discrete functions not only in the IF area, but also in the front end and mixer areas using totally digital downconverters.

Extremely low-noise amplifiers using GaAs FET, SiGe, indium phosphide, and new photonic technologies have also been recently introduced, allowing receivers for the first time to listen deeper into space for possible signs of extraterrestrial life. In today's technology, such a specialized communications receiver can provide useful reception of intelligent signals from a planet located some 15,000 light years away. Although this may seem as a great distance, it is only half way to the center of the Milky Way galaxy. (The Milky Way galaxy is 100,000 light years in diameter and 1,000 light years in thickness at the center.) It has been established that such a receiver will cover a preferred frequency range between 1,400 MHz and 1,727 MHz, the "water hole" frequency. This is shown in Figure 23.1.

The rationale behind this choice is based on the ideas that:

- Life in the universe is a function of water.
- An extraterrestrial civilization attempting to communicate with us would transmit in the water hole frequency band, which is located between the spectral line of hydrogen (1.420 GHz) and the spectral lines of hydroxyl (1.612 GHz to 1.727 GHz).

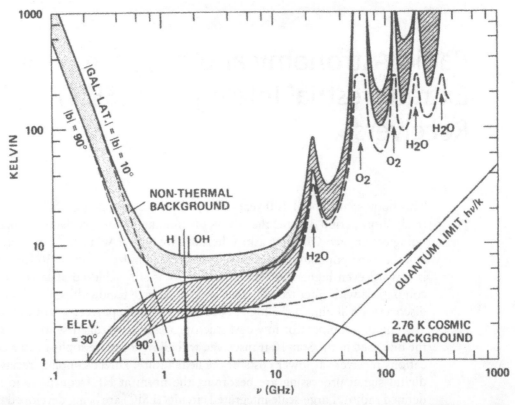

Figure 23.1 The "water hole" window provides the best overall noise performance for an interstellar communications receiver. This frequency band was also chosen as a meeting point for extraterrestrial civilizations, which base their existence on water. The signal shown represents the hypothetical strength of a transmitter similar to that of the Arecibo Observatory, located some 60 light years away and aimed at us.

Although other frequencies have been previously suggested, it has been discovered that this band of frequencies is the quietest in the spectrum as shown in Figure 23.1.

Existing radio observatories such as Arecibo are already searching the universe at the mentioned frequencies and more. An organized worldwide effort has been initiated to systematically listen for extraterrestrial signals. This is the search for extraterrestrial intelligence (SETI). In addition, radio noise studies are performed by several radio astronomy receivers located around the globe. Among these receivers, the Arecibo observatory receiver in Puerto Rico is a key receiver installation, which is worth mentioning. This is shown in Figure 23.2.

The Arecibo Observatory is part of the National Astronomy and Ionosphere Center (NAIC), a national research center operated by Cornell University under a cooperative agreement with the National Science Foundation (NSF). The NSF is an independent federal agency whose aim is to promote scientific and engineering progress in the United States

The "dish" antenna is 305m (1,000 feet) in diameter, 167 feet deep, and covers an area of about 20 acres. The surface is made of almost 40,000 perforated aluminum panels, each measuring about 3 feet by 6 feet, supported by a network of steel cables strung across the underlying sinkhole. It is a spherical (not parabolic) reflector.

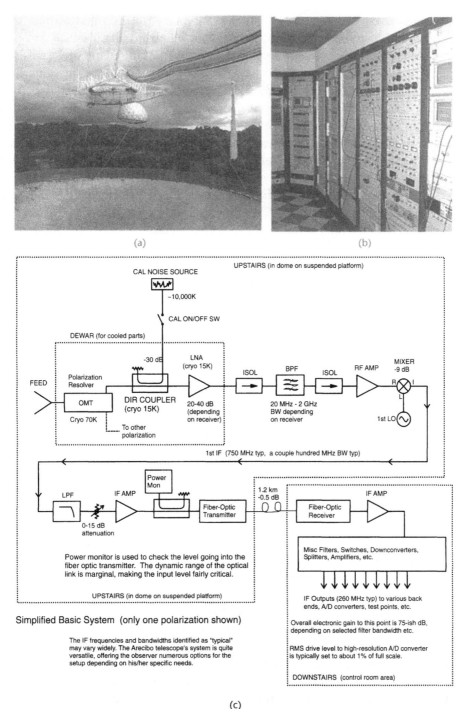

(c)

Figure 23.2 (a–c) At the time of this writing, the Arecibo Observatory is the world's largest radio telescope. Its receiver comprises a front end, an IF/LO section, and a back end. It covers the frequency range of ~300 MHz to 10 GHz in several bands. The front end is suspended above the 305-m diameter dish antenna, which provides approximately 70 dB of gain at 2.380 GHz. The first conversion (located inside the dome) is common to an IF of 750 MHz, and subsequent conversions in the control room area bring the signal to baseband. The overall receiver gain (between the antenna feed and the A/D input) is typically about 75 dB, but this can vary widely. The above pictures depict the suspended Gregorian dome where corrective optics, the front end, and cryogenics are located, and the IF/LO room on the ground. The two are linked by ~1.2 km of analog optical fiber. (Figure courtesy of the Arecibo Observatory, National Astronomy and Ionosphere Center, operated by Cornell University under a cooperative agreement with the National Science Foundation.)

Suspended 450 feet above the reflector is the 900-ton platform as shown in Figure 23.2. Similar in design to a bridge, it hangs in midair on 18 cables, which are strung from three reinforced concrete towers. One of the towers is 365 feet high, and the other two are 265 feet high. All three tops are at the same elevation. The combined volume of reinforced concrete in all three towers is 9,100 cubic yards. Each tower is back-guyed to ground anchors with seven 3.25-inch diameter steel bridge cables. Another system of three pairs of cables runs from each corner of the platform to large concrete blocks under the reflector. They are attached to giant jacks, which allow adjustment of the height of each corner with millimeter precision.

Just below the triangular frame of the upper platform is a circular track on which the azimuth arm turns. The azimuth arm is a bow shaped structure 328 feet long. The curved part of the arm is another track, on which a carriage house on one side and the Gregorian dome on the other side can be positioned anywhere up to twenty degrees from the vertical axle.

Inside the Gregorian dome two subreflectors (secondary and tertiary) focus radiation to a point in space where a set of horn antennae can be positioned to gather it.

Hanging below the carriage house are various linear antennas each tuned to a narrow band of frequencies. The long antenna sticking down is the 1-megawatt radar antenna.

The antennae point downward, and are designed specially for the Arecibo spherical reflector. By aiming a feed antenna at a certain point on the reflector, radio emissions originating from a very small area of the sky in line with the feed antenna will be focused on the feed antenna inside the Gregorian dome.

Attached to the antennas are very sensitive and highly complex radio receivers. These devices operate through conduction cooling at 15K using a closed-cycle helium gas cooling system to maintain a very low front end receiver temperature. At such cold temperatures the electron noise in the receivers is very small, and only the incoming radio signals, which are very weak, are amplified.

A total of 26 electric motors control the platform. These motors drive the azimuth and the Gregorian dome and carriage house to any position with precision.

The tertiary reflector can be moved to improve focusing. Front end receivers are moved into focus on a rotating floor inside the Gregorian dome and the dynamic vertical tie downs are activated as needed to maintain a precise platform position.

The 1-megawatt planetary radar transmitter, located in a special room inside the dome, directs radar waves to objects in our solar system. Analyzing the echoes provides information about surface properties and object dynamics.

The Arecibo radio receivers receive signals from the farthest reaches of the universe, from quasars, pulsars, and galaxies that emit radio waves, which arrive at Earth 100 million years later as signals so weak that they can only be detected by a giant eye like this. SETI is also part of the program.

The giant size of the reflector is what makes the Arecibo Observatory so special to scientists. It provides 70 dB of gain at 2.380 GHz ahead of the receiver electronics. It is the largest curved focusing antenna on the planet, which means it is the world's most sensitive radio receiver. Other radio telescopes may require several hours observing a given radio source to collect enough energy for analysis, whereas at Arecibo this may require just a few minutes of observation.

The Arecibo transmitter was used to send an extraterrestrial message on November 19, 1974, to the globular cluster M13 where possible intelligent life was suspected.

The Arecibo message was a simple graphic picture consisting of 73 rows of 23 bits (1,679 binary bits—eq. of 205 bytes) per row as shown in Figure 23.3. This number of rows and columns was chosen because each is a prime number. Prime numbers could be guessed by any intelligent recipients, and that would help them decode the graphics.

The message was sent by simple frequency-shift keying (FSK) between two frequencies (10 Hz) in the 2.380-GHz band and transmitted with a power of 20 terawatts (20 trillion watts, EIRP). This is the strongest manmade signal ever transmitted. The message contained the kind of information that any culture would want to learn about us: where we are located (within a depiction of our solar system); what we look like (a crude stick figure); a simple drawing of the transmitting antenna; and something about our biological construction (DNA and some of the building blocks of our biochemistry).

The IF frequencies and bandwidths identified as "typical" in the block diagram from Figure 23.2 may vary widely. The Arecibo telescope's back end IF/LO system is quite versatile, offering the observer numerous options for the setup depending on his or her specific needs. For example, the output of a wideband receiver may be split into several sub-bands appearing at different ports but converted to a common frequency range, to be sent to a multichannel spectrometer whose individual inputs have more limited bandwidths.

The term "back end" refers to a data acquisition and processing system such as a spectrometer. Usually, the selected back end is where the signal is digitized, with the object of providing further processing so that the data can be stored with minimal storage requirements. Some types of observations at the Arecibo radio telescope can generate data at rates significantly exceeding one terabyte per hour, but for most observations the rate can be reduced to a tiny fraction of this by suitable processing in the back end. The CAL injection scheme permits adding a small amount of noise in a controlled manner to the input signal. The resulting change in output can be used in processing to account for system gain variations over time and is an important part of making quantitative measurements. The input equivalent noise temperature of a cooled LNA in the system is usually only about 2 or 3K (noise figure about 0.04 dB or less). However, a variety of real-world factors conspire to make the overall noise performance of the telescope about an order of magnitude worse. These include:

- Feed-footprint spillover beyond the edge of the dish, admitting black body radiation from the terrain;
- Absorptive losses in the feed horn and room-temperature wave guide components;
- Blackbody radiation from the surrounding terrain and from the Sun being coupled into the beam by scattering off the feed platform and dome;
- Noise contributions from system components following the LNA;
- Radiation from other astronomical sources feeding into the antenna side lobes.

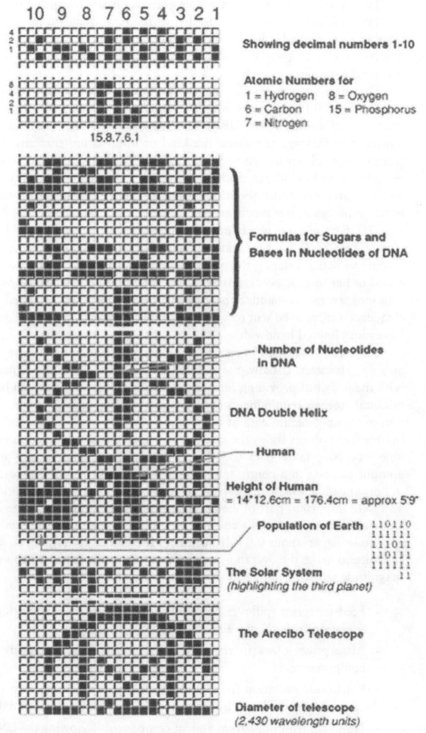

Figure 23.3 Content of the extraterrestrial data signal emitted by the Arecibo installation to the globular cluster M13 where possible intelligent life was suspected.

As a result, the overall system temperature of the Arecibo radio telescope is more like 25–30K, depending on which receiver is being used. It is worse for the UHF receivers where continuum radiation from the galaxy is stronger. On rainy days, it can be much worse at the highest frequencies (7 to 10 GHz) because of blackbody radiation from the rain itself and from a partially absorbing layer of water clinging to the dish surface during times of heavy rainfall.

In the absence of RFI, the dynamic range requirements of a radio telescope signal path are minimal. One is looking at noise of an almost constant level with the object of measuring rather small fractional changes in its amplitude level. However, RFI can easily be many orders of magnitude stronger, and it is highly desirable to accommodate the RFI signals without incurring nonlinear effects such as gain compression and intermodulation products.

While the primary dish at Arecibo is located in a well-shielded depression, the suspended feed platform is slightly above the surrounding hilltops and is line-of-sight to many RFI sources such as aviation radars, cell sites, broadcast stations, and so fourth. Energy striking the platform from these sources is in part scattered into the focusing beam and thus enters the receiver's input. Considerable effort is expended at the observatory towards ameliorating these interference sources.

CHAPTER 24

Digital Signal Processing (DSP) and Software-Defined Radio (SDR)

24.1 Introduction

In the previous chapters, we have discussed how RF signal processing can be performed using analog methods. However, RF signal processing can also be implemented in the digital domain using digital signal processing (DSP). In addition, both analog and digital signal processing can be used together in what is called mixed signal processing.

Digital signal processing (DSP) deals with transforming analog signals into sequential parallel digital numbers of binary format, manipulating them, using algorithms to extract information from noise, and translating them back into analog information for further interpretation [1–16]. In certain radar and EW receiver applications, the last operation is omitted as digital information is used directly for analysis. It is possible today to design entire receivers using DSP, as we will see later in this chapter. However, this is only possible up to a certain frequency. Unfortunately, advances in analog-to-digital (A/D) converters and digital-to-analog (D/A) converters have been relatively slow. According to the well-known Walden curves [1], the progress in performance of A/D converters has followed a general slope of one bit per octave of the sample rate. In addition, the SNR improvement occurs at the rate of one bit every 5.3 years, a fact that does not track with Moore's law (the number of transistors on a chip doubles every 18–24 months) if we equate doubling the number of transistors to a one-bit improvement.

State-of-the-art data conversion technologies are just now allowing digital processing at frequencies of up to 3 to 5 GHz, with the majority of applications at HF or relatively low intermediate frequency (IF) and/or at baseband. Aperture jitter, transistor matching, and increasing drain-bulk capacitance also remain major problems in the development of high-speed, low-power A/D converters. In general, the high bandwidth and high dynamic range performance of A/D converters have been directly proportional with power consumption. In spite of these difficulties, steady progress in speed and power has been made in recent years. Following these trends, a six-bit A/D converter with a 10-GHz effective resolution bandwidth has been recently available, but with a projected power consumption of around 4 watts.

All this makes the realization of the ideal software-defined radio (SDR) a more distant goal. The ideal SDR architecture would allow the digitization of an entire band of frequencies at the carrier and at any frequency. Then all processing, de-modulation, and so forth would take place in the digital domain. One of the receivers presented later in this chapter does this at HF. However, most higher-frequency receivers today using DSP are simply digitized superheterodynes.

Although DSP is a highly mathematical topic, which can occupy an entire book, our discussion will deal mainly with the major processes as well as the hardware necessary to perform DSP on RF signals. For a detailed treatise of the subject, you are directed to the many works listed at the end of this chapter.

The advantages offered by DSP are many. Among them are:

- Less sensitivity to component tolerance and environmental impacts;
- Elimination of bulky analog parts such as inductors or capacitors by dealing directly with mathematical models rather than physical parts, which would be too large for integration;
- Full integration and consistent performance and quality reproduction;
- Increased dynamic range by increasing word length (impacted by operational frequency of parts, increased parts count, and associated cost);
- Ability to use floating point mathematical manipulation of data in addition to fixed point manipulation (in analog processing dynamic range is limited by power supply rails);
- Ability to multiplex two or more signals at once if using additional sampling between primary samples;
- Ability to maintain exact linear phase in processing (no delay distortion);
- Sampling rate can be different in the various DSP sections of the processors to adapt to various applications (multirate);
- Ability to make constant changes in the process characteristics by simply changing coefficients in the software.

24.2   Time-Domain and Frequency-Domain Representation of Discrete Time Signals

The most important concept in digital signal processing is the process of continuous sampling of RF information in order to provide binary numbers, which in turn are representative of input analog signals. The information is quantized in both time and amplitude by using a sample-and-hold (S/H) and/or an analog-to-digital (A/D) converter. A high-quality, low phase noise, low-jitter sampling clock (master reference oscillator) is used to control the sampling.

The sample-and-hold mechanism retains the instantaneous sampled information until the A/D translates it into a binary number to the nearest integer number with an error of half the LSB value. Shown in Figure 24.1 is a basic DSP system block diagram. It is comprised of a sampler quantizer block, which outputs parallel binary numbers to the DSP section. This section, in turn, performs the mathematical operations intended to improve the signal-to-noise characteristics of a

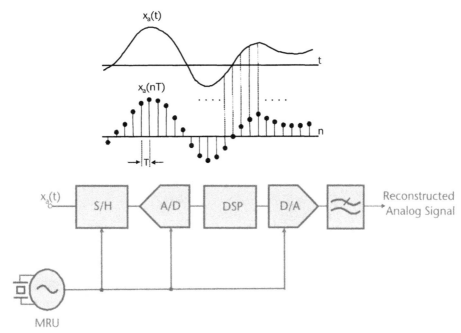

Figure 24.1   Block diagram of a DSP. The system performs continuous sampling of RF information in order to provide binary numbers, which in turn are representative of input analog signals. The information is quantized in time and amplitude by using an S/H and/or an A/D converter. A high-quality, low phase noise, low-jitter sampling clock (master reference) is used to perform the sampling. The A/D outputs parallel binary numbers to the DSP section that performs the mathematical operations intended to improve the signal-to-noise characteristics of a received signal. This, in turn, provides arithmetically modified parallel binary numbers to a digital-to-analog (D/A) converter followed by a lowpass filter intended to smooth out the reconstructed staircase-like analog signals.

received signal. This provides arithmetically modified parallel binary numbers to a digital-to-analog converter followed by a lowpass filter as we discussed.

Throughout a DSP system, the sampling frequency can be different for different stages of a processor in order to take advantage of either speeding up or slowing down the data to benefit a particular application. This is known as multirate digital signal processing. The process of converting from one sampling rate to another is known as decimation.

The sampling process is similar to the analog RF mixing process with the exception that the output is a series of parallel digital words, which can be manipulated via simple arithmetic operations. These operations are simple binary digital addition, multiplication (which uses multiple addition), subtraction (which uses inverse addition), and delay processes (recalling information). They are performed sequentially by using two binary values at any given time. A graphical representation of such process is shown in Figure 24.2. There is usually no integration or differentiation, since the process is performed using only the above arithmetic operations. However, it is possible to approximate integration and differentiation digitally, if required by the processing as shown in some of the later diagrams in this chapter.

After processing, the modified sequence of parallel numbers is presented sequentially to a D/A converter, which outputs a staircase-like signal.

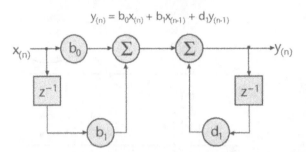

Figure 24.2 Graphic representation of a digital sampling process, which uses arithmetic operations such as multiple addition, subtraction (inverse addition), and delay (z^{-1}) processes (recalling information). These operations are performed sequentially (indicated by the arrows) using only two binary values at any given time.

Because this output presents sharp rise times in its composition, a lowpass filter is used to smooth out these problems and an improved reconstructed analog signal results.

If we consider a continuous function from the input information being sampled as $x_c(t)$, $-\infty < t < \infty$, where x_c is a continuous function of the continuous variable t, then the set of samples is $-\infty < n < \infty$, where the relation between t and n is defined by the sampling process $n = q(t)$.

24.3 Baseband Sampling Theory

We will discuss next the sampling operation. As shown in Figure 24.3, the analog signal is sampled once every T seconds resulting in a sampled data sequence, which is usually normalized to 1 second. Sampling is done for both the positive side of a waveform as well as the negative side of the waveform, and the mathematical representation of the signals is normalized to a point $n = 0$ as shown. Sampling is very important in reproducing the data because it has to be done fast enough to be representative of the data by obeying certain rules. These rules belong to the Shannon sampling theorem, or the Nyquist sampling theorem, named after these authors'

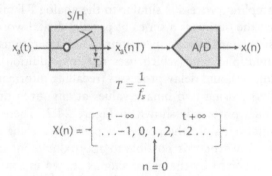

Figure 24.3 The process of sampling. An analog signal is sampled once every t seconds resulting in a sampled data sequence, which is normalized to 1 second. The sampling is done for both the positive side of a waveform as well as the negative side of the waveform, and mathematical representation of the signals is normalized to a point $n = 0$. The sample-and-hold circuit holds the data until it is transferred to the A/D.

papers [4, 5] on the subject. The sampling theorem indicates that a continuous signal can be sampled without loss only if it does not contain frequency components above one-half of the sampling rate. For instance, a sampling rate of 4 kHz requires the analog signal to be composed of frequencies below 2 kHz. If frequencies are above this limit, they will be aliased with frequencies between 0 and 2 kHz, combining with the data that was originally there. Thus, the Nyquist rate is $f_s = 2W$. This law applies to baseband as well as broadband signals, where W is the bandwidth of the information being sampled.

Subsampling in DSP can also be utilized where interim bands in between the successive multiples of $f_s/2$ can be used. In these applications, the sampling of the RF signal is much less than the Nyquist rate by a factor of up to, say, 16, or even more.

Oversampling can also be used. This is defined as $OSR = f_S/(2BW)$. In such case, the roll-off effect is minimized or eliminated. The quantization noise of the system is also improved because it is spread over a much wider frequency band. In addition, antialiasing filters can be more relaxed making them easier to integrate.

24.4   Bandpass Sampling Theory

As previously discussed, sampling is similar to RF mixing, where the LO is replaced by a sampling reference clock. The input to the A/D is also analog, and its output is a binary parallel digital word, which is presented to the D/A. The input to the D/A is a digital word and the output of the D/A is a staircase signal, which is smoothed out by a lowpass filter. The reference clock is an infinite set of impulses with period T, which mixes with the input signal, $x_a(t)$. Consequently:

$$s(t) = \sum_{n=-\infty}^{\infty} \delta(t - nT)$$

As we discussed earlier in the Nyquist theory definition, the input signal is limited to less than half the sample frequency, f_s, where $f_s = 1/T$, as defined by the theory. The spectrum of the output, $y_a(t) = x_a(t)s(t)$, has a double-sideband spectra for every integer multiple of the reference clock. By applying the Fourier transform of $y_a(t)$:

$$Y(j\omega) = \frac{1}{T} \sum_{k=-\infty}^{\infty} X(j\omega - kj\omega_s)$$

where $\omega_s = 2\omega/T$. If the baseband signal, $x_a(t)$, contains frequency components greater than the Nyquist frequency, then its signal content overlaps or causes aliasing. Looking at Figure 24.4, we can see how the sampling process takes place including the aliasing phenomenon. The baseband Fourier transform $X(j\omega)$ of $x_a(t)$ is shown in Figure 24.4(a). The spectra of $S(j\omega)$ is shown in Figure 24.4(b), and the

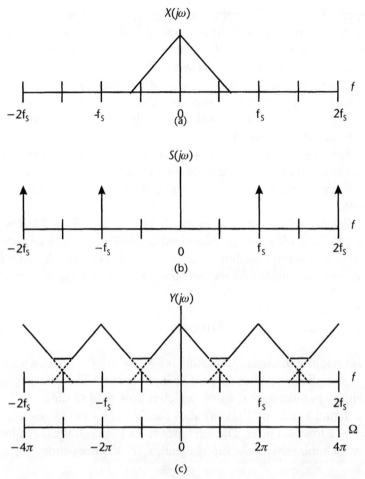

Figure 24.4 (a) Baseband spectrum $X(j\omega)$. (b) Sampling spectrum $S(j\omega)$. (c) Convolution of the two spectra $Y(j\omega)$.

convolution of the two $(Y(j\omega) = X(j\omega) * S(j\omega))$ is shown in Figure 24.4(c). In addition, the frequency axis in the digital domain, Ω is normalized by F_s, such that $\Omega = \omega/F_s$, as shown in Figure 24.4(c).

Subsampling can also be done if using a bandpass signal at a frequency higher than $F_s/2$. Considering that the bandpass spectrum is kept between integer multiples of the Nyquist frequency, an example of a properly subsampled bandpass signal showing no aliasing is shown in Figure 24.5. The Fourier transform $X(j\omega)$ of the bandpass signal $Xa(t)$, is shown in Figure 24.5(a). The convolution of $X(j\omega)$ with $S(j\omega)$ from Figure 24.4(b) results in $Y(j\omega)$, which is shown in Figure 24.5(b) along with both sets of axis, f and Ω [2, 3].

The spectrum shown in Figure 24.5(b) resembles the output of a D/A. As we have mentioned before, in order to reproduce the original baseband signal, a low-pass filter, $G(j\omega)$, is used after the D/A. This and the output spectrum, $S(j\omega)$, are shown in Figure 24.6(a). In order to reconstruct a bandpass signal, a bandpass filter, $K(j\omega)$, is used after the D/A. The resulting spectrum, $T(j\omega)$, is shown in Figure 24.6(b).

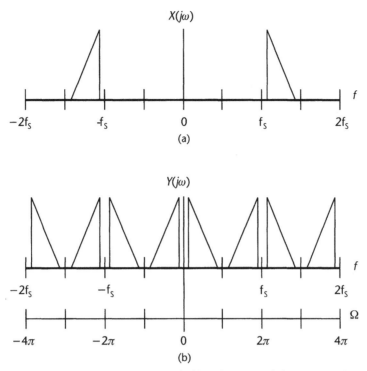

Figure 24.5 An example of a properly subsampled bandpass signal showing no aliasing. (a) Bandpass spectrum $X(j\omega)$. (b) Sampled spectrum, $Y(j\omega)$.

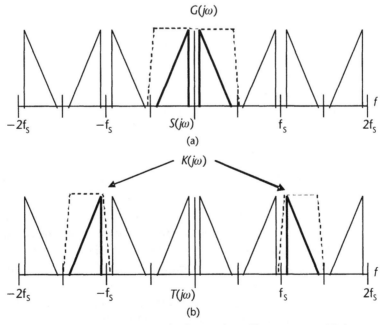

Figure 24.6 (a) Lowpass filter spectrum $S(j\omega)$. (b) Bandpass filter spectrum, $T(j\omega)$.

24.5 Analog-to-Digital (A/D) Conversion

The process of translating analog signals into parallel digital words is performed using A/D converters. These integrated circuit devices are classified by their type and by the number of bits available, which in turn dictates their signal-to-noise ratios (SNR) or their linear dynamic range equivalents. The maximum theoretical signal-to-noise ratio for a static A/D converter is expressed by the equation:

$$SNR_{max} = 6.02N + 1.76 \, dB \tag{24.1}$$

where N is the number of A/D bits.

This equation is derived from the process of quantizing a dynamic signal where the error voltage in the A/D is representative of a noise source that corrupts the digital representation of the input signal. The RMS amplitude of this error can then be derived from the following standard equations:

$$V_{in} RMS = \frac{2^N q}{2\sqrt{2}}$$

$$SNR = 20 \log \frac{2^N q / 2\sqrt{2}}{q / \sqrt{12}}$$

$$= 20 \left[N \log 2 + \log \sqrt{12} - \log \left(2\sqrt{2} \right) \right]$$

$$= 6.02N + 10.79 - 9.03$$

$$SNR_{RMS} = 6.02N + 1.76$$

It can be seen from these equations that an increase in the number of bits reduces the error amplitude by a factor of 2 for each additional bit, which in turn results in an SNR improvement of 6 dB per bit.

In reality, the actual SNR is less than the ideal SNR. If the SNR is less than the theoretical maximum, then the "effective" number of bits for an A/D converter is said to be less than the nominal number of bits as shown in the following equation:

$$N_{effective} = SNR_{actual} - 1.76 \, dB / 6.02 \, dB \tag{24.2}$$

The "effective" number of bits parameter is very important in using analog-to-digital converters. However, the number of bits in an A/D converter does not limit the dynamic range of a receiver. Intervention by AGC can keep large input dynamic ranges focused usually one third of the way down from the top of the A/D's linear dynamic range regardless of where the input RF signal is on the receiver's front end composite dynamic range. In addition, maintaining at least half the LSB of noise in the A/D will preserve linearity below the level of the LSB. Consequently, signals that are below the half LSB noise floor will exhibit a small SNR. High processing gain is then required after the A/D section (in the DSP section) to extract these low-level signals.

Depending on their frequency of operation and technology, the number of bits in A/D converters varies between one for a delta-sigma ($\Delta\Sigma$) modulator to as many as twenty four or even thirty two in some high fidelity cases. Most common static A/Ds have resolutions of 8, 10, 12, or 16 bits. Up to 32 bits A/Ds have been realized.

There are several types of A/Ds. Among them are the successive approximation, dual-slope, flash, and delta-sigma modulator. We will discuss some of the most popular types next.

24.6 Successive Approximation A/D

One of the most popular A/D converter type is the successive approximation. It uses three interconnected functional blocks, an analog comparator, a successive approximation register (SAR) and a D/A converter. Initially, the SAR is reset to zero. Then each bit in the SAR is set to 1 in sequence and starting with the most significant bit (MSB). At this point, the D/A converts this binary number into a representative voltage, which in turn is compared in the analog comparator with the input voltage and a representation of the signal is realized after several cycles are achieved. The digital result is then latched at the output.

24.7 Dual-Slope A/D

Dual slope A/Ds are very popular when high resolution is desired such as in measurement instrumentation. The main element in a dual-slope A/D is a capacitor, which starts out fully discharged, and is then charged for a certain amount of time by the source to be measured. It is at this point that the capacitor is switched to a known negative reference to be discharged to a zero value. A digital counter is then used to measure the discharge time. The elapsed time value obtained by the counter is then proportional to the input voltage. Increasing the counterresolution by adding additional counters achieves increased resolution.

24.8 Flash A/D

The most popular A/D converter is the parallel or flash architecture. It provides the fastest approach to quantizing analog signals. This approach uses the input signal voltage to compare with a set of fixed voltages produced by a ladder of precision resistors corresponding to binary weights. This brute-force approach allows the input level to be compared simultaneously with all preset levels. A typical flash A/D block diagram is shown in Figure 24.7.

Points on the input sine wave signal can be quantized based on the Nyquist sampling theory discussed previously, by simply being compared with the precision resistor network feeding the comparators. The outputs are strobbed at $\Phi 1$ and are then latched in what is called a *thermometer code decoder*, because of the resemblance with a thermometer reading of the data. In order for the analog signal to be properly quantized, its voltage level has to be within the end points of the precision reference voltage divider resistors ($R/2$) located at V_{ref} and ground. The data is further strobbed at $\Phi 2$ in order to be latched at the output by the N-

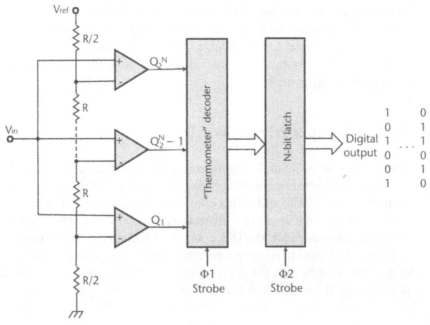

Figure 24.7 Block diagram of a typical flash A/D.

bit latch as shown. The digital output is a series of strobbed parallel binary words representative of the input signal.

Theoretically, flash A/Ds can operate at very high frequencies without sample-and-hold circuits. In order to do this, the response of the comparators has to be compatible with the transient frequency components at up to half the Nyquist limit to ensure high-frequency operation. These requirements and the limitations in resolution due to additional comparators and laser-trimmed resistors limit the practical number of bits being used as the frequency is increased, imposing additional power requirements as well. Up to 24-bit flash A/Ds can be manufactured at frequencies of up to 150 MHz. As frequency goes up, the effective number of bits is progressively reduced to 10, 8, or 6 bits as the frequency is increased to 2 GHz and beyond. Consequently, signal-to-noise (SNR) ratios are also reduced accordingly. As previously mentioned, progress in higher-frequency flash A/Ds has not followed the Moore law. This trend will continue for some time. However, up to 3 GHz and above flash A/Ds have been manufactured and are used in specialized receiver systems.

Most commercially available A/Ds today have balanced inputs exhibiting high impedance. This is intentional to provide immunity to common-mode problems. In most cases, adapting a single-ended IF output to the balanced input of an A/D is performed using an unbalanced to balanced differential transformation as shown in Figure 24.8.

24.9 Delta-Sigma ($\Delta\Sigma$) Modulator A/D

A good example of oversampling A/D converters is the delta-sigma ($\Delta\Sigma$) modulator. In this approach, the quantization noise is shaped around a desired passband

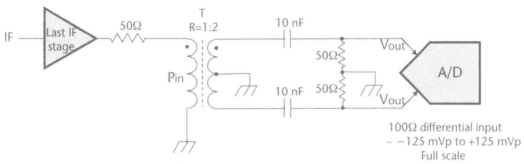

Figure 24.8   Method of feeding a balanced differential input A/D from a single-ended IF strip.

(baseband or bandpass) for greater improvements in SNR. The noise-shaping properties of the $\Delta\Sigma$ modulator have been compared with a notch filter function combined with a peak function. Baseband $\Delta\Sigma$ signal processing techniques are well-established for audio applications. However, bandpass $\Delta\Sigma$ modulators have only recently evolved, and are less known.

$\Delta\Sigma$ A/Ds are not always 1-bit. Multibit $\Delta\Sigma$ A/Ds overcome many of the limitations of 1-bit $\Delta\Sigma$ A/Ds but also introduce new issues. The main characteristic of $\Delta\Sigma$ is that they use oversampling and shaping of the quantization noise to achieve high resolution (ENOB, say, 16, 18, or even 24 bits) with low-resolution quantizers. The burden of achieving a high-dynamic range is shifted from requiring high-precision analog components to more complexity in the digital domain with digital filtering. $\Delta\Sigma$ A/D modulators can achieve an additional $20 \log 10 \, (N \times OSR)$ (dB) improvement in SNR beyond oversampled A/D converters, where N is the order of the modulator.

24.10   Delta-Sigma, Quantizing, and Noise Shaping

The basic concept of $\Delta\Sigma$ A/D modulators is that of using a coarse quantizer utilized within a feedback loop such that the power density of the resulting quantization error is suppressed within a specific frequency band of interest. This technique is known as quantization noise shaping or $\Delta\Sigma$ modulation. An example of a $\Delta\Sigma$ A/D is shown in Figure 24.9.

Using oversampling and the feedback loop, $\Delta\Sigma$ modulators allow the quantization errors of a coarse quantizer to be suppressed in a narrowband output. This is useful in narrowband IFs in receivers, and also in phase-locked loop outputs to improve on the phase noise performance. Bandpass $\Delta\Sigma$ modulators have been recently used in IFs at up to several GHz. An example of a receiver using a $\Delta\Sigma$ modulator is shown in Figure 24.10.

24.11   Digital-to-Analog (D/A) Conversion

Digital-to-analog (D/A) converters are used to reconstruct binary information such as obtained from DSP engines into analog signals. D/A conversion is easier

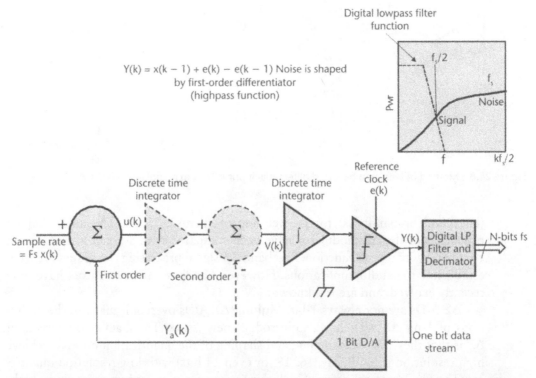

Figure 24.9 Recent advances in circuit speed enable bandpass RF ΔΣ modulators to be used at IF frequencies beyond 1 GHz. A model of a first-order ΔΣ converter (the second order shown in dotted lines).

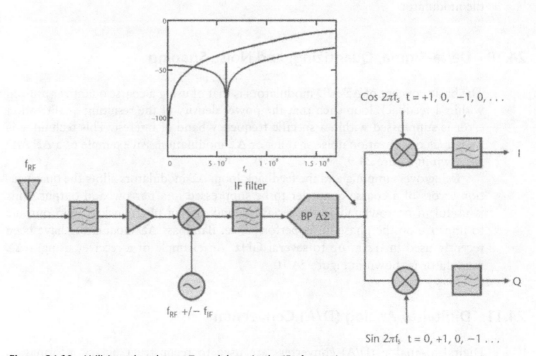

Figure 24.10 Utilizing a bandpass ΔΣ modulator in the IF of a receiver.

to implement than A/D conversion. Simple D/A converters can be designed using only resistor networks. A digital binary word can be converted to an analog output, which may be current or voltage. There are basically three types of D/As:

- Binary-weighed resistor network;
- R-2R ladder resistor network;
- Multiplying and bit stream.

The binary-weighed resistors network D/A simply adds binary-weighed resistors values to form the weighed sum of all nonzero bits from a binary parallel input word. This is shown in Figure 24.11(a). It can be seen that as the number of bits goes up, the resistors value increases tremendously making it very hard to maintain proper accuracy in production. For instance, if a 16-bit D/A is required, the ratio between the largest resistor value to the lowest is 2^{16} or 65,536. If the smallest resistor is 1 KΩ, then the largest resistor value is 65 MΩ. Maintaining a 1% precision for all values involved becomes a manufacturing impossibility. Consequently, binary-weighed resistor network D/A lends only to simple (up to 4 bits) devices.

The most effective type of digital-to-analog converter is the well known R-2R binary ladder network as shown in Figure 24.11(b).

In the multiplying D/A the output voltage (or current) is equal to a reference voltage times a constant determined by the digital input code divided by 2^n. More sophisticated D/A converters use precision constant current generators, switching circuits using FET switches, trimmed resistive circuits, and operational amplifiers fabricated on single substrates.

One of the most critical parameters of D/A converters is the conversion time or the settling time of the device. This is the time it takes a D/A from the initial data presented to the inputs to obtain a stable quantized value representing the input code. R-2R D/A converters excel in this regard.

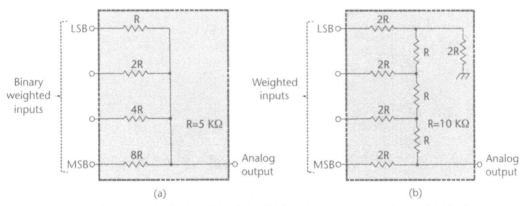

Figure 24.11   (a) Binary-weighed D/A. Maintaining high-resistance accuracy beyond 4 bits becomes an impossibility. (b) R-2R resistor ladder D/A converter uses only two resistor values, R and 2R to achieve binary-weighed inputs. Typical resistance values are 10 KΩ (R) and 20 KΩ (2R).

24.12   Staircase Reconstruction

A 6-bit R-2R D/A is shown in Figure 24.12. Its output is a staircase-like signal, which follows the number of bits used. A lowpass filter is usually used to smooth out the output in order to obtain an analog signal. A similar D/A converter application was presented in Chapter 18. It was used as the output of a digital AGC for a communications receiver. In this application, a cascaded up/down binary counter using two 74193 integrated circuits was interfaced to an 8-bit D/A converter to produce precision AGC signals for a receiver.

24.13   Bit Stream D/A

Multiplying D/A have disadvantages insofar as the accuracy of their most significant bit staying constant over a given temperature range. In an 8-bit D/A, for instance, the MSB must be accurate to one part in 2^8 (Figure 24.13). Obtaining and maintaining this accuracy over time and a temperature range is not an easy job. To remedy this problem, a bit-stream conversion technique similar to that used in $\Delta\Sigma$ modulators has been used. A higher sampling frequency is also used.

The advantages of this type of converter are high linearity combined with low cost, owed to the fact that most of the processing takes place in the digital domain and requirements for the analog antialiasing filter after the output can therefore be relaxed. For these reasons, this design is very popular in digital consumer electronics.

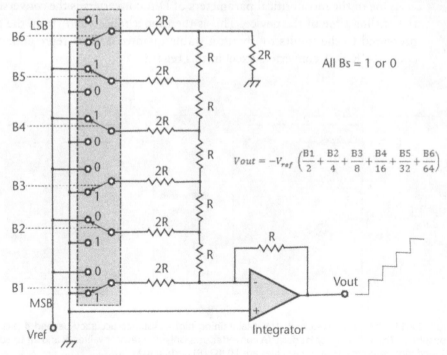

$$Vout = -V_{ref}\left(\frac{B1}{2} + \frac{B2}{4} + \frac{B3}{8} + \frac{B4}{16} + \frac{B5}{32} + \frac{B6}{64}\right)$$

Figure 24.12   A six-bit R-2R D/A implementation.

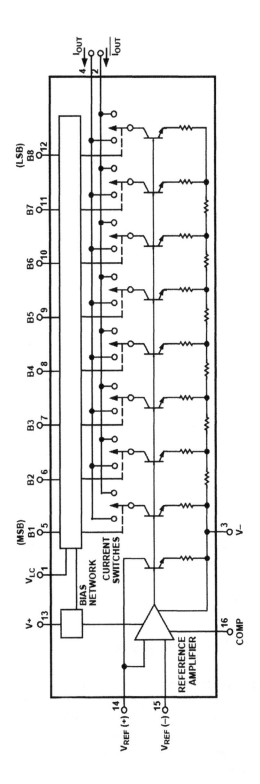

Figure 24.13 Example of an actual 8-bit multiplying D/A, DAC08. (Courtesy of Analog Devices.)

24.14   The Fourier Transform

We have discussed the key hardware and sampling elements involved in digital signal processing (DSP). We will now discuss the actual process involved in DSP.

Signal processing can be mathematically understood in either the time domain or the frequency domain. However, in DSP, most of the work is done primarily in the frequency domain. The two methods are equal in their analytical ability. However, in certain applications, one could describe signal processing better than the other. More often than not, problems that are hard to solve in the time domain can be easily resolved in the frequency domain and vice versa.

In order to be able to work efficiently with DSP in the frequency domain, we first need to have a basic understanding of the Fourier transform. The Fourier transform was named after the eighteenth-century mathematician Jean Baptiste Joseph Fourier (1768–1830) who, in 1812, was awarded a prize for scientific accomplishment for his complete explanation of the motion of heat in solid bodies. Today, this mathematical explanation is at the basis of explaining digital signal processing.

The Fourier transform as applied to DSP describes the relation between time and frequency through the following equations.

For the frequency-domain to the time-domain conversion:

$$x_{(t)} = \frac{1}{2\pi} \int_{-\infty}^{\infty} X(j\omega) e^{j\omega t} \, dt \tag{24.3}$$

For the time-domain to the frequency-domain conversion:

$$X_{(j\omega)} = \int_{-\infty}^{\infty} x(t) e^{-j\omega t} \, dt \tag{24.4}$$

where $j = \sqrt{-1}$ and $\omega = 2\pi f$.

It can be noticed from the above equations that the two transformations integrate over the $-\infty$ to $+\infty$, resulting in the concept of negative frequencies, which are not considered in the practical world. In practice, the complex exponential $e \pm j\omega t$ is replaced with the trigonometric expression $e - j\omega t = \cos \omega t \pm j \sin \omega t$.

24.15   Discrete and Fast Fourier Transforms

Although the Fourier transform equations are very useful, in reality they are not used directly in digital signal processing because they assume continuous data transformation, which is unrealistic. Analog data in DSP is sampled periodically according to the Nyquist criteria established earlier in this chapter. In order to allow DSP engines to process frequency-time domain transformations, the original Fourier transform equations have been modified into the discrete Fourier transform (DFT) equations as follows.

For a frequency-domain to a time-domain conversion:

$$x_{(n)} = \sum_{k=0}^{N-1} X_k \exp\left[j\frac{2\pi nk}{N}\right] \tag{24.5}$$

For a time-domain to a frequency-domain conversion:

$$x_{(n)} = \frac{1}{N}\sum_{k=0}^{N-1} x_k \exp\left[-j\frac{2\pi nk}{N}\right] \tag{24.6}$$

These DFT equations assume a quantized continuous signal or a finite duration signal. It can be seen that the integrals have been replaced with summation over an N discrete data points. Also, the frequency-domain to time-domain equation (24.5) differs from the time-domain to the frequency-domain equation (24.6) as a result of a change in the sign of the imaginary j and the scaling of the multiplier $1/N$. Using DFTs in computer programs has been further modified for increased efficiency into what is known as the fast Fourier transform (FFT). The FFT allows computing DFT functions more efficiently. They sort data by using data permutation algorithms known as the "butterfly" permutations because of the graphic resemblance of the flow diagrams are derived. The resulting Fourier transform calculations are speeded up as a result. Other even faster transforms such as the fast Hartley transform can be used for even more efficiency.

24.16   Digital Filters

The most important attribute of DSP in receivers is the ability to create filters (IF filters), which are very versatile as far as changing bandwidth on command and exhibiting linear phase response at the same time. While analog filters are generally designed for a single bandwidth response per filter, DSP filters can be tailored to adapt instantly to many bandwidth requirements, which is of great advantage when used in IF receivers, or directly in direct sampling receivers. Consequently, a receiver can be designed using a single analog preconditioning IF filter followed by a DSP section, which can provide many reconfigurable software filters, thus eliminating the complex switching mechanisms discussed in Chapter 17. Furthermore, digital filters can be implemented using the classical analog approximations discussed earlier in detail in Chapter 4, such as the Butterworth, Chebyshev, or elliptical designs as mathematically superimposed on infinite and finite response (IIR, FIR) algorithms as we will discuss next.

Additional functions such as noise canceling and suppression of unwanted carriers are also possible with DSP. An entire receiver can be implemented using DSP without using the traditional analog RF mixers involved in superheterodyne receivers. We will discuss a direct sampling HF receiver implementation next.

24.17   Infinite Impulse Response (IIR) Filters

IIR filters are fundamental parts of DSP systems. However, they are only one of the types of digital filters used in these systems. Their main characteristic is the presence of feedback, which gives them the infinite response characteristic, unlike their FIR counterparts. This actually means that if we input an impulse (a "1" followed by "0" samples) to an IIR filter, an infinite number of nonzero values would be theoretically outputted. The main advantage of IIR filters is their ability to perform functions using less computational power and less memory than when using similar FIR filters. However, there is a price. They are more sensitive to overflow problems and are not unconditionally stable such as their counterparts, FIR filters. They are also subject to noise generated by calculations, limit cycles, and low-level oscillations, which are a direct consequence of the feedback process. They are also harder to implement using fixed-point arithmetic and they do not match the computational advantages of FIR filters in multirate applications. Although all analog filter approximation types mentioned in Chapter 4 can be implemented with IIR designs, only low-order filter implementations are being used.

24.18   Finite Impulse Response (FIR) Filters

DSP can also be implemented using finite-impulse response (FIR) filters. In an FIR filter, the impulse response is said to be finite because there is no feedback, which by definition guarantees a finite response. Mathematically, FIR filters are represented by linear discrete-time functions in which the output sequence is related to the input and the impulse response of the filter by the convolution sum:

$$y_{(n)} = \sum_{m=0}^{M} x_{(m)} h(n-m) \qquad (24.7)$$

FIR filters operate as follows: An input sample is inputted to a delay line. Each sample accumulated in the delay line is multiplied by a coefficient. Then, each accumulated result is shifted to allow for a new similar operation. This actually means that if we input an impulse (a "1" followed by "0" samples) to an FIR filter, a finite number of zeroes will come out after the "1" sample has been outputted by the filter. The frequency response of an Nth order FIR filter is given by:

$$H_{(\omega)} = \sum_{n=0}^{N-1} h(n) e^{-j\omega n} \qquad (24.8)$$

One of the big advantages of FIR filters is the fact that they are simple to implement and that linear phase response can be achieved easily. Most FIR filter calculations can be implemented by looping a single software instruction. In addition, and unlike their IIR counterparts, FIR filters can be designed in multirate applications exhibiting decimation or interpolation, or both. This further allows better efficiency in software implementation. In addition, FIR filters can use fractional (less than

1) coefficients unlike the IIR filters. This makes the implementation more efficient, especially when using fixed-point arithmetic operations.

24.19 Smoothing Windows—Hanning/Hamming, Blackman, and Kaiser Bessel

When selecting a set of time-domain samples in designing a digital filter, the process introduces some high-frequency components that were not present originally in the continuously sampled signal. These artifacts show up as peaks in the FIR filter response and are undesirable. They are referred to as "leakage." We can reduce their amplitude by tapering off the amplitude of the samples towards each end of the sampling set, using smoothing windows. In effect, windowing can be thought of as sets of fudge factors intended to tweak coefficients for improved performance.

Windowing functions are typically applied, using a point-by-point multiplication process to the input of an FFT, in order to control the level of adjacent spectral artifacts that appear in the spectral response. Windowing manipulates the input data using values that smooth out or decrease toward zero the impulse response at each end of the data cycle.

There are many types of windows. Among them are the triangular, Bartlett, exact Blackman, Blackman, Blackman-Harris, Hanning, Hamming, Kaiser-Bessel, Poisson, Reimann, and several more. We will discuss some of them next.

The triangular window is a simple and fast window function, where the multiplication factor decreases linearly and symmetrically toward the ends of a sample. Another window is the Hanning window. The Hanning window is considered the best for attenuating the artifacts. Closely related to it, the Hamming window is probably the most popular method used today because it produces less objectionable sampling for many FIR filters applications. In a Hamming window, the samples about the center of the set get multiplied by the largest factor, and this decreases in a smooth sinusoidal fashion as we go away from the center, with the samples at each end of the window being multiplied by zero, and producing a smooth filter response. The commonly used Hanning and Hamming windows can be expressed by the following equations:

$$w\,Hanning(n) = 0.5 - 0.5\cos\left(\frac{2\pi n}{N}\right)$$ (24.9)

and

$$w\,Hamming(n) = 0.54 - 0.46\cos\left(\frac{2\pi n}{N}\right)$$ (24.10)

where N is the length of the window and w is the window value. They are both derived from the general cosine function form of:

$$w(n) = \alpha - \beta \cos\left(\frac{2\pi}{N}\right) \qquad (24.11)$$

Hanning, Hamming, and Blackman windows are shown in Figure 24.14. It can be seen from this figure that the first two types of windows are similar. However, they are different because the Hamming window does not get as close to zero near the edges as the Hanning window. Hamming windows are probably the most popular functions used in FIR filter designs today.

Another type of windows is the exact Blackman and its derivative, the Blackman window. The basic equation for the exact Blackman is:

$$y_i = x_i \left[a_0 - \cos(w) + a_2 \cos(2w) \right] \qquad (24.12)$$

for $i = 0, 1, 2, \ldots, n-1$ and $w = \frac{2\pi i}{n}$, where n is the window length and $a_0 = 7{,}938/18{,}608$, $a_1 = 9{,}240/18{,}608$, and $a_2 = 1{,}430/18{,}608$.

The Blackman window is a modified version of the exact Blackman window. The equation for the Blackman window is:

$$y_i = x_i \left[0.42 - 0.5\cos(w) + 0.8\cos(2w) \right] \qquad (24.13)$$

for $i = 0, 1, 2, \ldots, n-1$, where n is the window length.

However, other types have been used. Among them is the Blackman-Harris window, which when used in a FIR filter, it is believed to give the best out-of-band rejection. However, this is at a cost. The transition band from full response (gain 1) to the first zero response widens to accommodate a better sidelobe rejection in the final filter response.

Another type of window is the Kaiser-Bessel window. This is a flexible smoothing function whose shape can be modified on demand by adjusting its β input. Depending on the application, we can change the shape of the window such as to

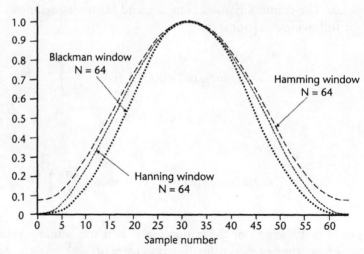

Figure 24.14 Hanning, Hamming, and Blackman ($N = 64$ samples) windows examples. The Hanning and Hamming windows are similar but differ in that the Hamming window does not get as close to zero near the edges as the Hanning window. The Blackman is narrower and goes down to zero.

control the spectral leakage amount. The Kaiser-Bessel window is used mainly for detecting two signals in close frequency proximity and of different amplitude. A list of some of the most popular smoothing windows is given in Table 24.1. For additional information on smoothing windows, you are directed to the References and Selected Bibliography at the end of this chapter.

24.20   Phase Noise and Jitter Considerations: Choosing Offsets in Bandpass Digital Signal Processing

Jitter is very important in the performance of DSP. The meaning of phase noise and jitter in receiver systems has previously been discussed in detail. Design of high performance master reference oscillators and mathematical methods of translating the phase noise into jitter have also been presented in this chapter. This information applies equally to DSP reference sampling oscillators.

Sampling jitter is distortion introduced by the variation in the sampling interval of an A/D by the reference oscillator. This, in turn, increases the noise floor of the signal being processed. Just as in superheterodyne mixing systems, the phase noise and resulting jitter produced by the MRU should not spoil the MDS of a receiver using DSP. Often, this MRU is one and the same as the receiver's synthesizer MRU, which may serve as reference for other local oscillators (LO) in the receiver system.

If DSP is used as a last IF, the phase noise quality of this reference should be equal to the previous conversion stages. If the receiver is a true sampling type, the phase noise (jitter) performance should be compatible with the projected MDS of the receiver. The reader is directed to Chapter 16 for the mathematical relationships of transforming phase noise into jitter.

In addition, much care should be taken in choosing LO frequencies in bandpass DSP systems in order to avoid alias problems introduced by sampling. A graphic method of choosing the sampling frequency and the center frequency/bandwidth of the related IF preceding a DSP section is shown in Figure 24.15.

24.21   Practical Software-Defined Radios (SDR)

As high-speed A/D and D/A converters migrate into new digital receivers, performance specifications in view of the new processing methodology become very

Table 24.1   Characteristics of Several Smoothing Filters

Smoothing Window	−3-dB Main Lobe Width (Bins)	−6-dB Lobe Width (Bins)	Maximum Side Lobe Level (dB)	Sidelobe Roll-Off Rate (Decibels Per Decade)
Hanning	1.44	2.0	−32	60
Hamming	1.3	1.81	−43	20
Blackman-Harris	1.62	2.27	−71	20
Exact Blackman	1.61	2.25	−67	20
Blackman	1.64	2.3	−58	60

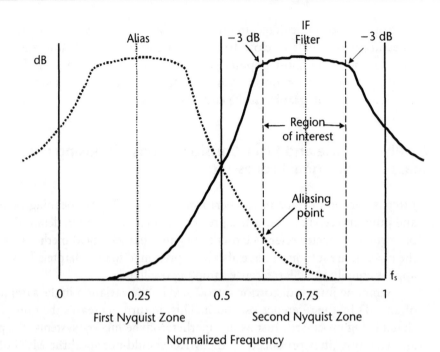

Figure 24.15 Graphic method of determining IF filter requirements preceding a bandpass DSP system in a receiver.

important to the system receiver designer. The meaning of dynamic range as explained in Chapter 13 changes somewhat to accommodate for the new digital terminology. However, it needs to be understood through the classic dynamic range concepts, which we have already explained. Specifications like spurious free dynamic range (SFDR) in a digital signal processor have slightly different explanations than the standard linear SFDR explanation discussed earlier because digital devices introduce a new and different set of variables.

In digital signal processing, spurious free dynamic range means the usable dynamic range before spurious noise spikes appear near a desired fundamental signal. SFDR is then the measure of the difference in amplitude between the fundamental and the largest harmonically or nonharmonically related spur from DC to the full Nyquist (f_S) bandwidth (or sometimes half Nyquist, or $f_S/2$). SFDR is usually specified in dBc.

In addition, for a given digital converter, a reasonable SFDR specification is a direct result of the number of bits of the device and can reach values up to 100 dBFS (FS = full scale) for modern AD converters. In an ideal system, the worst-case signal-to-noise ratio is calculated by the previously explained (24.1) where N is the number of bits of the converter. For instance, the worst-case signal-to-noise ratio for a 12-bit ideal digital system is 74 dB. This would be with the assumption that the quantization noise is uniformly distributed. In reality, this number is actually less than the ideal case by about 10% as we previously discussed.

Today, the classical receiving system is changing rapidly, due to advances in many technologies. The communications receiver of tomorrow will be increasingly called upon to assume demanding tasks of signal processing in a complex environment. Receivers using digital sampling signal processing techniques have been implemented to about 3 GHz. Their realization depends clearly on the progress of A/D converter technology, which has its design limitations as frequency goes up. The higher the frequency is, the fewer the number of bits; there is also less dynamic range. As we mentioned before, the exponential progress experienced in other areas of electronics has not been the case in digital signal processing at microwave frequencies. Because of this fact, receivers at microwave frequencies still use superheterodyne approaches followed by high IFs where limited dynamic range A/Ds can be used for digital signal processing in combination with complex AGCs. Consequently, the ideal digital receiver implemented at microwave frequencies remains a future goal.

24.22   The ADAT Software-Defined Radio

Complete digital sampling receivers are currently being implemented at the lower frequencies, primarily at HF and VHF/UHF frequencies. These digital receivers are known as software-defined radios (SDR) (other software-defined radio definitions exist).

As usual, new developments in the RF field are done by ham radio pioneers who are also talented design engineers. An example of an HF digital sampling receiver is the ADAT, ADT-200A transceiver. This is shown in Figure 24.16.

A block diagram of the ADT-200A receiver is shown in Figure 24.17. The receiving section starts with a high dynamic range preamplifier (14-dB gain) proceeded by a 0- to 35-dB programmable attenuator, which can be either operator controlled or automatically controlled by the AGC system and a half-octave filter bank. The ADT-200A uses the latest generation of high-performance DSP manufactured by Analog Devices Inc., capable of up to 2 billion instructions per second, leaving room for future options.

The front end is followed by an alias lowpass filter, which is matched to the higher impedance of a 14-bit A/D converter through a 1:4 RF transformer. The A/D is an Analog Devices part, the AD-6645, which offers an S/N ratio of 74 dB over

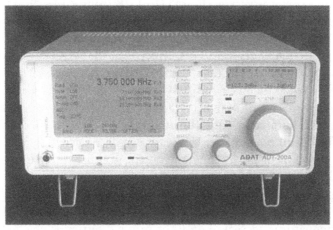

Figure 24.16   The ADAT, ADT-200A is an HF transceiver that uses an advanced digital sampling approach for both the receiver section and the transmitter section. (Courtesy of ADAT Corporation, Switzerland.)

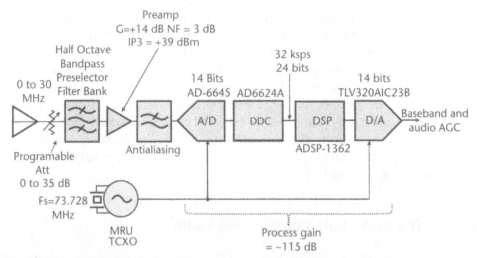

Figure 24.17   Block diagram of the ADAT, ADT-200A digital sampling receiver.

the half Nyquist bandwidth of 36.86 MHz. After the subsequent decimation, an instantaneous linear dynamic range of 120 dB is achieved.

Because of the almost total digital nature of the approach, there is no gain variation in the whole receiver until to the last stage. In other words, the ideal SNR condition of $6.02 \times N + 1.76$ has been satisfied. Consequently, the traditional AGC is not used the way it is used in analog radios. This is possible due to the fact, that the dynamic range of the A/D is about 115 dB at a bandwidth of 2.4 kHz (to be correct: S/N at half Nyquist $+10*\log(f_S/2*B) = 73.2 + 41.8$ dB $= 115$ dB, $f_S = 73.728$ MHz). Therefore, besides the input attenuator's range of 0, 5, to 35 dB, there is no automatic gain control before the back end as shown in Figure 24.18. The attenuator in front of the receiver serves to complete the AGC application, adapting the dynamic range from about 117 dB to high-level signal situations. The effects of the programmable attenuator on the receiver's dynamic range (MDS and compression point) are shown in Figure 24.19.

The total digital implementation in the ADT-200A receiver has the advantage that the S-meter is very accurate (+/−1 dB from −148 dBm to +17 dBm RF input range, and there are no negative effects on the baseband information such as AGC "popping," which is sometimes experienced in analog radios due to signal delays through IF quartz crystal filters. Moreover, the AGC in the ADT200A, also has a digital preemptive part, that "warns" the AGC block of the strong transients before they reach the back end AGC control mechanism. Consequently, the receiver behaves as in a well-designed traditional analog receiver and the S-meter can be precisely calibrated in both dBm and dBμV.

The 14-dB preamplifier is always active, preceded by the attenuator. As previously mentioned, the attenuator control is either manual or automatic, using the broadband peak sum voltage on the A/D converter input. Attack and decay times are obtained digitally as well. If the voltage rises over −2 dB full-scale for 1 second, the attenuator is increased by 5 dB; if it rises for 5 seconds below −8 dB full-scale, the attenuator is decreased until the original setting is reached again.

As shown in Figure 24.19, one can find different noise figures for the MDS, depending on preamplifier, attenuator setting, and bandwidth. Generally, the noise

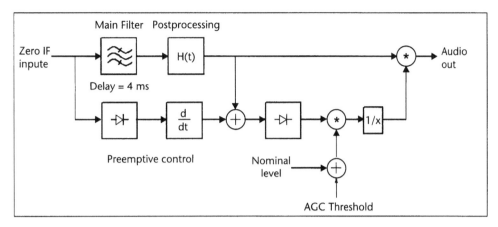

Figure 24.18   The ADT-200A receiver uses no AGC for the digital conversion section as its dynamic range is guaranteed at 117 dB. Implementation of the AGC is at baseband as shown. Additional AGC is obtained at the front end using the programmable 0- to 35-dB attenuator, which can be either manually or automatically operated in combination with the back end AGC. Because of the nature of the A/D dynamic range, there is no gain variation in the whole receiver until to the last stage. In other words, the ideal SNR of $6.02 \times N + 1.76$ has been satisfied (includes the attenuator effect). Consequently, traditional AGC is not used the way it is used in analog radios. This is possible due to the fact that the dynamic range of the A/D is about 115 dB at a bandwidth of 2.4 kHz (to be correct: S/N at half Nyquist +10 log $(f_S/2B) = 73.2 + 41.8$ dB = 115 dB, $f_S = 73.728$ MHz). (Courtesy of ADAT Corporation, Switzerland.)

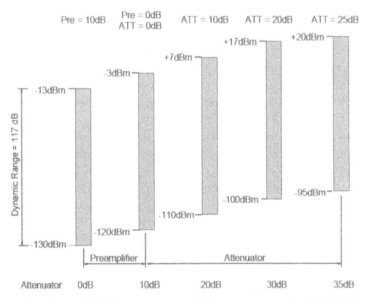

Figure 24.19   The effects of the input programmable attenuator on the ADT-200A receiver MDS and compression point. The bottom numbers represent the MDS while the top numbers are the respective compression points followed by IP3 numbers (not shown). When a strong signal is present, which exceeds the digital down converter's SFDR and its back end AGC, the front end AGC (programmable attenuator) is inserted to compensate for it. (Courtesy of the ADAT Corporation, Switzerland.)

figure of the ADT-200A is 10 dB (+/−1 dB) from 1 MHz to 30 MHz. This means that with the preamplifier on (which is always the case) and an ultimate bandwidth of 2.4 kHz, the MDS is around −130 dBm, and with an ultimate bandwidth of 100 Hz, the MDS is −144 dBm. Due to the precise S-meter, these figures can be checked directly with the ADT-200A system without any external equipment.

As we discussed in Chapter 13, the IP3 is not a direct physical measure. Without being included in the dynamic range equations presented in Chapter 13, IP3 is actually meaningless as it can be showed at many levels of the two tones. In addition, an A/D converter usually shows the best IP3 near full scale.

In addition, in some A/Ds, the third-order product does not necessarily follow the classic 3-to-1 curve range.

In A/D converters, performance can usually be improved using dither. The ADT-200A receiver uses digital dithering to improve on its performance. Usually, this is a very small amount of random noise (white noise), which is added to the input before conversion. Its amplitude is set to be about half of the least significant bit. Its effect is to cause the state of the LSB to randomly oscillate between 0 and 1 in the presence of very low levels of input, instead of sticking to a fixed value. Rather than the signal simply being cut off altogether at this low level (which is only being quantized to a resolution of 1 bit), it extends the effective range of signals that the A/D converter can convert, at the expense of a slight increase in noise. Effectively, the quantization error is diffused across a series of noise values, which is far less objectionable than a hard cutoff. The result is an accurate representation of the signal over time. A suitable FIR filter at the output of the system can thus recover small signal variations as it is the case in the ADAT-200A receiver.

An audio signal of a very low level (with respect to the bit depth of the A/D) sampled without dither usually sounds distorted and unpleasant. Without dither the low level always yields a "1" from the A/D. With dithering, the true level of the audio is still reproduced as a series of values over time, rather than a series of separate bits at one instance in time. The IP3 of the ADT-200A receiver is shown in Figure 24.20.

The sideband suppression (or image rejection in this case) in the digital down-converter of the ADT-200A is a mixer function driven by a numerical oscillator, which shifts the desired spectrum into a zero IF band. This mixing and all subsequent processing is a complex scheme of splitting the I and a Q channels as shown in Figure 24.21.

As this processing is done in the digital domain, the phase shifts and amplitude imbalances are assumed to be ideal. This means that the suppression of the unwanted sideband in this system is greater than 110 dB as shown in Figure 24.22. Shown is the simulation of the direct conversion receiver using the quadrature signal processing for sideband image suppression. This example assumes a signal above the LO frequency and the undesired sideband (image) below the LO frequency. The sideband selection, only changes the sign in the summing device.

One of the big advantages of this system is an almost ideal digital signal processor in which mathematical algorithms can be exactly reproduced during a long time without any aging effects. Consequently, there is no need for additional correction techniques as discussed in Chapter 7. The only limiting factor here is the numerical resolution of the processors. In the ADT-200A case, this is 40 bits for filters and 32 bits for other parts of the design. For instance, in the RIT (receiver incremental tuning) of the ADT-200A, a mathematical IF of 1.3 kHz is used to allow the passband tuning function to move the filter curve over the carrier without

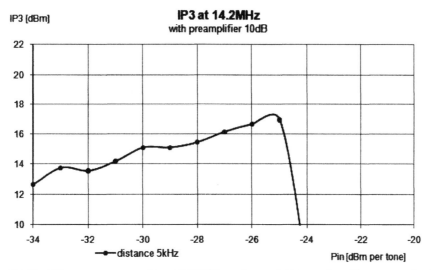

Figure 24.20   IP3 measurements of the ADT-200A receiver at 5-kHz spacing. (Courtesy of the ADAT Corporation, Switzerland.)

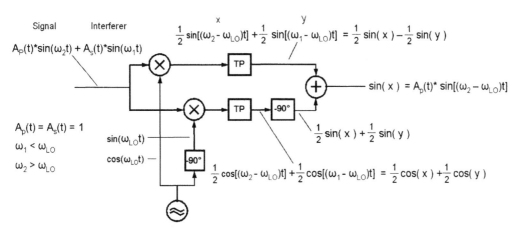

Figure 24.21   Simulation of the direct conversion sampling receiver using quadrature signal processing for sideband image suppression. This example assumes a signal above the LO frequency and the undesired sideband (image) below the LO frequency. The sideband selection, only changes the sign in the summing device.

having any problems with the image. After that, the shifted spectrum is mixed back to its original frequencies. This is only possible with a complex signal representation and the practically ideal mathematical signal treatment.

This is why, in production, the whole receiver needs only one adjustment: setting the compensation for the insertion loss of the analog preselector filters in the half-octave bank.

Another interesting phenomenon attributed to the digital nature of the receiver is that the 90° phase shifter is implemented into the filter function, so that the Q of the filter shows a perfect linear phase of +90° for negative frequencies and −90° for positive frequencies. The reproduction of this phase lends to a jitter, which is better than 0.0001°. This gives a clear disadvantage of the digital approach over

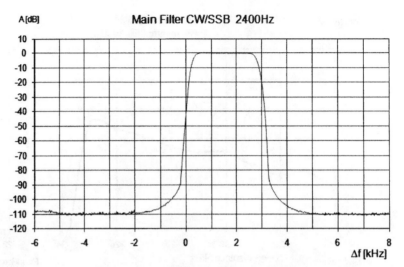

Figure 24.22   FIR filter response of the ADT-200A receiver showing an almost perfect suppression of the lower sideband (image). (Courtesy of ADAT Corporation, Switzerland.)

analog direct conversion receiver approaches, which have to use one of the image calibration methods discussed in Chapter 7.

Another advantage of the digital approach is that there is only one oscillator [the master reference unit (MRU)] in the entire receiver, feeding all mathematical sections of the digital signal processor so that the system is fully coherent. With the only oscillator in the ADT-200A being a crystal oscillator with a very low phase noise/jitter performance (typically < 0.5 ps), one can assume, that the whole receiver has a low phase noise performance of −140 dBc at 1 kHz and beyond. There is no direct measurement of the phase noise possible with this system, but it can be determined by the reciprocal mixing effect as shown in Figure 24.23.

Figure 24.24 shows MATLAB simulations of the ADT-200A FIR filters. To protect from out-of-band signals (not the image), the ADT-200A receiver uses a

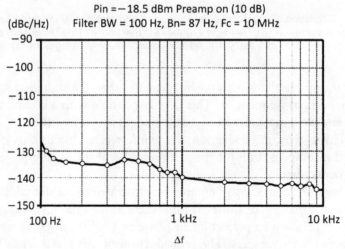

Figure 24.23   Phase noise performance of the ADT-200A receiver. (Courtesy of ADAT Corporation, Switzerland.)

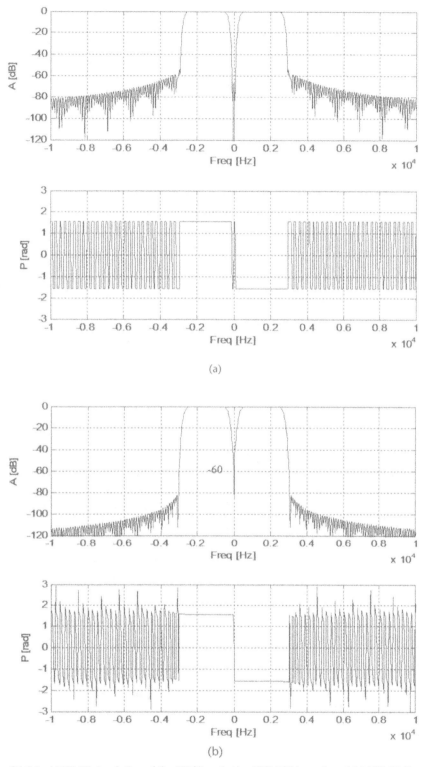

Figure 24.24 MATLAB simulation of the FIR filters in the ADT-200A receiver. (a) MATLAB Simulation of a FIR bandpass filter, 256 taps, using a Hamming window ($\omega Ham(n) = 0.54 + 0.46 \cos(\pi m/N)$ where N is order/2 and m is 0, 1, ..., $N-1$). (b) MATLAB simulation of a FIR bandpass filter, 256 taps, using a Kaiser window. (c) MATLAB simulation of a FIR bandpass filter, 256 taps, using a Blackman-Harris window. (Courtesy of ADAT Corporation, Switzerland.)

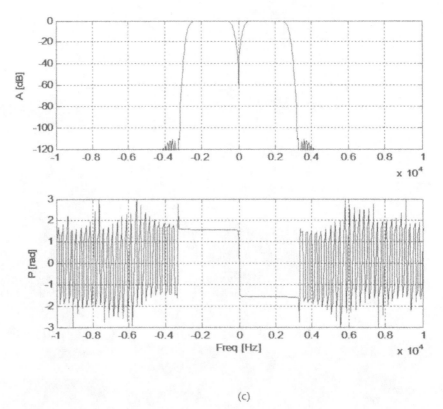

(c)

Figure 24.24   (continued)

half-octave filter bank similar to that used in the Star-10 transceiver, which we discussed earlier in Chapter 14. In the ADT-200A, because of the zero IF conversion, these filters are not utilized to reject the image, which is on the opposite side band of the FIR filter as we previously discussed. The composite response of the ADT-200A receiver is shown in Figure 24.25. In addition, the ADT-200A receiver can receive concurrently on two separate frequencies by combining filters in the half-octave bank. This is shown in Figures 24.26. Actual implementation of the half-octave filter bank in the ADT-200A is shown in Figure 24.27. Figure 24.28 shows the actual implementation of the entire ADT-200A transceiver.

24.23   Other Software-Defined Radios (SDR)

So far, we have discussed only one form of SDR. However, SDRs have many definitions. Among them is the example shown in Figure 24.29.

24.24   Defining Software-Defined Radios (SDR)

Although seemingly simple to define at first glance, software defined radio (SDR) is nothing but simple to define. According to [6], SDRs are described as "radio equipment, where one or more functionality-defining features are accomplished in

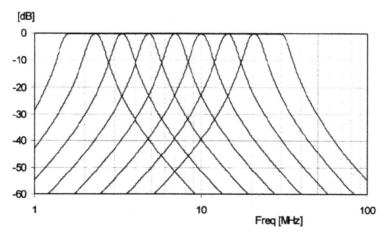

Figure 24.25 Composite frequency response of the front end half-octave filter bank used in the ADT-200A receiver. (Courtesy of the ADAT Corporation, Switzerland.)

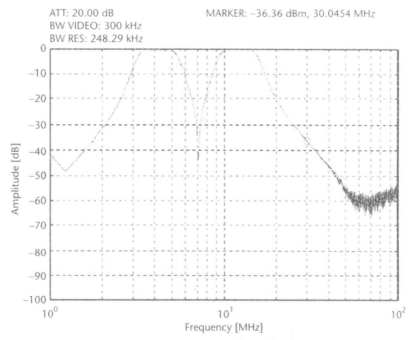

Figure 24.26 Combining filters in the half-octave filter bank to provide concurrent operation in different portions of the HF spectrum. (Courtesy of the ADAT Corporation, Switzerland.)

a software algorithm rather than in hardware." This functionality usually addresses the lower physical layers of the seven-layer ISO-OSI (International Standardization Organization–Open Systems Interconnection) model described in Chapter 4, but can go up into the higher levels, as well, depending on the application. A further differentiation of SDR is provided by the various de facto SDR tiers as outlined in Table 24.2.

An interesting cross-section of practical application examples of SDRs can be found in the product line of Radixon Group, an Australia-based company known

The Preselector

Attenuator, 0 … 35 dB
in 5 dB Steps

Half octave filters,
switched by high
curren FETs

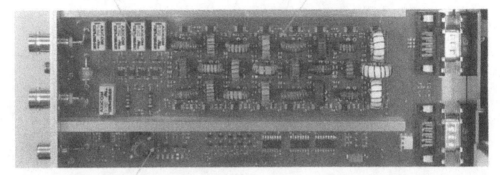

VLF-Front end, for 60, 75,
77.5 and 137 kHz

Figure 24.27 Implementation of the ADT-200A preselector using eight half-octave filters similar to those discussed in Chapter 14. (Courtesy of the ADAT Corporation, Switzerland.)

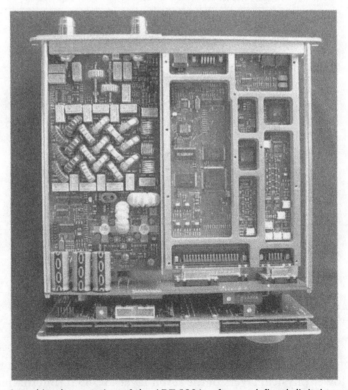

Figure 24.28 Actual implementation of the ADT-200A software defined digital transceiver. (Courtesy of the ADAT Corporation, Switzerland.)

through its brand name WiNRADiO. The evolution of WiNRADiO's receiver products seems to reflect the evolution of SDRs themselves: The first product, the WR-1000i, is an ISA-bus card for the personal computer (PC). The WR-1000i is

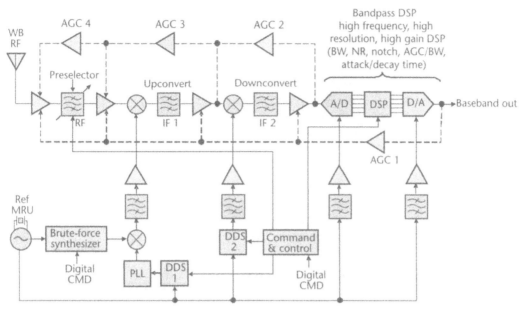

Figure 24.29 A different concept of a software-defined receiver uses a dual conversion hybrid superhetero-dyne approach, with the last IF fed to a bandpass digital signal processor.

Table 24.2 SDR De Facto Classifications

Tier 1	An SDR where limited functions such as power levels are being software controlled. Frequency or mode is not software controlled.
Tier 2	An SDR where a significant amount of functions are software configurable. Such radio can also be called a software-controlled radio (SCR). Functions such as frequency, modulation, waveform generation, bandwidth, and encryption may be addressed. However, the RF front end is implemented in analog hardware and is not software reconfigurable.
Tier 3	An SDR where reconfigurability extends as close to the antenna as possible. The front end is also software configurable. This radio has full functional programmability.
Tier 4	An SDR where in addition to the functionality mentioned in tier 3, additional functions can be achieved via software. Among them are multiple and concurrent frequencies operability, bandwidths, and modes operation. This radio is a step ahead of the ISR radio and is called the ultimate software radio (USR) because it can change its functionality on demand.

Source: [6].

a typical tier 1 SDR, where the role of the computer is confined to a mere receiver control via a simple graphical user interface resembling a conventional receiver, while the actual signal processing remains entirely conventional.

In designing this product, the task was to prove that a broadband receiver can coexist with the electrically noisy PC environment. This product represented an important milestone because it was the first commercially available communications receiver on a personal computer platform, paving the way for more advanced future SDR products. The WR-1000i receiver and its graphical user interface (GUI) are shown in Figure 24.30.

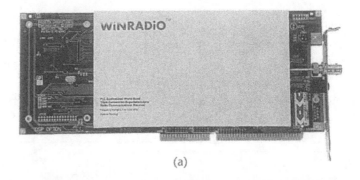

(a)

(b)

Figure 24.30 (a) WiNRADiO WR-1000i receiver. (b) GUI for the WR-1000i receiver. (Courtesy of WiNRADiO, Australia.)

Contemporary successors of this product are already tier 2 SDR products. They still rely on a conventional analog downconverting front end, but the analog processing ends at the last intermediate frequency, which is low enough to be subsequently processed (i.e., filtered and demodulated) by the host PC. An example of this type of receiver is the WiNRADiO WR-G305. This is a software-defined PC-based wideband scanning receiver covering a frequency range from 9 kHz to 1.8 GHz (expandable to 3.5 GHz with an optional converter). The block diagram of the WR-G305 receiver is shown in Figure 24.31.

The WR-G305 SDR is available in two versions: the WR-G305e USB model and the WR-G305i internal PCI card model. While the USB model contains its own A/D, the PCI model relies on the PC sound card for the IF signal digitization. Both units, their internal implementation, and the GUI interface are shown in Figure 24.32.

The receiver is based on a double-conversion process, where the incoming frequency is downconverted to the output frequency, which is provided as a filtered output ready to be digitized. The paths are switchable in order to minimize images and spurious mixing products. Filters are specially selected to ensure flat phase response and digitally controlled attenuators are employed in the front end as well as in the path.

The entire system features an excellent phase stability and flatness throughout the entire frequency range, with minimum amplitude and phase distortion, as well as minimum amplitude and phase mismatch between channels. This also includes digitally controlled phase adjustment of the output signal for precise phase

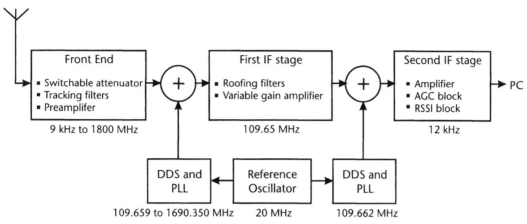

Figure 24.31   WR-G305 block diagram. (Courtesy of WiNRADiO, Australia.)

calibration in multichannel phase-coherent applications. Figure 24.32(d) shows the actual display of the WR-G305i receiver.

These SDRs also provide a "study" feature as part of the receiver application software, which makes it possible to observe, in real time, the inner workings on the receiver demodulators, using interactive block diagrams that show their structure. This is shown in Figure 24.33.

In this example, note that the input of this AM demodulator is a digitized IF signal and not already hardware-separated I/Q signals as might be expected. Such an approach is in fact often preferable for narrowband SDRs, because the I/Q separation can be performed more comfortably and accurately in software by simply multiplying the signal with sin and cos functions at the IF center frequency.

Below the interactive block diagram, there are two color-coded spectrum analyzers. Left or right clicking a mouse on any of the various "test points" inside the block diagram will connect either the left or right spectrum analyzer to the associated point, and the color of the point will change accordingly.

There is also a vector voltmeter showing amplitude and phase differences between the two selected points, at a frequency indicated by the user-controllable cursor. This makes it possible to click around the block diagram to see the actual signal at the various test points in the SDR receiver being shaped by the signal processing and finally demodulated. Each demodulation type (AM, FM, SSB, and so forth) has its own interactive block diagram.

Another interesting representative of the WiNRADiO SDR family, which also demonstrates a completely different SDR concept in practice, is the WR-G31DDC receiver, also known as Excalibur. As its model number suggests, this model relies on a digital downconverter to baseband. The block diagram of this receiver is shown in Figure 24.34.

The G31DDC is an HF receiver, covering a range of 9 kHz to 50 MHz. An interesting feature of this receiver is the ability to display the entire 50-MHz operational bandwidth of the receiver in a real-time spectrum analyzer. This functionality is provided by one portion of the FPGA, while the remaining portion is dedicated to the digital downconverter with a processing bandwidth variable up to 2 MHz. The GUI of this product has been improved from that of the WR-1000i receiver

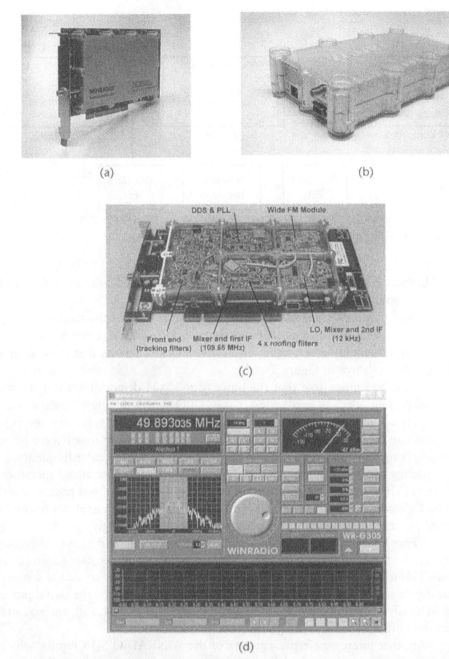

(a)   (b)

(c)

(d)

Figure 24.32   (a) The WR-G305i is a PC-based SDR manufactured by WiNRADiO. It covers a frequency range of 9 kHz to 1.8 GHz and can be expanded to 3.5 GHz with an optional converter. (b) The WR-G305e is the USB version of the same receiver. (c) Actual implementation. (d) GUI for both receivers. (Courtesy of WiNRADiO, Australia.)

and demonstrates how far the consumer-applied SDR technology has advanced in just 15 years. While the early products offered not much more than simple graphical equivalents of conventional receiver controls, the GUI of the WR-G31DDC receiver includes three interactive real-time spectrum analyzers (one showing the

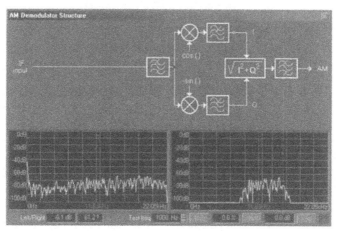

Figure 24.33 WR-G305 display of interactive demodulator structure (IDS). (Courtesy of WiNRA-DiO, Australia.)

entire received spectrum, one showing the DDC bandwidth, and one covering the selected demodulator bandwidth) and many other features.

This receiver can demodulate and record three independent channels at the same time, as long as they fall within the current DDC bandwidth (i.e., as long as they are less than 2 MHz apart). The GUI interface of this receiver is shown in Figure 24.35.

In addition to the above receivers, Radixon also makes SDR building blocks designed for a wide range of professional applications, such as fast search SDRs or phase-coherent receiver arrays for direction-finding applications.

One such product range is the WR-G526e product family. This family of modular products includes both analog front-end and digital back-end modules that can be used to construct an SDR to suit a particular application.

The available analog front-end modules cover a very wide frequency range from LF to SHF frequencies, employing conventional upconversion superheterodyne technology. These receivers are also produced with separate local oscillators in order to provide phase coherence for multiple receiver arrays. These local oscillators work with 0.01-ppm accuracy. The WR-526e SDR modules are shown in Figure 24.36(a), while their implementation in a rack-mountable receiver platform using eight receivers is shown in Figure 24.36(b).

The digital back-end modules are constructed around the dual core Blackfin DSP (Analog Devices BF 561) and an FPGA. Output data can be either via a 60 MSPS DSP daisy chain, which can accommodate several back-ends working in a phase-coherent arrangement, or via an LVDS output port operating at 2 Gbps, shown in Figure 24.37.

The internal FPGA structure contains four DDC channels, an interface for a snapshot memory, and some "glue" interface logic. The internal FPGA configuration for the WR-G526e/DSPS module is shown in Figure 24.38.

A typical WR-G526e rack-mountable receiver platform can contain up to eight WR-G526e front-end and back-end modules in a phase-coherent arrangement, master reference oscillators, interface circuitry, and power supplies.

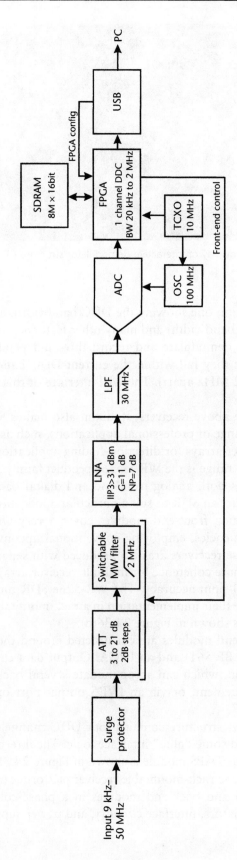

Figure 24.34 WR-G31DDC block diagram. (Courtesy of WiNRADiO, Australia.)

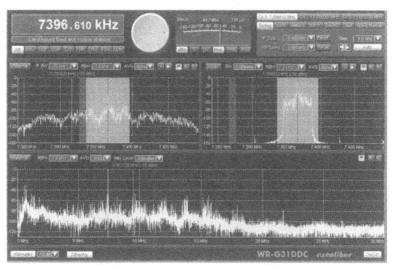

Figure 24.35   WR-G31DDC GUI. (Courtesy of WiNRADiO, Australia.)

24.25   Cognitive Radio

Cognitive radio is a new field in radio communications and wireless data networking, which is based on adaptive system configurations allowing user requirements to be configured on the fly as a function of the application. This is possible in the context of flexible tier 4 SDRs, which can provide configuration resources most appropriate for these needs. The idea of cognitive radio was first introduced in the early 1990s. It was thought of as the ideal goal offered by the new generations of SDR platforms just evolving. It was believed that by providing a flexible adaptive communications means, the frequency spectrum could be managed much more efficiently using on-demand dynamically reconfigurable radios that would automatically change their variables in response to new needs. This included reconfigurable spectrum utilization as a function of time and application, and adaptive modulation and bandwidth assignments. New functionality was recently added to the cognitive radio requirements. One of these requirements is dynamically sensing spectrum densities in real time and reallocating entire frequency bands on demand in addition to frequency hopping and spread spectrum modulation schemes. Because of this, regulatory bodies such as the Federal Communications Commission in the United States along with other regulatory bodies worldwide have taken a new position in frequency spectrum assignments specifically directed toward providing this flexibility from a regulatory point of view.

Although initially, cognitive radio was thought of as an extension of software-defined radio (SDR) design, the entire field was found to be much broader than initially thought. Cognitive radio today extends to total spectrum management, which is performed dynamically and in real time. Much of the work today is directed toward the design of sophisticated spectrum sensing apparatus, algorithms, and networking protocols for exchanging sensing data between various nodes in communications networks such as APCO-25, the Joint Tactical Radio System (JTRS), and others. Spectrum polling networks using orthogonal frequency-division multiple access (OFDMA) modulation have been envisioned in emergency and WLAN

(a)

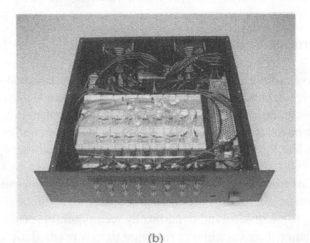

(b)

Figure 24.36   (a) Example of WR-G526e SDR coherent receiver modules. (b) Rack-mountable platform using eight WR-G526e phase-coherent receivers. (Courtesy of WiNRADiO, Australia.)

applications where higher link throughput and transmission path extensions can be provided on demand.

24.26 Conclusions

The field of SDR is constantly developing with new configurations emerging along with the A/D and D/A technology advances. Although it is hard to predict the future of SDR, it will most definitely play an important role in the future of wireless communications. The radio of the future will most definitely blend together with the computer and the Internet technologies in unimaginable ways.

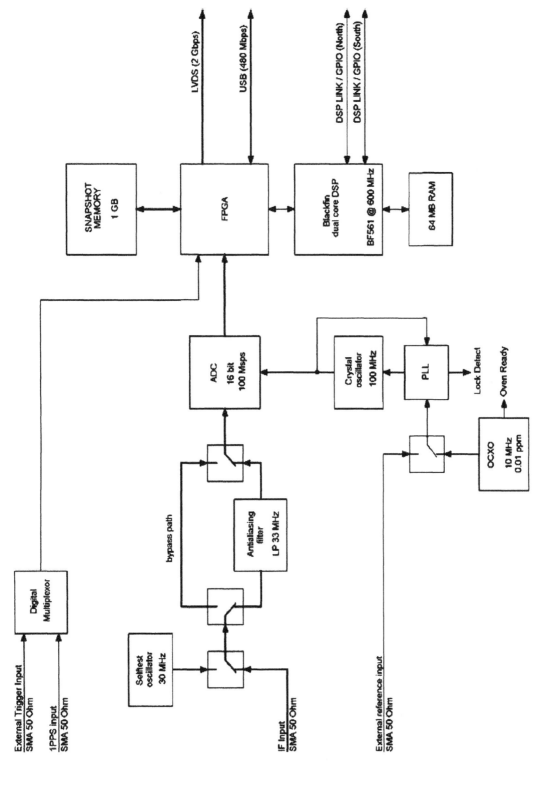

Figure 24.37 WR-G526e/DSPS back-end module—block diagram. (Courtesy of WiNRADiO, Australia.)

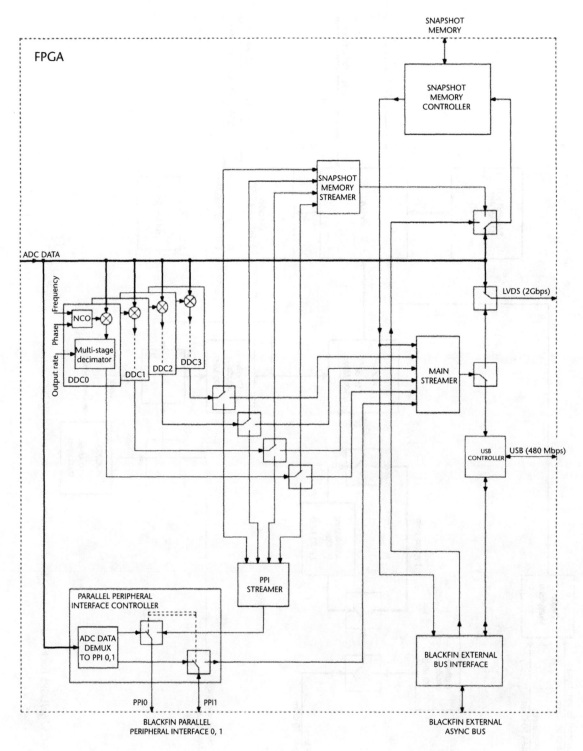

Figure 24.38   WR-G526e/DSPS module—internal structure of the FPGA. (Courtesy of WiNRADiO, Australia.)

References

[1] Walden, R. H., "Performance Trends for Analog-to-Digital Converters," *IEEE Communications Magazine*, Vol. 37, No. 2, February 1999.

[2] Hentschel, T., M. Henker, and G. Fettweis, "The Digital Front-End of Software Radio Terminals," *IEEE Personal Communications*, August 1999.

[3] Lai, E., *Practical Digital Signal Processing for Engineers and Technicians*, Burlington, MA: Newnes, 2004.

[4] Nyquist, H., "Certain Topics in Telegraph Transmission Theory," *AIEE Trans.*, 1928.

[5] Shannon, C. E., "Communication in the Presence of Noise," *Proc. IRE*, January 1949.

[6] "Definition of SDR," *Radio & Electronics*, http://www.radio-electronics.com/info/receivers/sdr/software-defined-radios-tutorial.php.

Selected Bibliography

Aziz, P. M., H. V. Sorensen, and J. van der Spiegel, "An Overview of Sigma-Delta Converters," *Signal Processing Magazine*, Vol. 13, No. 1, January 1996.

Alkin, O., *PC-DSP*, Upper Saddle River, NJ: Prentice-Hall, 1990.

Ashley, S., "Cognitive Radio," *Scientific American*, March 2006.

Banu, M., et al., "A BiCMOS Double Low Receiver for GSM," *Proceedings of the CICC 97 IEEE*, 1997.

Baringer, C., et al., "3-Bit, 8 GSPS Flash ADC," *Proceedings of the International Conference on Indium Phosphide and Related Materials*, April 1996.

Bergeron, B., "Digital Signal Processing," *Communications Quarterly*, Fall 1990.

Blackman, R. B., and J. W. Tukey, *The Measurement of Power Spectra*, New York: Dover Publications, 1958.

Boute, R., *The Geometry of Bandpass Sampling*.

Brigham, E. O., *The Fast Fourier Transform*, Upper Saddle River, NJ: Prentice-Hall, 1974.

Broesch, J. D., *Digital Signal Processing Demystified*, Solana Beach, CA: High Text Publications, 1997.

Buck, J., M. Daniel, and A. Singer, *Computer Explorations in Signals and Systems Using MATLAB*, Upper Saddle River, NJ: Prentice Hall, 1997.

Burrus, C. S., et al., *Computer-Based Exercises for Signal Processing Using MATLAB*, Upper Saddle River, NJ: Prentice-Hall, 1994.

Cherry, J. A., and W. M. Snelgrove, "Clock Jitter and Quantizer Metastability in Continuous-Time Delta-Sigma Modulators," *IEEE Trans. on Circuits and Systems II*, Vol. 46, No. 6, June 1999.

Cherry, J. A., and W. M. Snelgrove, *Continuous-Time Delta-Sigma Modulators for High-Speed A/D Conversion*, Boston, MA: Kluwer Academic Publishers, 2000.

Cherry, J. A., W. M. Snelgrove, and W. Gao, "On the Design of a Fourth-Order Continuous-Time-LC Delta-Sigma Modulator for UHF A/D Conversion," *IEEE Trans. on Circuits and Systems II*, Vol. 47, No. 6, 2000.

Cho, T. B., and P. R. Gray, "A 10 b 20 Msample/s 35 mW Pipeline A/D Converter," *IEEE Journal of Solid-State Circuits*, March 1995.

Choi, M., and A. A. Abidi, "A 6-b 1.3 Gsample/s A/D Converter in 0.35-μm CMOS," *IEEE Journal of Solid-State Circuits*, December 2001.

Cooley, J. W., "How the FFT Gained Acceptance," *Signal Processing Magazine*, Vol. 9, No. 1, January 1992.

Cosand, A. E., et al., "IF-Sampling Fourth-Order Bandpass DS Modulator for Digital Receiver Applications," *Solid-State Circuits*, Vol. 39, No. 10, October 2004.

Crochiere, R. E., and L. R. Rabiner, *Multirate Digital Signal Processing*, Upper Saddle River, NJ: Prentice-Hall, 1983.

da Fonte Dias, V., "Signal Processing in the Sigma-Delta Domain," *Microelectronics Journal*, Vol. 26, 1995.

Dalton, D., et al., "A 200-MSPS 6-bit Flash ADC in 0.6-μm CMOS," *IEEE Trans. on Circuits and Systems, Analog and Digital Signal Processing*, November 1998.

Demler, M. J., *High-Speed Analog-to-Digital Conversion*, San Diego, CA: Academic Press, 1991.

Drentea, C., "The Star-10 Transceiver," *QEX*, November/December 2007.

Dudgeon, D. E., R. M. Mersereau, and R. M. Merser, *Multidimensional Digital Signal Processing*, Upper Saddle River, NJ: Prentice-Hall, 1984.

Dutta Roy, S. C., "Digital Signal Processsing Introduction," Video Course, Department of Electrical Engineering University of Delhi, http://www.youtube.com/watch?v=y1BHUlpnfIQ.

Fairchild Semiconductors, "SPT7760 8-bit, 1 GSPS, Flash A/D Converter," datasheet, 1995.

Flynn, M., and B. Sheahan, "A 400-Msample/s, 6-b CMOS Folding and Interpolating ADC," *IEEE Journal of Solid-State Circuits*, December 1998.

Galton, I., "Delta-Sigma Data Conversion in Wireless Transceivers," *IEEE Trans. on Microwave Theory and Techniques*, Vol. 50, No. 1, 2002.

Geelen, G., "1.1 GSample/s CMOS A/D Converter," *IEEE International Solid-State Circuits Conference, Digest of Technical Papers*, February 2001.

Glas, J. P. F., "Digital I/Q Imbalance Compensation in a Receiver," *GLOBECOM 1998 IEEE*, Vol. 3, 1998.

Gold, B., and C. M. Rader, *Digital Processing of Signals*, New York: McGraw-Hill, 1969.

Goodenough, F., "Analog Technology of All Varieties Dominate ISSCC," *Electronic Design*, Vol. 44, February 19, 1996.

Goyal, V.K., "Theoretical Foundations of Transform Coding," *Signal Processing Magazine*, Vol. 18, No. 5, September 2001.

Hamming, R. W., *Digital Filters*, Upper Saddle River, NJ: Prentice-Hall, 1977.

Harris, F. J., "On the Use of Windows for Harmonic Analysis with the Discrete Fourier Transform," *Proc. IEEE*, Vol. 66, 1978, pp. 51–83.

Hayes, M. H., *Statistical Digital Signal Processing and Modeling*, New York: John Wiley & Sons, 1996.

Heideman, M., D. Johnson, and C. Burrus, "Gauss and the History of the Fast Fourier Transform," *ASSP Magazine*, Vol. 1, No. 4, October 1984.

Hein, S., and A. Zakhor, *Sigma Delta Modulators: Nonlinear Decoding Algorithms and Stability Analysis*, Boston, MA: Kluwer Academic, 1993.

Horowitz, P., and H. Winfield, *The Art of Electronics*, New York: Cambridge University Press, 1980.

Ingle, V. K., and J. G. Proakis, *Digital Signal Processing Using MATLAB*, Pacific Grove, CA: Brooks/Cole Publishing, 2000.

Inglis, A. F., *Electronic Communications Handbook*, New York: McGraw-Hill, 1988.

Irie, K., et al., "An 8b 500MS/s Full Nyquist Cascade A/D Converter," *Symposium on VLSI Circuits, Digest of Technical Papers*, June 1999.

James, J. F., *A Student's Guide to Fourier Transforms: With Applications in Physics and Engineering*, 2nd ed., New York: Cambridge University Press, 2002.

Jensen, J. F., A. E. Cosand, and H. C. Choe, "IF Sampling 4th Order Bandpass DS Modulator for Digital Receiver Applications," *2003 IEEE GaAs IC Symposium Technical Digest*, San Diego, CA, 2003.

Jensen, J. F., et al., "A 3.2-GHz Second-Order Delta-Sigma Modulator Implemented in InP HBT Technology," *IEEE Journal of Solid-State Circuits*, Vol. 30, No. 10, October 1995.

Johnson, D. H., and D. E. Dudgeon, *Array Signal Processing: Concepts and Techniques*, Upper Saddle River, NJ: Prentice-Hall, 1993.

Jurgen, A. E. P., et al., "A Sixth-Order Continuous-Time Bandpass Sigma-Delta Modulator for Digital Radio IF," *IEEE Solid State Circuits*, Vol. 34, No. 12, December 1999.

Kaplan, T., et al., "A 1.3-GHz IF Digitizer Using a 4th-Order Continuous-Time Bandpass SD Modulator," *Proceedings of the IEEE Custom Integrated Circuits Conference*, September 21–24, 2003.

Kaplan, T. S., et al., "A 2GS/s 3b/spl Delta/Sigma/-Modulated DAC with a Tunable Switched-Capacitor Bandpass DAC Mismatch Shaper," *Solid-State Circuits Conference, Digest of Technical Papers*, February 15–19, 2004.

Kaplan, T. S., et al., "A 2-GSPS 3-Bit DS Modulated DAC with Tunable Bandpass Mismatch Shaping," *IEEE Journal of Solid-State Circuits*, Vol. 40, No. 3, March 2005.

Kinget, P., and M. Steyaert, "Impact of Transistor Mismatch on the Speed-Accuracy—Power Trade-Off of Analog CMOS Circuits," *Proceedings of the IEEE Custom Integrated Circuits Conference*, May 1996.

Kroupa, V. F., *Direct Digital Frequency Synthesis*, New York: IEEE Press, 1999.

Lee, E. A., and P. Varaiya, *Structure and Interpretation of Signals and Systems*, Reading, MA: Addison-Wesley, 2003.

Lim, J. S., (ed.), *Speech Enhancement*, Upper Saddle River, NJ: Prentice-Hall, 1983.

Lim, J. S., *Two-Dimensional Signal and Image Processing*, Upper Saddle River, NJ: Prentice-Hall, 1990.

Lim, J. S., and A. V. Oppenheim, (eds.), *Advanced Topics in Signal Processing*, Upper Saddle River, NJ: Prentice-Hall, 1988.

Luh, L., et al., "A 4GHz 4th-Order Passive LC Bandpass Delta-Sigma Modulator with IF at 1.4GHz," *2006 IEEE Symposium on VLSI Circuits Digest of Papers*, Honolulu, HI, 2006.

Lynn, P. A., *An Introduction to the Analysis and Processing of Signals*, New York: John Wiley & Sons, 1973.

Lyons, R. G., *Understanding Digital Signal Processing*, Upper Saddle River, NJ: Prentice-Hall, 2004.

Mahajan, A., M. Agarwal, and A. K. Chaturvedi, "A Novel Method for Down-Conversion of Multiple Bandpass Signals," *IEEE Trans. on Wireless Comm.*, February 2006.

Maxim Integrated Products, "MAX108 +/- 5V, 1.5 GSPS, 8-Bit ADC with On-Chip 2.2GHz Track/Hold Amplifier," datasheet, 2001.

McClellan, J. H., et al., *Computer-Based Exercises for Signal Processing Using MATLAB*, Upper Saddle River, NJ: Prentice-Hall, 1998.

McClellan, J. H., and C. M. Rader, *Number Theory in Digital Signal Processing*, Upper Saddle River, NJ: Prentice-Hall, 1979.

McClellan, J. H., R. W. Schafer, and M. A. Yoder, *DSP First: A Multimedia Approach*, Upper Saddle River, NJ: Prentice-Hall, 1998.

Mehr, I., and D. Dalton, "A 500-MSample/s 6-Bit Nyquist-Rate ADC for Disk-Drive Read-Channel Applications," *IEEE Journal of Solid-State Circuits*, July 1999.

Mitra, S. K., *Digital Signal Processing: A Computer Based Approach,* 3rd ed., New York: McGraw-Hill, 2006.

Nagaraj, K., et al., "A Dual-Mode 700MSample/s 6-bit 200MSample/s 7-Bit A/D Converter in a 0.25 um Digital CMOS Process," *IEEE Journal of Solid-State Circuits*, December 2000.

Nary, K. R., et al., "An 8-Bit 2 Giga Sample Per Second Analog to Digital Converter," *IEEE GaAs IC Symposium, Technical Digest*, Vol. 17, October 1995.

Nezami, M. K., "Wireless Digital Radios, RF Architecture and Signal Processing," Raytheon, 2007.

Norsworthy, S. R., R. Schreier, and G. C. Temes, *Delta-Sigma Data Converters: Theory, Design, and Simulation*, New York: IEEE Press, 1997.

Nuttall, A. H., "Some Windows with Very Good Sidelobe Behavior," *IEEE Trans. on Acoustics, Speech, and Signal Processing*, Vol. 29, No. 1, February 1981.

Oliver, B. M., and J. Billingham, "Project Cyclops, A Design Study of a System for Detecting Extraterrestrial Intelligence," NASA CR 114445, 1972.

Oppenheim, A. V., (ed.), *Papers on Digital Signal Processing*, Cambridge, MA: MIT Press, 1969.

Oppenheim, A. V., and S. H. Nawab, (eds.), *Symbolic and Knowledge-Based Signal Processing*, Upper Saddle River, NJ: Prentice-Hall, 1992.

Oppenheim, A. V., and R. W. Schafer, *Discrete-Time Signal Processing*, Upper Saddle River, NJ: Prentice-Hall, 1989.

Oppenheim, A.V., and R.W. Schafer, "From Frequency to Quefrency: A History of the Cepstrum," *Signal Processing Magazine*, Vol. 21, No. 5, September 2004.

Oppenheim, A. V., A. S. Willsky, and S. H. Nawab, *Signals and Systems,* 2nd ed., Upper Saddle River, NJ: Prentice-Hall, 1997.

Parks, T. W., and C. S. Burrus, *Digital Filter Design*, New York: John Wiley & Sons, 1987.

Pelgrom, M. J. M., H. P. Tuinhout, and M. Vertregt, "Transistor Matching in Analog CMOS Applications," *International Electron Devices Meeting, Technical Digest*, December 1998.

Picard, R., *Effective Computing*, Cambridge, MA, MIT Press.

Pohlmann, K. C., *Principles of Digital Audio,* 3rd ed., New York: McGraw-Hill, 1995.

Poor, V., and G. W. Wornell, (eds.), *Wireless Communications: Signal Processing Perspectives*, Upper Saddle River, NJ: Prentice-Hall, 1998.

Poulton, K., et al., "A 6-b, 4 GSa/s GaAs HBT ADC," *IEEE Journal of Solid-State Circuits*, October 1995.

Poulton, K., et al., "An 8-GSa/s 8-Bit ADC System," *Symposium on VLSI Circuits, Digest of Technical Papers*, June 1997.

Poulton, K., et al., "A 4GSample/s 8b ADC in 0.35-μm CMOS," *IEEE International Solid-State Circuits Conference, Digest of Technical Papers*, February 2002.

Proakis, J. G., and D. G. Manolakis, *Digital Signal Processing: Principles, Algorithms, and Applications*, Upper Saddle River, NJ: Prentice-Hall.

Quarmby, D., *Signal Processor Chips*, Upper Saddle River, NJ: Prentice-Hall, 1985.

Quatieri, T., *Discrete-Time Speech Signal Processing: Principles and Practice*, Upper Saddle River, NJ: Prentice-Hall, 2002.

Rabiner, L., "The Chirp Z-Transform Algorithm—A Lesson in Serendipity," *Signal Processing Magazine*, Vol. 21, No. 2, March 2004.

Rabiner, L. R., and B. Gold, *Theory and Application of Digital Signal Processing*, Upper Saddle River, NJ: Prentice-Hall, 1975.

Rabiner, L. R., and R. W. Schafer, *Digital Processing of Speech Signals*, Upper Saddle River, NJ: Prentice-Hall, 1978.

Raghavan, G., et al., "Architecture, Design, and Test of Continuous-Time Tunable Intermediate-Frequency Bandpass Delta-Sigma Modulators," *IEEE Journal of Solid State Circuits*, Vol. 36, No. 1, January 2001.

Raghavan, G., et al., "A Bandpass ΣΔ Modulator with 92 dB SNR and Center Frequency Continuously Programmable from 0 to 70 MHz," *IEEE International Solid-State Circuits Conference Digest of Technical Papers*, Vol. 40, February 1997.

Ramirez, R. W., *The FFT: Fundamentals and Concepts*, Upper Saddle River, NJ: Prentice-Hall, 1985.

Russell, A., "Efficient Rational Sampling Rate Alteration Using IIR Filters," *Signal Processing Letters*, Vol. 7, No. 1, January 2000.

Salkintzis, A. K., H. Nie, and P. T. Mathiopoulos, "ADC and DSP Challenges in the Development of Software Radio Based Stations," *IEEE Personal Communications*, August 1999.

Sansen, W., "Analog Circuit Design in Scaled CMOS Technology," *Symposium on VLSI Circuits, Digest of Technical Papers*, June 1996.

Scholtens, P., "A 2.5 Volt 6 Bit 600MS/s Flash ADC in 0.25 μm CMOS," *Proceedings of the European Solid-State Circuits Conference*, September 2000.

Scholtens, P., and M. Vertregt, "A 1.6 GSample/s Flash ADC in 0.18 μm CMOS Using Averaging Termination," *IEEE International Solid-State Circuits Conference, Digest of Technical Papers*, February 2002.

Setty, P., et al., "A 5.75 b 350 MSample/s or 6.75b 150MSample/s Reconfigurable Flash ADC for a PRML Read Channel," *IEEE International Solid-State Circuits Conference, Digest of Technical Papers*, February 1998.

Sigmon, K., *MATLAB Primer*, 3rd ed., Boca Raton, FL: CRC Press.

Smith, D., *Digital Signal Processing Technology*, Newington, CT: ARRL, 2001.

Smith, D., "Signals, Samples, and Stuff: A DSP Tutorial," (Parts 1–4), *QEX*, May–October 1998.

Smith, S. W., *The Scientist and Engineer's Guide to Digital Signal Processing*, San Diego, CA: California Technical Publishing, 1997.

Steiglitz, K., *A Digital Signal Processing Primer*, Reading, MA: Addison-Wesley, 1996.

Steyaert, J. C. M., "A Single-Chip 900 MHz CMOS Receiver Front-End with a High Performance Low- Topology," *IEEE Solid-State Circuits*, Vol. 30, No. 12, December 1995.

Steyaert, M., et al., "Custom Analog Low Power Design: The Problem of Low Voltage and Mismatch," *Proceedings of the IEEE Custom Integrated Circuits Conference*, May 1997.

Sun, Y., "Generalized Bandpass Sampling Receivers for Software Defined Radio," Ph.D. thesis, Royal Institute of Technology, Sweden, May 2006.

Sushihara, K., et al., "A 6b 800 MSample/s CMOS A/D Converter," *IEEE International Solid-State Circuits Conference, Digest of Technical Papers*, February 2000.

Tamba, Y., and K. Yamakido, "A CMOS 6b 500MSample/s ADC for Hard Disk Drive Read Channel," *IEEE International Solid-State Circuits Conference, Digest of Technical Papers*, February 1999.

Terrell, T. J., *Introduction to Digital Filters*, New York: John Wiley & Sons, 1980.

Tribolet, J. M., *Seismic Applications of Homomorphic Signal Processing*, Upper Saddle River, NJ: Prentice-Hall, 1979.

Tsukamoto, S., W. G. Schofield, and T. Endo, "A CMOS 6-b, 400-MSample/s ADC with Error Correction," *IEEE Journal of Solid-State Circuits*, December 1998.

Turek, D. B., "Design of Efficient Digital Interpolation Filters for Integer Upsampling," Master's thesis, Massachusetts Institute of Technology.

Uyttenhove, K., and M. Steyaert, "Speed-Power-Accuracy Trade-Off in High-Speed ADC's: What About Nano-Electronics?" *Proceedings of the IEEE Custom Integrated Circuits Conference*, May 2001.

Vaidyanathan, P. P., *Multirate Systems and Filter Banks*, Upper Saddle River, NJ: Prentice-Hall, 1993.

Vaughan, R. G., N. L. Scott, and D. R. White, "The Theory of Bandpass Sampling," *IEEE Trans. on Signal Processing*, Vol. 39, No. 9, September 1991.

Vittoz, E. A., "Future of Analog in the VLSI Environment," *Proceedings of the IEEE International Symposium on Circuits and Systems*, May 1990.

Walden, R. H., "Analog-to-Digital Converter Survey and Analysis," *IEEE Journal on Selected Areas in Communication*, April 1999.

Walden, R. H., "Analog-to-Digital Converter Technology Comparison," *IEEE GaAs IC Symposium, Technical Digest*, October 1994.

Walden, R. H., "Performance Trends for Analog-to-Digital Converters," *IEEE Communications Magazine*, February 1999.

Walden, R. H., "Spreadsheet of Data for ADC Survey," Hughes Research Laboratories (HRL), http://www.hrl.com/TECHLABS/micro/ADC/adc.html.

Wornell, G. W., *Signal Processing with Fractals: A Wavelet-Based Approach*, Upper Saddle River, NJ: Prentice-Hall, 1996.

Yoon, K., S. Park, and W. Kim, "A 6b 500MSample/s CMOS Flash ADC with a Background Interpolated Auto-Zeroing Technique," *IEEE International Solid-State Circuits Conference, Digest of Technical Papers*, February 1999.

Zavrel, Jr., R. J., and M. Morin, "DSP for RF Designers," *Communication Design Conference*, San Jose, CA, 2001.

Zavrel, R. J., "Tomorrow's Receivers: What Will the Next 20 Years Bring?" *Ham Radio*, November 1987.

CHAPTER 25

Electronic Warfare (EW) Receivers

In addition to communications, radio astronomy, and monitoring, high probability of intercept receivers are used in electronic warfare (EW) from 2 GHz to 20 GHz and above. Electronic warfare is defined as the action by electronic means that exploits or reduces enemy use of the electromagnetic spectrum and retains its use for friendly forces. New developments in analog and digital filters should allow for the realization of single upconversion receivers, which can achieve the required final selectivity in the first IF. Extremely high dynamic ranges will be required to cope with the future's densely populated RF environment. The receiver of the future will be a hybridized approach involving several generic receiver types such as the superheterodyne, tuned radio frequency (TRF), and/or Bragg cell, providing an increased probability of intercept over the conventional types.

During peace time, HPOI receivers are used to characterize enemy defense abilities. During war time, HPOI receivers are used to judge enemy intentions, anticipate hostile actions, support command and control operations, and assess operational effectiveness. Additional applications include defense tactical electronic support measures (ESM), battle planning, and identifying hostile emitters.

Such multisensor systems, including millimeter wave (MMW) terminal seekers, use advanced digital antiradiation homing (ARH) receivers and GPS/INS sensors to rapidly engage traditional and advanced enemy air defense targets including time-sensitive "blinking" radar strike targets.

In these applications, the HPOI receiver system can become very complex and requires even more advanced system development and technology choices than previously discussed.

Today's HPOI receiver systems are confronted with dynamic range constraints, which are getting progressively more demanding in terms of signal densities, minimum time of analysis, format complexity, and more. The need for higher performance in smaller volumes is also imminent.

Specific receiver requirements are for high probability of intercept with wide instantaneous bandwidth, minimum scanning time, high throughput processing capacity, and accurate parameter measurements. These characteristics are necessary for adequate signature analysis in order to hand off accurate data for sensor cueing and power management.

25.1   Probability of Intercept (POI)

POI is the probability that a receiver system will not only detect, but process and identify an emitter within a minimum amount of time. More often than not, the emitter is a low probability of intercept source, which requires an even better receiver.

Although there is no clear methodology of defining probability of intercept (POI), there are several analytical methods of computing it. It is the job of the RF designer and system analyst to decide how to compute the POI, given a specific type of emitter and receiver used.

In addition to the scanning superheterodyne, several other types of receivers are used in concert in order to increase the probability of intercept. Many times, emitters are elusive in their transmissions. These are known as low probability of intercept emitters.

The primary criterion in detection considers the probability that an emitter will be detected on a single receiver scan or near instantaneously. This requires a combination of ultrawideband, ultrafast scanning superheterodyne and bulk detection receivers using special technologies and sometimes combining RF with photonic processing as we will see later on.

Alternative detection criterion considers the cumulative probability of detection meaning that an emitter has been detected at least once within a limited frequency range and time constraints. Probability of intercept of a receiver system depends on several factors:

- Pulse width of the emitter's signal;
- Instantaneous field-of-view;
- Instantaneous bandwidth;
- Receiver sensitivity;
- Receiver resolution;
- Receiver dwell time;
- Receiver scan time (if applicable);
- System throughput including data processing;
- Stored emitter parameter data validity;
- Other reaction time constraints;
- Number and type of emitters that must be tracked per unit of time.

Most of these properties are contradictory. Trade studies need to be performed to come up with a comprehensive receiver system solution for a specific application. These studies include but are not limited to:

- Bandwidth analysis;
- Dispersive delay line bandwidth;
- Scan rate;
- Scan width—step resolution;

- Time to scan total bandwidth;
- Total scan period with retrace;
- Ultimate frequency resolution;
- Maximum pulses per scan;
- Time period for output per scan;
- Output pulse width from delay lines;
- POI for shortest pulse.

The ideal digital receiver was briefly discussed in Chapter 10. The perfect electronic warfare (EW) receiver would be a totally digital receiver defined as a high-resolution, high dynamic range analog-to-digital converter (A/D) and signal processor connected directly to the antenna. Such a receiver would translate modulated or pulsed RF signals of any frequency, bandwidth, phase, and amplitude into digital words, which can be processed and manipulated near instantly using digital signal processing (DSP) techniques. The trend is to eliminate mixers altogether and go back to TRF radios that use new modern signal processing technologies at RF frequencies. An ideal digital EW receiver is shown in Figure 25.1.

However, there are limitations to the signal processing speed in both A/D and D/A technologies. In addition, to accommodate for the extremely wide RF coverage, a very high quality MRU would have to be switchable and tunable to allow for the best alias situation. Today, high-resolution sampling digital receivers can be implemented at the lower frequencies as previously explained in Chapter 10 (see patent 6,882,310). Their realization at higher frequencies has been impractical due to technological limitations of today's A/D technology. Digital signal processing (DSP) is also possible in software-defined radios at the intermediate frequencies (IF) or directly at baseband.

Some of the various receiver technologies used for wideband receivers are:

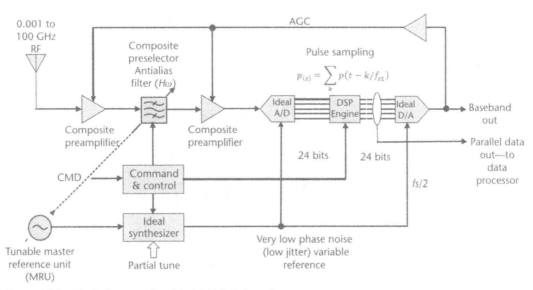

Figure 25.1 Block diagram of an ideal EW digital receiver.

- Crystal video—warning receiver;
- IF log video;
- Detected log video amplifier;
- Instantaneous frequency measurement (IFM);
- Phase detection—used for direction of arrival;
- Ultrawideband scanning superheterodyne;
- Channelized receiver—activity monitor;
- Bragg cell—activity monitor;
- Combinations of the above.

25.2 Crystal Video Receiver

A crystal video receiver is a form of a tuned radio frequency (TRF) receiver consisting of a frequency multiplexer or a filter bank, which splits a wide input frequency range into several broad contiguous bands, which, in turn, are filtered and logarithmically amplified before detection.

Amplitude and video detection occurs at baseband in each of the bands by using a "crystal" square law diode detector. This is the simplest form of an electronic counter measures (ECM) receiver.

The crystal video receiver is low cost and small. However, it has low sensitivity due to a large noise bandwidth, and is subject to blocking from strong in-band signals. It is usable only in singular emitter frequency environments and in near proximity to these emitters, primarily for detection of low duty cycle pulse or CW signals such as used by police radars. Its purpose is to give rapid warning responses in the presence of a radar signal. A block diagram of a crystal video receiver is shown in Figure 25.2.

Looking at Figure 25.2, it can be seen that each channelized gain stage is a limiting amplifier. The output of each channel is a video signal proportional to the

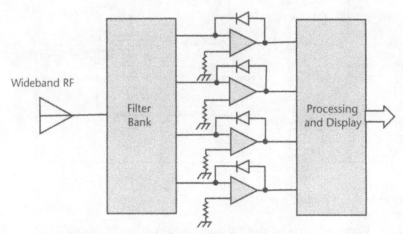

Figure 25.2 Block diagram of a crystal video receiver.

input power. Some designs use this limited IF to feed additional IFM receivers or phase discriminators.

Almost all detected log-video amplifiers in these receivers are usually followed by A/D and digital circuitry so that the output is a digital word. Part of the design difficulty is developing the timing for trigger pulse for the A/D and keeping log amps and diodes performance consistent over a wide temperature range. More recently, the logarithmic amplifier transfer functions have been implemented in a single DSP block, which manages such shifts in performance. More information about this can be found in [1].

Whether it is a detected log video design or an IF log amplifier, the output of this type of receiver is a video voltage proportional to the RF input power. All log amplifiers have good rise times for incoming RF pulses, but poor fall times, which limit the response time to a second incoming pulse. A typical set of performance characteristics for an crystal video receiver is shown in Table 25.1.

In a crystal video receiver, several microwave transmissions from hostile radars are seen concurrently by a wideband receiving antenna. From here, signals are fed into the crystal video receiver. The receiver sections are tuned to consecutive slices of the covered band, which allow simultaneous reception and discrimination of radars operating over a wide bandwidth.

The receiver output is a set of signals, which represents the envelope of the detected microwave signals. Crystal video receivers are usually used as warning radar receivers.

25.3 The Compressive (Microscan) Receiver

The compressive or microscan receiver is a form of superheterodyne receiver usually using double conversion to combine the properties of a wideband, fast sweeping receiver with the narrowband and pulse compression and decompression characteristics of a SAW delay line implemented in a relatively narrowband IF. A comprehensive receiver block diagram is shown in Figure 25.3. This example shows two CW signals as an input and two pulses delayed in time as an output. The output pulses can vary in amplitude indicating different signal strengths.

Such receiver can scan a frequency band several orders of magnitude faster (up to 10,000 times faster) than would be possible with conventional scanning superheterodyne receivers.

The main advantage of using a compressive receiver is that if more than one signal is present in the observation bandwidth, then this receiver will give a spectrum

Table 25.1 Typical Specifications of a Crystal Video Receiver

Total bandwidth (typical)	2 to 18 GHz
Instantaneous bandwidth	Octave
Ultimate resolution	10-MHz video
Linear dynamic range	>45 dB
MDS (typical)	<–60 dBm (with preamp)
POI	100% instantaneous

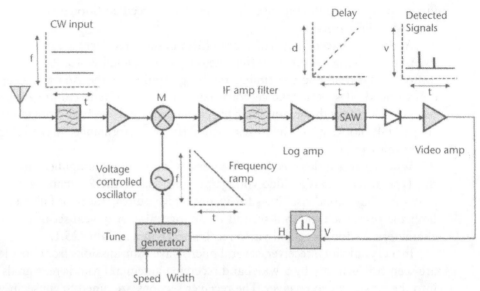

Figure 25.3 Block diagram of a comprehensive (microscan) receiver.

analyzer type of display where multiple signals are displayed as a function of chirp time.

This technique can also be used to observe pulse-on-pulse or CW signals that are time coincident. The compressive receiver can be used to deal with very dense signal environments. A set of typical specifications for a compressive receiver is shown in Table 25.2.

25.4 Instantaneous Frequency Measurement (IFM) Receiver and Digital Instantaneous Frequency Measurement Receiver (DIFM)

The IFM and DIFM receivers are a more complex form of the TRF receiver using bandpass/band reject front-end filters and a combination of delay lines and

Table 25.2 Compressive (Microscan) Receiver Typical Specifications

Total bandwidth	2 to 18 GHz
	Compatible with total bandwidth
Double conversion superheterodyne	Compatible with pulse rice time and SAW compression/dispersion line technology
Instantaneous bandwidth	1 GHz typical
Ultimate bandwidth	Narrow
Linear dynamic range	80 dB typical
MDS	−100 dBm
POI	100%

Note: First sidelobe dynamic range is a significant limiting factor for a compressive receiver (or any other sampling receiver for that matter).

phase detectors to allow for near instantaneous (within fractions of nsec) frequency measurement of single pulse signatures. This type of receiver is usually used for a jammer quick set-on mechanism or as an acquisition receiver to set up a slower, narrowband, high-resolution receiver. A block diagram of an IFM/DIFM receiver is shown in Figure 25.4. An IFM measurement mechanism is shown in Figure 25.5.

IFM receivers usually perform the IFM discrimination at some lower IF frequency because limiter amps and well-balanced mixers can be obtained at these frequencies. There are companies that manufacture a special set of mixers intended specifically for IFM or phase detection use. These designs have better isolations and diode-matching characteristics than standard performance mixers. IFM receiver

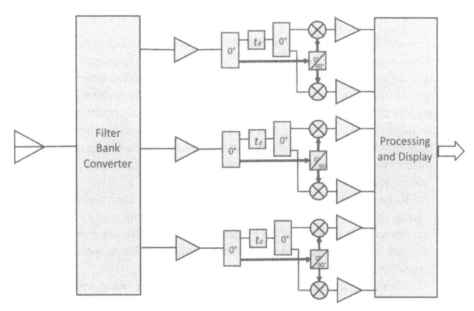

Figure 25.4   Block diagram of an IFM/DIFM measurement receiver.

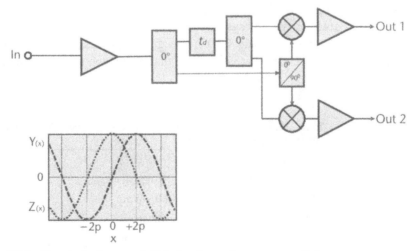

Figure 25.5   Near instantaneous (within fractions of a nanosecond) frequency measurement of single pulse signatures is possible using time delay discrimination.

designs tend to be expensive. It is common to use IFM circuits in RF channelizers. Typical performance of an IFM/DIFM receiver is shown in Table 25.3.

A deficiency of IFM receivers is that long pulses and CW signals can saturate an IFM circuit making it blind to short pulsed signals. It is a common practice to detect pulse widths of 100 μs or longer as continuous waves (CW). Another issue with IFM circuits is that they will not work with time coincident signals. A common approach used for IFM circuits is to power divide the limited signal and use a bank of IFM circuits utilizing different time delays, which in turn changes the effective bandwidth of the circuit.

25.5 Phase Detection in Interferometer Receivers

A phase detector is not considered a type of receiver but it is an important type of discriminator used in interferometer receivers. Interferometer systems are used for direction finding, which in turn is dependent on phase detection circuits to measure the relative phase difference between two or more channels. Phase differences greater than one cycle must be determined by using algorithms based on the dimensions of the antenna array and the signals frequency. The algorithms determine the number of cycle differences and the phase detection circuitry measures the phase difference within a cycle. A phase detector circuit gives sine and cosine outputs as shown in Figure 25.6.

In an interferometer receiver equipped with a phase detector circuit, the video output voltage is proportional to the difference in phase of two channels. Although one mixer will give phase difference, two are usually used so that only the linear portion of the response is used. One output is a sine response and the other is a cosine response.

These circuits are critical in the design of a microwave interferometer receiving system for determining the direction of arrival of a signal. This type of detection can also be used in radio astronomy receivers. The way such a system works bases itself on the fact that mixing two signals of identical frequency results in the second harmonic product and a DC component. This is typical of mixer phase detectors as we discussed previously in Chapter 15.

Consequently, the voltage of the DC component is proportional to the phase difference of the two signals. Care in design and processing corrections must be applied to reduce the inherent DC offsets that occur due to the problem of LO to RF

Table 25.3 Instantaneous Frequency Measurement (IFM/DIFM) Receiver Typical Performance

Total bandwidth	0.5 to 18 GHz
Instantaneous bandwidth	1 GHz typical
Ultimate resolution	1 MHz typical
Linear dynamic range	<50 dB typical
MDS	−70 dBm typical
POI	100% instantaneous (subnanoseconds)

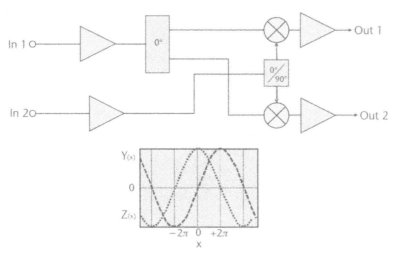

Figure 25.6 Typical phase detector arrangement used in an interferometer receiver.

isolation, caused by the lack of perfect diode matching in the mixers and changes caused by temperature variations.

25.6 Wideband Swept Superheterodyne Receivers

A wideband, microwave swept superheterodyne receiver is defined as a fast sweeping/hopping wide IF bandwidth (>500 MHz per frequency byte) receiver performing a fast Fourier transform directly at the IF and using a high-frequency (>1 GHz center frequency) bandpass A/D-DSP signal processor. Such systems are becoming the norm in signal analysis, but higher-frequency A/D technology, as mentioned before, is advancing more slowly than other signal processing technologies.

Super wideband swept superheterodyne receivers require complex banks of automatically switched half-octave front-end filters in order to avoid image problems over an extremely wide band set of RF frequencies, usually from 2 to 18 GHz. Upconverting above this range becomes a necessity in order to simplify front-end complexity and minimize receiver size. Typical performance for a wideband microwave swept superheterodyne receiver is shown in Table 25.4.

25.7 Narrowband Swept Superheterodyne Receivers

A narrowband, swept superheterodyne receiver uses a combination of a wideband, swept/switched type receiver (as discussed above) combined with an additional narrowband second or third IF sweeping through the first very wide bandwidth IF for increased resolution. It utilizes a lower-frequency second IF to take advantage of lower-frequency A/Ds with an increased number of bits (>10) for higher dynamic range implementation. This receiver uses multiple AGC loops and a complex synthesizer to achieve the highest possible dynamic range at the narrowest bandwidth and step resolution required. Requirements for such a receiver are shown in Table 25.5.

Table 25.4 Wideband Swept Superheterodyne Receiver Typical Performance

Total bandwidth	2 to 18 GHz
IF center frequency	1 GHz typical
Front end composite	8 half-octave filter bank
Instantaneous bandwidth	1 GHz typical
Ultimate resolution	Limited by A/D
Linear dynamic range	>100 dB typical
MDS	−110 dBm typical
POI	<100% near instantaneous

Table 25.5 Narrowband Microwave Swept Superheterodyne Receiver Typical Performance

Total bandwidth	2 to 20 GHz
Double conversion superheterodyne	First IF: 1.5 GHz typical, second IF: 0.5 GHz typical
Front end composite	7 half-octave filter bank
Instantaneous bandwidth	>0.5 GHz typical
Ultimate resolution	10/100 Hz typical
Linear dynamic range	>130 dB
MDS	−138 dBm typical
POI	<100% not instantaneous

Such a receiver is described in U.S. patent 7,139,545 [2]. A block diagram of this receiver as it appears in the patent is shown in Figure 25.7.

Looking at Figure 25.7, an ultrawideband receiver system includes an intelligent preselector stage to upconvert received signals to signals within a first IF frequency range at 26 GHz. It uses a synthesized stepped first local oscillator to downconvert the signals within the first IF frequency range to signals within a second IF frequency range using a fixed second oscillator signal. It also uses a second downconverting stage to downconvert the signals within the second IF frequency range to signals within a third IF frequency range using a high-resolution synthesizer. The center frequency of the first IF frequency range is at least 30% to 35% higher than a highest frequency in the receive frequency range to provide an image frequency far from (e.g., more than twice) the highest frequency received. The receiver includes a synthesizer to coherently generate the first, second, and third oscillator signals using half-integer frequency dividers (see [3]) based on a master reference signal. Finally, the receiver uses a synthesized stepped third oscillator to provide an ultimate resolution of 10 Hz.

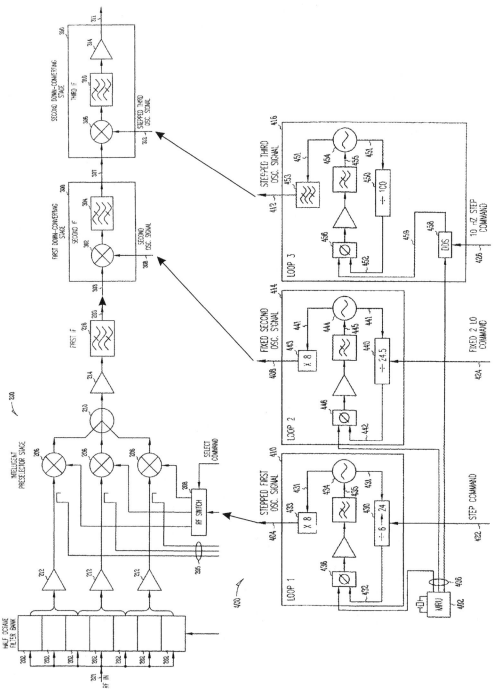

Figure 25.7 Block diagram of a wideband, microwave swept superheterodyne receiver. This receiver is described in [2].

25.8 Channelized Bulk Filter (Cued) Receiver

The channelized receiver uses a multiple superheterodyne approach that divides the frequency range into contiguous channels compatible with the final pulse resolution desired, usually less than 2 GHz wide. It uses a type of parallel receiver architecture with a wide input bandwidth and multiple narrowband outputs.

This technology allows much wider bandwidths to be monitored at each coarse frequency step. These bandwidths are typically in the 500- to 1,000-MHz range. Using a 500-MHz step bandwidth, a 2- to 20-GHz total bandwidth can be scanned in 36 steps instead of 1,800 steps for a 10-MHz IF bandwidth. This reduces the scanning time required and greatly increases the probability of intercept. A typical channelized bulk filter receiver is shown in Figure 25.8.

The classic approach to a channelized a bulk filter receiver uses various voltage comparator schemes (previously discussed) to form signals usable by the signal detection mechanism, in order to output useful information. Digital versions of this receiver approach use A/Ds to sample the signal bandwidth in order to obtain FFT samples of the frequency domain signals representation.

In the classical approach these signals have to be processed very quickly because the signals would be time delayed and phase measurements would be made after the channelized receiver has steered a main receiver. Typical performance for a channelized bulk filter receiver is shown in Table 25.6.

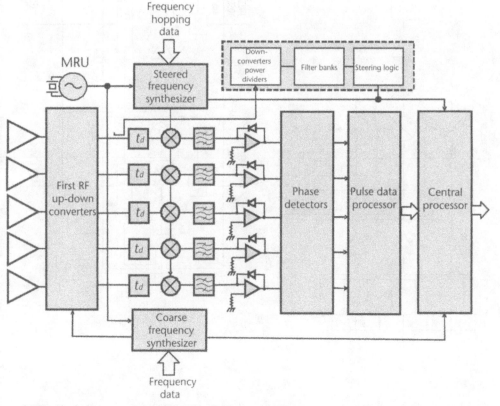

Figure 25.8 Block diagram of channelized bulk filter receiver.

25.9   The Bragg Cell or Acousto-Optic Receiver and Ultrawideband Instantaneous IFs

We will now look at another type of receiver and technology intended to capture large instantaneous microwave bandwidths. This is the Bragg cell receiver. This type of receiver is relatively new to EW, but has been used extensively in radio astronomy.

The Bragg cell receiver is a unique ultrawideband instantaneous receiver type, which blends together RF and photonic technologies. In radio astronomy, it is used to identify multiple radio sources concurrently. In EW, this technology has applications as a steering mechanism for much higher resolution receivers because of its instantaneous nature of receiving multiple signals concurrently and without frequency scanning. Some Bragg cell receiver implementations can provide at least four times shorter scanning and processing time than any other receiver or technology. In addition, Bragg cell technology can be used as a direct replacement for state-of-the art A/D converter technology at high IF frequencies and percentage bandwidths. This is because dynamic range offered using the Bragg cell processor technology, when used in an IF application, can provide greater bandwidth than comparable A/D technology as used at the same IF frequency.

A wideband instantaneous (acousto-optic) Bragg cell receiver is defined as a nonsweeping spectrum analyzer receiver system, which is laser driven and includes the Bragg cell transducer at a microwave frequency IF. It provides instantaneous signal processing of many RF signals over very large bandwidths. A Bragg cell receiver is a parallel processing superheterodyne radio receiver, which acts as an optical Fourier transformer that continuously converts the time-varying inputs into discrete frequency sets represented by separate instantaneous beam diffraction angles of a monochromatic laser light, corresponding to active input frequencies as processed through the Bragg cell transducer.

The Bragg cell is a block of very pure crystalline material, usually quartz or lithium niobate, approximately 1 cm × 1 cm × 10 cm in size with a transducer bonded to it. When excited by IF electrical signals within the passband of the transducer, the Bragg cell is internally exposed to sound waves corresponding to all the RF signals received. A Fourier transfer frequency decomposition takes place inside

Table 25.6   Channelized Bulk-Filter Receiver Typical Performance Specifications

Total bandwidth	2 to 20 GHz
Instantaneous bandwidth	0.5 GHz typical
MDS	−85 dBm typical
Pulse width resolution	100 ns typical
Signal resolution	10 MHz minimum, 30 MHz for a 50-dB signal separation
Frequency accuracy	+/− 2 MHz pulse-to-pulse
Data throughput	1 Mpps
POI	<100%

the Bragg cell and beams of information can be projected at diffent angles and viewed on a physical screen. These, in turn, can be detected in real time using photo diodes.

The interaction between sound and light was suggested by Louis Marcel Brillouin (a French physicist who lived between 1854 to 1948, famous for his ferromagnetic theory, perpetual motion opposition, and the Brillouin scattering phenomenon) some 80 years ago. The concept was proven in 1932 through an experiment involving a column of water in which sound was induced. Two fields (one broadband, the other monochromatic optical) were coupled via the photoelastic effect, wherein an induced strain produced a corresponding change in the index of refraction of the medium. This allowed for the first-order diffraction line to become more intense while providing cancellation of the other lines. This angle was later called the Bragg angle and the phenomenon constitutes the basis for today's acousto-optical (A/O) receiver, or as it is sometimes called, the Bragg cell receiver.

In a typical microwave Bragg cell receiver, several RF signals of interest present at the antenna are converted to a wideband IF. This is a typical superheterodyne process, and the signals contained within the IF are further amplified by a high-gain, high dynamic range IF amplifier, and are finally applied to the Bragg cell through the piezoelectric transducer, which is bonded to it as shown in Figure 25.9.

Modern Bragg cell receivers use a beam of monochromatic light such as that produced by a solid-state laser injected at the certain angle into the Bragg cell. An instantaneous deflection will occur for each of the sonic signals traveling through it. This happens because the acoustic energy in the material causes changes in the refractive index between the peaks and valleys of the wave. The resultant diffraction pattern is an instantaneous display of the signals, which can be viewed panoramically on a screen within the range of the signals applied to the transducer. The deflection angle and the intensity of the light beam are proportional respectively to the frequency and power of the signals being received.

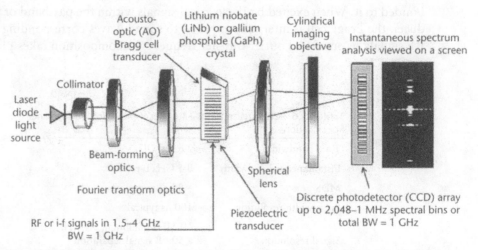

Figure 25.9 The Bragg cell receiver principle and typical display of received signals. The acousto-optic spectrum analyzer contains four essential components: the laser diode monochromatic light source, Fourier transform optics, acousto-optic Bragg cell, photodetector array, as well as the associated processing.

In modern Bragg cell receivers, an RF input signal produces an ultrasonic traveling phase grating within the device, whose interaction geometry defines the output optical field. AO devices are made of a suitable transparent optical medium onto which a piezoelectric transducer is bonded. Proper adjustment of the angle of incidence of a light source through the medium is required to allow for the first-order diffraction line to become more intense while canceling other lines.

The Bragg cell receiver can be viewed as a parallel-processing spectrum analyzer, which acts as an optical Fourier transformer that continuously converts the time-varying input into discrete frequency sets represented by separate instantaneous beam diffraction angles corresponding to each frequency.

The instantaneous nature of the Bragg cell receiver allows for a high POI since many signals can be "viewed" at the same time, as in the case of a radio telescope intended to capture and analyze large portions of the spectrum in a minimum amount of time.

Detection of the displayed signals is implemented by focusing the multiple light beams on a linear array of PIN photodiodes (or a CCD), which are spaced very closely together. The position of each detector corresponds to a specific frequency within the bandpass of the receiver allowing for parallel frequency detection of all signals within that band. As many as a thousand detectors can be incorporated into such a signal collection installation, allowing for a typical frequency cell resolution of 1 MHz or less.

Shown in Figures 25.10 and 25.11 are commercially available Bragg cells manufactured by AA Opto-Electronics Corporation.

The advantages of a wideband instantaneous (acousto-optic) Bragg cell receiver are:

- No variable LO required for resolution over the bandwidth of interest;
- Near instantaneous nature allows for high probability of intercept (POI) of many signals, all at the same time.

One of the disadvantages of a wideband instantaneous (acousto-optic) Bragg cell receiver is that limited linear spurious-free dynamic range of the Bragg processor

Figure 25.10 Commercially available Bragg cell (acousto-optic deflector) can provide 256 frequency spots resolution. (Courtesy of AA Opto-Electronic, France.)

Figure 25.11 Example of a multiple AO modulator (Bragg cell) assembly. The unit can deflect and modulate, simultaneously and independently, up to 16 laser channels operating at 180 MHz +/− 30 MHz per channel, for a total of approximately GHz composite bandwidth. For each of the 16 channels' devices, the number of resolvable spots (Rayleigh criterion) is 20 with an access time of 335 ns. Each channel requires +33 dBm (2 watts) of RF drive at an IF frequency of 180 MHz. Optical laser wavelength is 355 nm and active aperture per channel is 0.5×2 mm². Material is fused silica. (Figure courtesy of AA Opto-Electronic, France.)

(~50 dB) at microwave frequencies compares with that of low-resolution A/D processors, so multiple AGCs are necessary.

Implementation of a novel technology such as optical RF channelizers has always been a difficult sell even when there are clear advantages in terms of throughput (or high probability of intercept), ability to see spread spectrum signatures, small size, weight, and power characteristics.

Once the IF signal is digitized, the downstream processing (FFT) can be performed using conventional electronic means, and this, no doubt is attractive for a number of reasons.

If the RF power spectral density is monitored using one or more broadband IF Bragg processors, the acquired activity information can be further analyzed using a narrowband swept superheterodyne receiver such as the one described above. This, in turn, saves time and increases POI. In combination with the high-resolution receiver, the Bragg cell receiver provides the highest POI receiver system known to date.

Modern Bragg cells operate over portions of the spectrum between 40 MHz to 100 GHz. The use of Bragg cells for RF spectrum analysis has been well established since the 1980s. Relatively narrowband EW intercept receivers have been used along with wideband radio astronomy (RA) and SETI receivers applying the concept at very wide band IF frequencies, while hopping over even wider frequency ranges.

Receivers having larger near instantaneous bandwidths (500-MHz minimum) and exhibiting high resolution have been implemented over the frequency range of 2 GHz to 20 GHz using Bragg cell IFs. One such example is described in the U.S.

patent 7,324,797 [4]. The invention provides an improved wideband receiver and method. Such a receiver can be used to scan a wide frequency band, for a myriad of reasons, in an effort to receive and detect information. Typically, the scanning is accomplished in frequency steps, known as hops, corresponding to tuning a local oscillator in the process of frequency conversion.

In this receiver, large blocks of 1-GHz information bandwidth can be quickly scanned and analyzed over an ultrawide microwave range. A composite of as many as four ultrawideband Bragg cell processors can be used in the last IF of such a receiver extending the instantaneous bandwidth analyzed from 1 GHz to 4 GHz per hop. A block diagram of the receiver is shown in Figure 25.12. This shows the essence of the patent. It describes an integrated acousto-optical receiver using a Bragg cell processor at the second IF (center frequency of ~1.5 GHz). Six frequency blocks of 4 GHz each are scanned by the first converter and four blocks of 1 GHz are scanned within each 4-GHz block in the second conversion. The brute force synthesizer shown provides 6×4 hops or 24×1-GHz-wide scans for the entire 2- to 18-GHz RF range. By using four Bragg cells (instead of just one) in the second IF, the entire range of 2 to 18 GHz can be analyzed in only four hops. The solid-state laser, the optical elements, the Bragg cell, and the array of photo detectors can be incorporated on the same substrate, significantly reducing the size of the receiver. Consequently, the signal resolution is much better than that given by a crystal video receiver or any other methods presented earlier in this chapter, using wideband filters.

The typical performance of a receiver using a Bragg cell processor in the second IF is shown in Table 25.7.

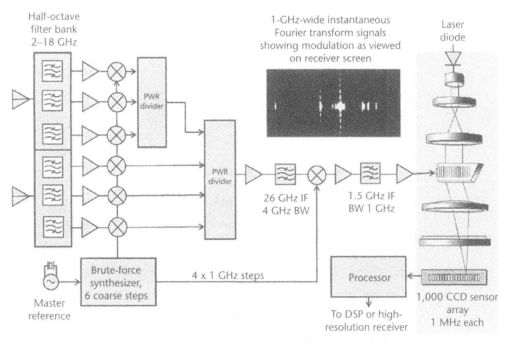

Figure 25.12   U.S. patent 7,324,797 [4] describes an integrated acousto-optical receiver using a Bragg cell processor at the second IF (center frequency of ~1.5 GHz).

25.10 Conclusions

In conclusion, acousto-optical signal processing technology has been emerging from the experimental laboratory development phase into an advanced process of signal collection with wideband applications. Bragg cells can now be used as Fourier transformers in the last IF of microwave wideband receivers, replacing A/D converters, and offering wider-band performance at higher IF frequencies. While still in its infancy, the Bragg cell receiver provides an improved signal processing tool in a dense RF spectrum.

Bragg cell is a disruptive technology. At the higher microwave frequencies, it can be implemented in lithium niobate (LiNb) or gallium phosphide (GaPh) as shown in Table 25.8.

Design of optical channelizers involves analysis of the optical wave interaction and propagation, which occurs when an input optical field is modulated by many signals.

For the instantaneous power spectral density of the input signal to be generated optically, the electrical signals must be converted to an equivalent optically modulated field.

Table 25.7 Typical Performance Specifications for an Advanced Ultrawideband Instantaneous Bragg Cell Receiver Using a Bragg Cell Processor in the Second IF

Total bandwidth	2 to 18 GHz
Double conversion superheterodyne	First IF: 1.5 GHz typical, second IF: 0.5 GHz typical
Front end composite	7 half-octave filter bank
Instantaneous bandwidth	>1 (up to 4) GHz typical
Ultimate resolution	1 MHz typical
Linear dynamic range	>100 dB with AGC
MDS	−100 dBm typical
POI	~100% Near instantaneous

Table 25.8 Bragg Cell Technology Trade-Offs of Lithium Niobate (LiNb) and Gallium Phosphide (GaPh)

Lithium Niobate (LiNb)	Gallium Phosphide (GaPh)
Cf <= 4 GHz	Cf <= 1 GHz (typical)
BW <= 2 GHz	BW <= 500 MHz
Laser wavelength = 550 nm	Laser wavelength = 850 nm
Ultimate resolution = 1 MHz/200 kHz	Ultimate resolution = 10 MHz
Fourier delay = fempto seconds	Fourier delay = fempto seconds
Physical size, butter brick	Physical size, butter brick

Bragg cells, because of the nature of the interaction that occurs between the optical field and the signals of interest, operate much like single-sideband (SSB) mixers, but at optical frequencies. The process is associated with the complex amplitude of the optical sidebands produced under small signal conditions. As with conventional mixers, the signal is translated by the carrier, in this case optical, in the upper (USB) or lower (LSB) sideband. What is different is the information contained in the stored time window of the signal.

References

[1] Willis, J. M., and C. C. W. McGuire, "Digital Crystal Video Receiver," U.S. Patent 7,113,229.

[2] Drentea, C., "Ultra-Wide Band Fully Synthesized High-Resolution Receiver and Method," U.S. Patent 7,139,545.

[3] Young, S. S., and K. Arnold, "Low Noise Fine Frequency Step Synthesizer," U.S. Patent 5,150,078, European Patent EP0545232.

[4] Drentea, C., "Bragg-Cell Application to High Probability of Intercept Receiver," U.S. Patent 7,324,797.

Selected Bibliography

Adamy, D. L., *Introduction to Electronic Warfare Modeling and Simulation*, Norwood, MA: Artech House, 2003.

Drentea, C., *The Art of RF System Design*, Raytheon, 2005.

Drentea, C., "Ultra Wide Band (20 MHz to 5 GHz) Analog to Digital Signal Processor," U.S. Patent 6,492,925.

Drentea, C., and T. Ashburn, *High Probability of Intercept: EW Receivers, Design and Applications*, Raytheon.

Guan, X., and A. Hajimiri, "A 24-GHz CMOS Front-End," *IEEE J. Solid-State Circuits*, Vol. 39, 2004.

Hossein-Zadeh, M., and A. F. J. Levi, "14.6-GHz LiNbO3 Microdisk Photonic Self-Homodyne RF Receiver," *IEEE MTT*, Vol. 54, 2006, pp. 821–831.

Hossein-Zadeh, M., and A. F. J. Levi, "Self-Homodyne RF-Optical LiNbO3 Microdisk Receiver," *Solid State Electron.*, Vol. 49, 2005, pp. 1428–1434.

Ilchenko, V. S., et al., "Coherent Resonant Ka-Band Photonic Microwave Receiver."

Ilchenko, V. S., et al., "Ka-Band All-Resonant Photonic Microwave Receiver," *IEEE Photon. Tech. Lett.*, Vol. 20, 2008, pp. 1600–1603.

Ilchenko, V. S., et al., "Photonic Front-End for Millimeter Wave Applications," *IEEE Photonic Proceedings*.

Ilchenko, V. S., et al., "SubmicroWatt Photonic Microwave Receiver," *IEEE Photon. Tech. Lett.*, Vol. 14, 2002, pp. 1602–1604.

Ilchenko, V. S., et al., "Whispering Gallery Mode Electro-Optic Modulator and Photonic Microwave Receiver," *J. Opt. Soc.*, 2003.

Matsko, A. B., et al., "RF Photonic Receiver Front-End Based on Crystalline Whispering Gallery Mode Resonators," *IEEE Photonics*.

Olbrich, M., et al., "A 3 GHz Instantaneous Bandwidth Acousto-Optical Spectrometer with 1 MHz Resolution," Physikalisches Institut, Universität zu Köln.

Pace, P. E., *Advanced Techniques for Digital Receivers*, Norwood, MA: Artech House, 2000.

Pace, P. E., *Detecting and Classifying Low Probability of Intercept Radar*, Norwood, MA: Artech House, 2004.

Schleher, D. C., *Introduction to Electronic Warfare*, Norwood, MA: Artech House, 1986.

"Sir Lawrence Bragg, Mathematician, Physicist, Nobel Laureate, 1890–1971," http://www.whitehat.com.au/australia/People/Bragg.asp.

Van Brunt, L. B., "Applied ECM," EW Engineering, Inc., Dunn Loring, VA, 1978.

CHAPTER 26

Conclusions

This book spans about 100 years of radio technology. While this is a relatively short period of time, the advent of wireless radio technology has had an enormous technological impact. In retrospect, it is hard to believe that only 100 years ago the first intelligent signal was heard with a coherer-type receiver.

Today, smart radios and new wireless devices avoid frequency traffic bottlenecks by sensing traffic densities, and communicating and switching automatically to liberate frequencies and avoid RF traffic jams.

Although frequency hopping and spread spectrum have been used to a great extent in the past, the new methodology goes beyond just modulation schemes. Engineers and regulatory bodies are now working together on new ways to allow wireless devices to share the frequency spectrum by listening first, analyzing spectrum activity against regulatory designations, using intelligent queuing algorithms, and reasoning and communicating between themselves, in order to reassign the spectrum usage for maximum throughput for all parties involved.

Although it is hard to predict what the next 100 years will bring, it is clear that we are on the right path, technologically. This book has made a forceful attempt to summarize all factors entering the design and development of high probability of intercept receivers. It is hoped that after reading it, the reader has a good understanding about how to achieve high dynamic range in designing communications receivers.

About the Author

Cornell Drentea is a technical consultant with over 40 years of hands-on experience in the aerospace, telecommunications, and electronics industries. He was previously a senior engineer/scientist at Hughes/Raytheon Systems and was a technology leader at the Avionics and Corporate Divisions of Honeywell. He was involved in the design and development of complex terrestrial and satellite communications networks and has made significant contributions in the design and development of RF, radar, guidance, and communications systems at frequencies of up to 100 GHz. Mr. Drentea has developed several state-of-the-art RF products including ultrawideband-high probability of intercept (HPOI) receivers, complex synthesizers, multimodulation transmitters, and agile deep-space transceivers. He has presented on RF design topics at prestigious technical forums such as IEEE, RF-Expo, and Sensors-Expo and has given comprehensive professional courses in RF receiver, synthesizer, and system design to some of the largest companies. He holds five patents and has published over 80 professional technical papers and articles in national and international magazines.

Index

The Artech House Intelligence and Information Operations Series

For further information on these and other Artech House titles, including previously considered out-of-print books now available through our In-Print-Forever® (IPF®) program, contact:

Artech House
685 Canton Street
Norwood, MA 02062
Phone: 781-769-9750
Fax: 781-769-6334
e-mail: artech@artechhouse.com

Artech House
16 Sussex Street
London SW1V 14RW UK
Phone: +44 (0)20-7596-8750
Fax: +44 (0)20-7630-0166
e-mail: artech-uk@artechhouse.com

Find us on the World Wide Web at: www.artechhouse.com

CPSIA information can be obtained
at www.ICGtesting.com
Printed in the USA
BVHW01*0345120618
518784BV00013B/46/P